Tropical Forests

The symposium on Tropical Forests: Botanical dynamics, speciation, and diversity was held at Aarhus University on the occasion of the twenty-fifth Anniversary of the Botanical Institute. The Danish Natural Science Research Council, the Research Foundation of Aarhus University, Aarhus Olie, and DANIDA, are hereby acknowleged for making the event possible.

Tropical Forests

Botanical Dynamics, Speciation and Diversity

Edited by

L. B. Holm-Nielsen, I. C. Nielsen and H. Balslev
Botanical Institute
Aarhus University

ACADEMIC PRESS
Harcourt Brace Jovanovich, Publishers
London □ San Diego □ New York
Boston □ Sydney □ Tokyo □ Toronto

ACADEMIC PRESS LIMITED
24-28 Oval Road
London NW1 7DX

United States Edition published by
ACADEMIC PRESS INC
San Diego, CA 92101

**British Library Cataloguing in Publication Data
is available**

This book is printed on acid-free paper

ISBN 0–12–353550–6

Printed in Great Britain by T. J. Press (Padstow) Ltd.,
Padstow, Cornwall

Foreword

Last year, I was greatly disappointed to be unable to attend the Tropical Forest Symposium at Aarhus University. Now that I have the papers in front of me, I am even more sorry that the Symposium coincided with the time of my international move! This is a wonderful, up-to-date assessment of dynamics, speciation and diversity in tropical forests. It is also a worthy tribute to the long and distinguished career of Professor Larsen, whose vision built the Botanical Institute that celebrated its twenty-fifth anniversary with this Symposium. Already, in the comparatively short life of the Botanical Institute of Aarhus University, its students have become leaders in the international botanical community and have produced a large number of useful and significant publications.

Tropical forests are disappearing at an alarming rate. To provide a rationale for their conservation and sustainable utilisation, we must understand their complexity, interactions and dynamics. This is not just another volume that proclaims the destruction of forests; it provides the data on which to base conservation and utilisation programmes.

The section on forest dynamics discusses both the physical features such as soil, landscape and water, and the patterns of tree distribution in the forest. These chapters show that, over the last 25 years, we have really begun to understand forest structure and composition because we have data from so many fields such as tree architecture, gap-phase dynamics, canopy structure, *etc*. The forest is now known to be much more dynamic than was thought by earlier workers who emphasized long-term stability.

The sections on diversity and speciation are largely written by systematists. It is most encouraging to see the many ecological and theoretical aspect that are emerging from their work. The combination of an ecological and taxonomic approach has led to many useful ideas about species diversity and speciation. This volume is a fine example of the way in which existing, but still far from complete, data are being synthesised to enable us to understand, not just the defination of each of the many tropical forest species, but also to understand how this remarkable species diversity is put together into the dynamic and evolving ecosystem that we know as tropical rain forest. It presents a strong case for the health and value of contemporary tropical systematics. We just need a larger workforce. This underlines the increased importance of the few remaining institutes that are training tropical systematists. This volume will continue to keep the Botanical Institute of Aarhus University in the forefront of tropical forest studies.

Ghillean T. Prance, Royal Botanic Gardens, Kew.

Organizing Committee

L. B. Holm-Nielsen, I. C. Nielsen, and H. Balslev

Symposium on

Tropical Forests: Botanical dynamics, speciation and diversity.

held at Aarhus University, 8-10 August 1988

Session 1. Dynamics - Chairmen P. S. Ashton and P. Windisch

Session 2. Speciation - Chairmen B. ter Welle and Hu Chi-ming

Session 3. Diversity - Chairmen O. Hamann and B. Nordenstam

Contributors

Andersson, L. Institut för Systematisk Botanik, University of Göteborg, Carl Skottsbergsgata 22, S-413 19 Göteborg, Sweden (p. 173)

Ashton, P. S. The Arnold Arboretum of Harvard University, 22 Divinity Avenue, Cambridge, MA 02 138, USA (p. 239)

Balslev, H. Botanisk Institut, Aarhus University, Nordlandsvej 68, DK-8240 Risskov, Denmark (p. 287)

Barthélémy, D. Laboratoire de Botanique, Institut Botanique, 163 rue A. Broussonet, F-34000 Montpellier, France (p.89)

Bruenig, E. F. Institut für Weltforstwirtschaft, University of Hamburg, Leuschnerstrasse 91, D-2050 Hamburg 80, Federal Republic of Germany (p. 75)

Castroviejo, S. Real Jardin Botánico, Plaza de Murillo 2, 28014 Madrid, Spain (p. 347)

Chen Z-y. South China Institute of Botany, Academia Sinica, Guangzhou, Wushan, Peoples Republic of China (p.185)

Dransfield, J. Herbarium, Royal Botanic Gardens, Kew, Richmond, Surrey, TW9 3AB England, UK (p. 153)

Edelin, C. Laboratoire de Botanique, Institut Botanique, 163 rue A. Broussonet, F-34000, Montpellier, France (p. 89)

Fcuillct, C. Centre ORSTOM de Cayenne, B. P. 165, 97323 Cayenne Cédex, French Guiana (p. 311)

Geesink, R. Rijksherbarium, Rapenburg 70-74, 2311 EZ Leiden, The Netherlands (p. 135)

Gentry, A. H. The Missouri Botanical Garden, P.O. Box 299, St. Louis, MO 63166-0299, USA (p. 113)

Hallé, F. Laboratoire de Botanique, Institut Botanique, 163 rue A. Broussonet, F-34000 Montpellier, France (p. 89)

Hansen, B. Botanisk Museum, University of Copenhagen, Gothersgade 130, DK-1123 Copenhagen K, Denmark (p. 201)

Hartshorn, G. S. World Wildlife Fund, 1250 24th Street, NW, Washington DC 20037, USA (p. 65)

Haynes, R. R. Department of Biology, University of Alabama, Tuscaloosa, AL 35486, USA (p. 211)

Holm-Nielsen, L. B. Botanisk Institut, Aarhus University, Nordlandsvej 68, DK-8240 Risskov, Denmark (p. 211)

Huang, Y-w. Institut für Weltforstwirtschaft, University of Hamburg, Leuschnerstrasse 91, D-2050 Hamburg 80, Federal Republic of Germany (p. 75)

Huber, O. C. V. G. and Instituto Venezolano de Investigaciones Cientificas (I. V. I. C.), Apartado 80405, Caracas 1080-A, Venezuela (p. 271)

Irion, G. Forschungsinstitut Senckenberg, Abteilung für Meeresgeologie und Meeresbiologie, Schleussenstrasse 39a, D-2940 Wilhelmshaven, Federal Republic of Germany (p. 23)

Iwatsuki, K. Botanical Gardens, University of Tokyo, Hakusan, Tokyo 112, Japan (p. 193)

Junk, W. J. Max Planck-Institut für Limnologie, Arbeitsgruppe für tropischer Ökologie, P. O. Box 165, D-2320 Plön, Federal Republic of Germany (p. 47)

Kornet, D. J. Instituut voor Teoretische Biologie, University of Leiden, Groenhovenstraat 5, 2311 BT Leiden, The Netherlands (p. 135)

Larsen, K. Botanisk Institut, Aarhus University, Nordlandsvej 68, DK-8240 Risskov, Denmark (p. 339)

Luteyn, J. L. The New York Botanical Garden, Bronx, NY 10458-5126, USA (p. 297)

Mori, S. A. The New York Botanical Garden, Bronx, NY 10458-5126, USA (p. 319)

Nielsen, I. C. Botanisk Institut, Aarhus University, Nordlandsvej 68, DK-8240 Risskov, Denmark (p. 355)

Oldeman, R. A. A. Institute of Silviculture and Forest Ecology, Agricultural University, P.O. Box 342, 6700 AN, Wageningen, The Netherlands (p. 3)

Ortiz-Crespo, F. US-AID, American Embassy, Quito, Ecuador (p. 335)

Polhill, R. M. Herbarium, Royal Botanic Gardens, Kew, Richmond, Surrey, TW9 3AB, England, UK (p. 221)

Ramella, L. Conservatoire et Jardin Botaniques, Case postale 60, CH-1192 Chambésy/GE/Switzerland (p. 259)

Räsänen, M. Department of Biology, University of Turku, SF-20500 Turku, Finland (p. 35)

Raven, P. H. The Missouri Botanical Garden, P.O. Box 299, St. Louis, MO 63166, USA (p. 365)

Renner, S. S. Botanisk Institut, Aarhus University, Nordlandsvej 68, DK-8240 Risskov, Denmark (p. 287)

Salo, J. Department of Biology, University of Turku, SF-20500 Turku, Finland (p. 35)

Shukla, V. K. S. Analytical Research and Development, Aarhus Oliefabrik A/S, P.O. Box 50, DK-8100 Aarhus C, Denmark (p. 355)

Spichiger, R. Conservatoire et Jardin Botaniques, Case postale 60, CH-1292 Chambésy/GE/Switzerland (p. 259)

Sumithraarachchi, D. B. Royal Botanic Gardens, Peradeniya, Kandy, Sri Lanka (p.253)

Swaine, M. D. Department of Botany, University of Aberdeen, St. Machar Drive, Aberdeen AB9 2UD, Scotland, UK (p. 101)

Preface

The botany of tropical forest has been a focal point of the research and teaching of the Botanical Institute since its establishment at Aarhus University 25 years ago. We have often felt the need for more knowledge about the diverse tropical ecosystems, for example when we saw the same meager data used to redundency in the argument for the conservation of some of these areas. The temperate areas of the world have not been developed in a sustainable way but we may be learning the lesson. The tropical forest still houses an unmatched biological richness. In order to prevent the total destruction of these areas the scientific community must be urged to produce and communicate further knowledge about them.

As Raven points out in the concluding remarks of this volume 150 000 species of higher plants in the tropics are being studied by only a few hundred botanists. These species and their potential uses are not likely to be studied before many of them become extinct. Training of taxonomists has been neglected in many countries, perhaps because many schools of taxonomists have isolated themselves from the adjacent disciplines. We have tried to avoid this in Aarhus and therefore we wanted to expand the theme of the symposium on Tropical Forest held in Aarhus in August 1988 to include dynamics and diversity aspects. It is our opinion that some of the major tasks for biologists in years to come will be to achieve better understanding of the importance of such concepts as dynamics and diversity of ecosystems. It must be stressed, however, that without further basic studies of the speciation and the taxonomy of the organisms of the systems such new understanding is not likely to be achieved. The research at the Botanical Institute has been directed both towards the taxonomy of tropical plants, the floristics of selected areas and, more recently, towards the tropical ecosystems. We have furthermore developed applied research in some related areas (cf. Shukla and Nielsen, this volume) and have been engaged in development projects such as reforestation of the Andean highlands with local species.

The Symposium was arranged to bring together a comprehensive but not too large group of people from forestry institutes, ecological institutes, private industry, botanical gardens, museums and universities on the occasion of the twenty-fifth anniversary of the Botanical Institute. The 144 participants represented 24 countries. Papers were presented orally and as posters. The 56 abstracts are published in Skov and Barfod, (1988), AAU-Reports **18**, 1-46. The present volume includes the written version of the oral presentations.

We wish to acknowledge the help of the session chairmen P. S.

Ashton, O. Hamann, Hu Chi-ming, B. Nordenstam, B. ter Welle, and Paulo Windisch who made the program move smoothly, and to F. Ortiz-Crespo who gave a stimulating morning appeal (Ortiz, this volume) about the situation of the tropical forests as seen from the point of view of the people living in the tropical countries.

The papers were rewied by a number of colleauges with special knowledge of the topics treated. Their promt and critical comments were most helpful to our editing, and we are most grateful to them: L. Andersson, P. S. Ashton, A. Barfod, B. Boom, V. A. Funk, A. H. Gentry, R. Haase, O. Hamann, R. R. Haynes, N. Jacobsen, J. A. Korstgård, J. Kress, J. Kuijt, K. Larsen, J. L. Luteyn, T. V. Madsen, U. Molau, S. A. Mori, R. A. A. Oldeman, K. Rahn, T. Ray, S. S. Renner, M. Richardson, K. Sand-Jensen, W. G. Sombroek, K. Torssell, D. Wasshausen, C. Westerkamp, and B. Øllgaard. Special thanks go to S. Churchill who read through all the papers.

The scientific and technical editing of this book was carried out at the Botanical Institute, Aarhus University. Without the outstanding help of the staff the completion of this task within few months would not have been possible. The editors appreciate the fine artwork by A. Sloth and K. Tind, and the word processesing and layout by A. Thygesen and B. Højstrøm. F. Nørgaard assisted in the computing and A. Boyd was linguistic consultant.

We wish to emphasize that although we have placed the individual papers in a context, the full story about the dynamics, speciation and diversity of the tropical forests is not told. We have found that each of the published papers in its own way provides us with further ideas about the tropical forests. We have intended to bring people and thoughts together which in the further work with these important and exciting forests will fertilize the creation and transmission of new knowledge to the benefit of science and the tropical forests.

Lauritz B. Holm-Nielsen, Ivan Nielsen and Henrik Balslev

Contents

Dynamics

Speciation

Diversity

Past, Present and Future

Introduction

It used to be agreeable to many scientists that at least some areas of tropical forests existed under relatively stable climatic conditions during the geologic history in which the Angiosperms evolved.

This myth of the forest ecosystems being stable climaxes has now been challenged. The present knowledge about the dynamics of these systems and the evolution of the landscapes in which they occur now justifies the richness in niches and diversity. The understanding of the tropical forests as high-diversity systems help us explain the apparent stability often observed. Further studies will without doubt provide science with new understanding of the theory and philosophy of diversity as a concept, and of dynamics as a stabilizing element in nature

In his introductory paper to the section on dynamics in tropical forests Oldeman presents the hypothesis of ecological interference and the linkage to the architecture of forest mosaic, the building and the dynamics of forest eco-units, and the distribution patterns of habitats. The geology and geomorphology of some regions are about to be better understood (Irion and Salo), and theories on the history of the Amazon Basin, such as the refugium theory, presented by Haffer in 1974 must be revised accordingly. The ecological dynamics caused by rythmic flooding of extended areas in the Amazon Basin (Junk), is an example of natural dynamic conditions which may have influenced diversification of the forests and the speciation processes during the Pleistocene. The small scale dynamics, of natural and man-made gap structures (Hartshorn) and of canopy structure as seen during the regeneration or succession resulting from sporadic events such as forest burning caused by lightning strikes (Bruenig and Huang) are important features for the understanding of the forests. The bioarchitectural concept employed by Hallé and collaborators (Barthélémy *et al.)* may prove to be a useful tool in the further and detailed studies of the structure of tropical forests. The concluding paper by Swaine draws the attention to some of the research which is needed in order to get closer to an understanding of the dynamics of tropical forests.

The second chapter on the speciation of tropical plant groups is introduced by Gentry´s review. Though his theory of explosive evolution, examplified from Cerro Centinela in western Ecuador, where he suggests that species evolved during 15 years of isolation, may be considered only a challenging theory, it provides us with new ideas, and reminds us that our own prejudices may be the major obstacle in the search for further understanding. The theory of entropy explained and examplified from Malesian legumes (Geesink and Kornet) provides a philosophical background for the understanding of speciation in the tropics. A series of papers describe the speciation of plant groups which represent different growth forms in the tropics. The woody monocotyledons are examplified by the palms of Madagascar (Dransfield). A theory on the evolution of *Heliconia* (Anderssons) and studies on Zingiberaceae (Chen) represent the giant hebs. Evolution in tropical forest ground herbs is demonstrated by examples in *Asplenium* (Iwatsuki), Acanthaceae (Hansen), Alismatidae (Haynes and Holm-Nielsen) and in Loranthaceous pseudoparasites in (Polhill). These studies all give us the monographers ideas of how speciation contributed to the diversification of the tropical forests. Several authors remind us of how incomplete the available data sets are. Even experts have difficulties in disentangling the taxonomy. They have to deal with insufficient hypotheses on the geological and geomorphological history, they do not have representative collections at hand, the ecological requirements of the plants are unknown, the biology, cytology *etc.* are only studied in certain cases, and the studies are only representative for fractions of the populations dealt with. In spite of these and many other difficulties much more comprehensive studies than formerly are carried out during these years as the examples of this volume indicate.

The third chapter on the diversity of tropical forests is introducced by Ashton's paper on the species richness of tropical forest based on his detailed studies of the forests of Sarawak. The diversity of tropical forests is on one hand a question of the diversity of vegetation types as shown for the Sri Lankan forests (Sumithraarachchi), for the forests of the Paraguayan Chaco (Spichiger), and for the Guayanan shrublands (Huber). On the other hand data on the diversity in terms of species numbers is demonstrated from tropical rain forests of Amazonean Ecuador (Balslev and Renner). How a single plant group such as the Ericaceae adapted and contributed to the diversity of the Neotropical Montane forests is discussed by Luteyn. It is futhermore shown how diversity of the Passifloraceae in French Guiana may be faciliated by the drastic ecological changes and dynamics of tropical forests (Feuillet), whereas the study of the Lecythidaceae provides an exellent example of how the diversity of tropical trees may depend on intricate and dynamic equilibria among species in very diverse systems (Mori). Such diversity may well have arisen through the dynamics of tropical forests, but the failure of the Lecythidaceae to be reastablished in secondary forests on cleared land suggests that the overall system need to be relatively stable and dynamics may rather occur in subsystems.

The combination of factors such as long history, tropical climate, dynamics and competition gave rise to the richness in niches and species diversity. The evolution of plant species in the tropical forest have doubtlessly followed same patterns as known elsewhere. Although the principles and methods of the molecular genetics are highly elaborated today, they have hardly been introduced to the studies of tropical forest plants. Such techniques may eventually provide much better knowledge in terms of species numbers in terms of species numbersabout population structures and provide the proof to the speculative theories of speciation based on similarities among morphological species. But for a long time to come such sophisticated studies, including biosystematic studies of tropical forest species must be limited to few examples. The few taxonomists will have more than enough to do in keeping in touch with the modern methods without abandoning the necessary descriptive work. A much better understanding of the speciation processes is essential to the knowledge of the diversity of the tropical forests.

Dynamics

Dynamics in tropical rain forests

R. A. A. Oldeman

Agricultural University, Wageningen, The Netherlands

The tropical forest ecologist, to some extent is still in limbo or rather in a green hell. According to some fantastic tales, hell would be a place where normal dimensions lack their normal constancy, leaving one in a permanent state of confusion. In tropical rain forests, such a state of the mind in ecologists is at least partly due to the use of temperate forest models. These may be efficient in temperate forests, but in tropical rain forests they too often act as confusing rather than guiding devices.

One of the main causes is the sheer species richness of the tropical rain forests. This has been known since 1956, when master biologists of that time met at the Unesco symposium at Kandy, Sri Lanka (UNESCO, 1958). Even if counts are limited to woody plants, species numbers per hectare in tropical rain forests are high. They often are so high that they exceed an ill-defined but very real limit, beyond which it becomes impossible to see the forest for its trees. Such trouble is indicated by the difficulty in establishing floristic minimum areas for tropical rain forest types occuring on certain sites (for sites, see *e.g.* Junk or Bruenig and Huang or Gentry, this volume). As recently as 1987, Hommel succeeded in defining rhinoceros habitats in Ujung Kulon, Java, Indonesia by combining land evaluation techniques with phytosociological methods. Still, this region lies outside the strict rain forest zones which are richest in species.

In the methodological background of species richness confusions, there are the inheritances of agricultural and pharmaceutical thinking. For instance, in the original agricultural meaning a "stand" is a timber stand. Its surface is measured in squares such as acres or hectares, and its vegetation is assessed in terms of numbers of merchantable trees per species per square. Timber volumes and prices are then calculated with age, size and shape of these trees. The build-up of the stands themselves most often was a consequence of management. Pure stands, containing one tree species, consist of a population of trees that is kept very homogeneous. Variation then can be considered as a statistical error. On the other hand, a frequent forest utilization type is the rural village forest called "coppice with standards." It contains man-made layers, *i.e.*, the coppice layer ("shrub layer") and the layer of standards or older left-over trees ("tree layer"). Below there is a "herb layer."

TROPICAL FORESTS
ISBN 0–12–353550–6

The dynamics of these stands were recognized early. Foresters and farmers have distinguished development phases for at least two centuries, and named them after the state of the crop tree or the products that could be harvested in such a phase. Seedling phases yield nothing, pole phases contain sapling trees yielding poles, mature phases yield sawn wood and later phases have depreciating names if they occur at all, *e.g.* degradation or decay phases. In all these forest stands, trees were there to be harvested. Therefore, dead fallen trees were rare and if present were due to a "disturbance" or "catastrophe," both terms referring of course to a management plan or intention rather than to the ecosystem itself.

The stands, their structure, architecture and (wood) productivity were assessed according to the productivity of the soil, defining together with other environmental factors a series of site classes or productivity classes. The dynamics of pure stands were expressed in yield tables, showing the statistical averages of tree number, stand height, mean diameter and average volume over time. This approach was mainly of German and French origin. It was brought to the tropics in colonial times, for instance by Dutch foresters educated in Germany or German foresters like Sir Dietrich Brandis who worked in the British Empire. Many of the actual production models are computerized versions of improved yield tables.

The second inheritance, from the pharmacists, came to our days mainly through taxonomy from its ancestral apothecary monks. In the pharmaceutical profession it is not the crop yield that matters, but the quality and quantity of a medical drug to be applied. This is closely linked to the precise identity of the drug procuring organisms. In its turn, the identity is established by a combination of *gestalt* perception and analysis of form. This led to a preoccupation with "natural" shapes, forms and patterns quite different from the agricultural squares and grids and unburdened by quantitative aspects required with yield calculations.

This short historical reflection is indispensable in order to understand the present book, particularly the section on forest dynamics. All authors, as all scientists in this field, are still wrestling with at least some of the inherited concepts. These sometimes are so implicit that they can only be traced through the axioms or basic rules that are tacitly assumed.

It may seem useless to question assumptions such as the usefulness of grids and squares in calculating important ecological processes or patterns, or the utilization of stochastic models for population shifts. Still, in declaring such assumptions indispensable, one adopts a method and a model. And this adoption means that problem statements must be adapted to the scientific method, instead of doing it the other way round. It is very important to pursue this topic in the present Symposium, where taxonomy and ecology meet.

Form, numbers and production

The basic scientific image of an organism still is its taxonomical description. This is needed in order to place in the natural realm all observations at lower levels, like chromosome counts or phenomena in tissue physiology, and at higher levels than the organism, such as species counts or interaction studies. A taxonomic description depends on morphological exactitude. The observation and measuring of forms extends from biochemistry (*e.g.* L- and R- molecules), to whole organisms (for trees, see Barthélémy *et al.*, this volume).

In combination with physiology, the morphological approach yields explanatory models for the organisms. For instance, the carbon processing in a plant as a system can be explained in terms of the functioning of its leaf-bearing branches and roots bearing root-hairs, as subsystems. As the terms indicate, this is a case of system analysis. The usual sequence of systems and subsystems used is the hierarchy of genes, chromosomes, nuclei, cells, tissues, organs and organisms. In architectural analyses of trees, at least three system levels are added (Barthélémy *et al.*, this volume), *i.e.*, organs, organ complexes (*e.g.* organ-bearing shoots), branched complexes (*e.g.* conforming to an architectural unit), and reiterated or metamorphic complexes (*e.g.* a very large reiterated tree).

As for systems analyses, this addition shows that some levels may be left out if inappropriate. For instance, in the case of a palm, the levels of branching and reiteration may go unused, because the species shows only one organ bearing stipe. This exemplifies the general rule that individualization of organisms may have taken place at all levels, from the genetic one (viruses), through the cellular (unicellular organisms) and following levels to the reiteration level (large dicotyledonous forest trees). With this in mind, the speciation chapter by Gentry (this volume) and the considerations on co-evolution by Ashton (this volume) may be looked at in the context of natural system hierarchies (Grene, 1987).

In particular, the "environment" exerting selection pressure upon organisms and which sometimes has prematurely been defined as an ecological niche, expressed as a multidimensional vector, can be pinpointed in a much more precise manner. For instance, the environment of a cell is either determined by the surrounding cells in a tissue to which it belongs, or is codetermined by interactions with other cells in case of free-living unicellulars. These other cells can be free-living or belong to a tissue in a larger organism without changing the above statement. More generally, the direct environment of an organism is the system immediately above its level of individualization (Oldeman, in press). This is another way of expressing one of the principles of systems analysis, *i.e.* a system is explained in terms of interaction among its subsystems.

Morphology and physiology are the basis of explanatory systems models at or below the organism level in the hierarchy of living systems. In the usual approach, floristic/faunistics and production ecology are the fundamentals of the final environment of the largest organism, which is either a biological community or an abiotic site, or a combination of both.

Swaine (this volume) shows how the first step usually is made, by counting species populations and their behavior pattern per hectare or per year. Like Alexandre (*e.g.* 1977, 1978, 1986), he carefully investigated the behavior of these trees with references to what is generally thought to be the most important factor, *i.e.* light. Tree behavior with reference to environmental factors has been called *temperament* in continental European forestry for centuries. For the tropics, the issue is being hotly debated nowadays, whether rain forest trees have a temperament or not. From a simple dichotomy in pioneer and non-pioneer trees (*e.g.* Whitmore, 1988), through more elaborate schemes (*e.g.* Oldeman and van Dijk, 1988) to the idea of a continuum of temperaments from sun-loving to shade tolerant trees (*e.g.* Bongers and Popma, 1988), the reaction of trees to their abiotic environment is both recognized as important and still remains unclear.

Two important aspects of this approach must be emphasized: (i) counts are often limited to trees or woody species, in the humid tropics, this is necessary in order to keep the species numbers to be processed manageable, at least in ecosystem models that rest on population counts; (ii) species are grouped in larger compartments like pioneers or shade tolerant trees. This simplifies the models still further, but mainly allows us to link population dynamics and light dynamics in the forest. If there is evidence of heavy climatic or biotic impacts upon the forest, resulting in regular partial elimination of its crown canopy, the proportion of pioneer tree populations is higher than under a lower impact regime.

This leads further to the impacts, their regularity and effect. For reasons mentioned above, the terms "disturbance" and "catastrophe" are avoided here because ecologically speaking such impacts are nothing else but perfectly regular and natural dynamic processes. Bruenig and Huang (this volume) as well as Ashton (this volume) considered the effects of such impacts upon the forest and its species richness. One of the important things their studies prove is that the consequences of the impact regime are closely linked to the "hospitality" of the abiotical site, as postulated by Oldeman (1983a). More fertile soils without too many hydrological (drought) or topographical (slopes) stress factors show an impact pattern with less large-sized gaps than elsewhere and higher average species richness. The diverse indicators for diversity emphasize this.

Notwithstanding the kaleidoscopic character of the impact-caused mosaic that characterizes the tropical rain forests according to Oldeman

(1974a) and Gentry (this volume), these patterns are most often considered as the heterogeneity of one stand (Bruenig and Huang, this volume). This corresponds to the classical Central European view of forests which are harvested in a "nature-conforming way," tree-wise selection cutting, after which the forest regeneration in the one-tree gaps is carefully tended (see Hartshorn, this volume, for tropical America; or de Graaf, 1986; Poels, 1987; Jonkers 1987). Stands treated in this way have been interpreted usually as mixed stands, with mixtures of species and ages, and not as mosaics of small stands.

In Amazonia, the hospitality of sites is linked to the flooding pattern, as is well established by Junk (this volume), and this in turn sheds light on long-term site developments which are linked to evolutionary selection (Irion, this volume). These long-term changes and movements in the hydrological state of the lowlands, linked to river and sea dynamics, should perhaps not be regarded as evidence against the Pleistocene refugium theory (*cf.* Gentry, this volume) but rather as a supplement to that theory, for which new evidence has recently been found in the central parts of Amazonia (van der Hammen, pers. comm., 1988).

In both evolutionary hypotheses, the character of the selecting environment is not considered. If annual floods cause the development of annual biorhythms (Junk, this volume), a direct selective impact of abiotic factors on perennial species is assumed, without considering the interactions within the biological community which may easily lead to rhythms being cancelled amongst mutually interacting species. The role of the direct environment, mentioned above, is left out. On the contrary, Ashton (this volume) has concentrated upon these intermediary environments as have Bruenig and Huang (this volume).

Intermediate environments are all linked to the pattern of impacts and mortality upon a forest carpet, covering a region.

The architecture of forests: eco-units and silvatic mosaics

The numerical data on species populations and the emphasis on the importance of site characteristics, which are a prerequisite for ecological production models, lead to an image of large, heterogeneous, mixed stands in which species richness, diversity and, to some extent, distribution can be expressed mathematically with the aid of statistical models. It is by this approach that Ashton (this volume) seeks to answer his two important questions about the deterministic character of floristic structure and specialization of component species of tropical rain forests.

Since 1974 (Oldeman 1974a,b) another approach adding architectural criteria to the numerical and production ones has been tried.

Traditional forest transect drawings are sketch-maps, illustrating the image of the forest from beside and from above as a crown projection. They are said to represent forest physiognomy (cf. Bruenig and Huang, this volume), which is a property that is not liable to be analysed. Figure 1 shows an analytical type of transect sketch that summarizes many tropical data (e.g. Oldeman in Jacobs, 1988; in press).

Barthélémy et al. (this volume) defined the means to analyse this transect. The trees form the skeleton of the forest. Therefore, a first diagnosis of forest architecture rests on an explanation in terms of tree architecture. According to this architecture one can define (i) *potential trees* (or trees of the future) which have still a potential for height growth, crown expansion or both, but which may also die young; (ii) *trees of the present* which have a potential for extension or expansion left, their size being the maximal one at the given site, but their sustenance being very durable because of an ability to replace lost parts; (iii) *trees of the past* which have been damaged or died back beyond the point of no return, and which are dead, dying, decaying or a combination of some of these states.

On the long but little detailed transect (Fig. 1A), tropical rain forest is shown as a mosaic consisting of bits and pieces of forest of different sizes and in different stages of development. As explained elsewhere (Oldeman 1983b), these bits and pieces are not gaps (*chablis* - Oldeman 1978) because only some of them display this destroyed canopy state, whereas the term patch is too little specific. If the inclusion of form criteria in the analysis is taken seriously and not treated as a corollary of numerical data, figure 1B shows three growth phases of an authentical living system, characterized by its form like a plant illustrated in a flora.

Figure 1. Tropical rain forest dynamics. **A** - Silvatic mosaic, composed by eco-units of different sizes and shapes, built by different species groups. *Chablis* denotes the innovation phase ("gap-phase") of an eco-unit, when the propagule bank is mobilized. **B** - Development phases of one eco-unit. The *innovation phase* is not shown, the sequence here considered showing the *aggradation phase* (growing, Ba), the *biostatic phase* (long stable, mature, Bb) and the *degradation phase* in which reorganisation is taking place (no innovation here). Note that diagnosis of the phase rests on the state of subsystems: trees of the past (stippled), of the present (cross hatched) and potential trees (blank). Note associated organisms (epiphytes, climbers), and thick border zone around eco-unit.

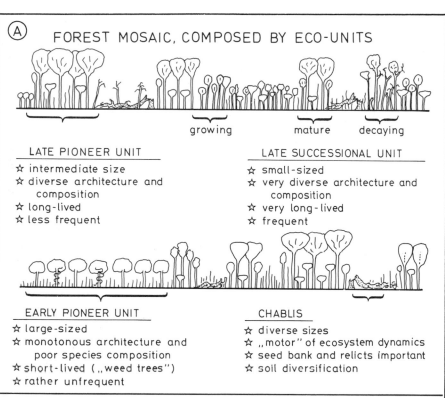

A FOREST MOSAIC, COMPOSED BY ECO-UNITS

growing   mature   decaying

LATE PIONEER UNIT
- ☆ intermediate size
- ☆ diverse architecture and composition
- ☆ long-lived
- ☆ less frequent

LATE SUCCESSIONAL UNIT
- ☆ small-sized
- ☆ very diverse architecture and composition
- ☆ very long-lived
- ☆ frequent

EARLY PIONEER UNIT
- ☆ large-sized
- ☆ monotonous architecture and poor species composition
- ☆ short-lived (,,weed trees")
- ☆ rather unfrequent

CHABLIS
- ☆ diverse sizes
- ☆ ,,motor" of ecosystem dynamics
- ☆ seed bank and relicts important
- ☆ soil diversification

B. One eco-unit in time

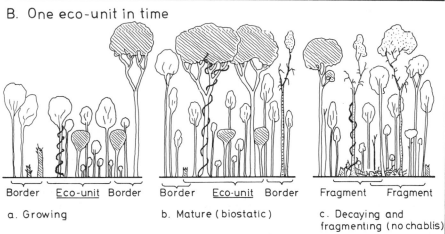

Border | Eco-unit | Border   Border | Eco-unit | Border   Fragment | Fragment

a. Growing

b. Mature (biostatic)

c. Decaying and fragmenting (no chablis)

The growth processes can be diagnosed and explained by the state of the tree components of such a system. If there are mainly potential trees, they compete and grow rapidly towards the canopy (Fig. 1Ba). When some trees have reached the present state, these dominate the architecture of the whole system and the remaining potential trees are suppressed risking premature mortality (Fig. 1Bb). When the present trees decay and become trees of the past, the whole system will be replaced by one or more others (Fig. 1Bc). The sequence of such a system is to pass through an *innovation phase*, the trees being mainly seedlings, an *aggradation phase* (fig. 1Ba), a *biostatic phase* (Fig. 1Bb) and a *degradation phase* (Fig. 1Bc).

Like Edelin coined the term *architectural unit* as a generic term for the basic architectural building block in tree architecture (Barthélémy *et al.*, this volume), Oldeman (1983a,b) coined the term *eco-unit* for the living system represented in figure 1B. Forest eco-units are circumscribed as: *"all surfaces on which at one moment in time a vegetation development has begun, of which the architecture, ecophysiological functioning and species composition are ordained by one set of trees until the end"* (Oldeman, in press). The replacement of these trees therefore automatically provokes the death of the eco-unit.

Although both the architecture and the functioning of a forest eco-unit is determined by its trees, its functions depend on all the organisms which live within its limits or which pass through it and serve as carriers of inputs and outputs. Forest eco-units can be divided according to size. Large eco-units are often built by pioneer trees, but they are less frequent than small eco-units, which are built by slow growing shade tolerant tree species. The size of the eco-unit is important and so is the form. Oldeman (1980, 1983b) gave some illustrations of large and small eco-units, including a small one in the undergrowth without any opening of the forest canopy (*cf.* also Gentry, Ashton, this volume).

If a forest is considered from an agricultural point of view, it is a collection of trees. In that case, the heterogeneous stand model would be very convenient. But a forest contains countless non-arborescent species of organisms against a limited number of tree species. Viewed from the side of the smaller-sized and most numerous species, an eco-unit is a type of ecosystem that is more liable to be understood as a collection of coherent habitats than some arbitrary square like an acre or a hectare of forest. Now can an eco-unit, as distinguished by its architecture or *form*, be considered as a living system being something more than a figment of the imagination?

Because form is the prime criterium, the spatial limits of an eco-unit have to be established first. These limits are a border layer which is between the inside of the system and the outside. The upper limit is formed by the crown canopy, in particular the leaf mass. The lateral border layers are transition zones towards the neighbouring eco-units, which are generally clearest in the innovation, aggradation and degradation phases (Fig. 1B). The lower border is hidden in the ground, formed by roots and their associated organisms, and unknown as to its precise architecture up to now. In studies of energy and nutrient budgets, these border zones are the ones where gradients, *e.g.* light or humidity gradients, are steep.

Four other attributes than the border layer have been distinguished as essential properties in a living system (also see Boyce, 1978). These are the transfer of information within the system, the orderly transfer of nutrients and energy in the system, its adjustment to changes in the direct environment and its multiplication. These attributes are not properties of the component organisms but of the eco-unit itself. Therefore they consist of interactive processes which are properties neither of one interacting organism, nor of the other.

These attributes have been extensively dealt with by Oldeman (in press). Information in eco-units is processed through regulation of physical factors, *e.g.* light and water distribution, and through chemicals playing a role roughly but not closely comparable to hormones in organisms. These information-carrying chemicals in ecosystems are called *impellors* (latin for "drivers") and include humus decomposition products like polyterpenes or polyphenols, plant attractants, pheromones and other substances including "poisons." Energy and nutrient transfer takes place in an ascending direction within living plants, in a descending way it passes through the litter cycle with its myriads of associated organisms, and laterally it is often carried by animals (inputs and outputs of the eco-unit).

Adjustment of eco-units to changing environmental factors also is a question of interactions among the component organisms. For adjustment, it nearly always leads to changed resource allocation patterns to different organisms. Information transfer leads to changing patterns in energy and nutrient transfer, so that some organisms profit more, others less than in an earlier state of the eco-unit. Examples are "pests," when production capacity is shifted from crop plants to insects, or grass proliferation under stands with impaired tree crowns letting through much light, so that more energy is allocated to grass and less to trees.

Finally, eco-units reproduce. If a new eco-unit is opened (a "gap" is created) in a certain forest type, everyone knows empirically that the forest growing back in that volume belongs to the forest type in question and can be recognized as such. This is due to an information-carrier containing the instructions needed to build such an eco-unit. The information-carrier is present in the form of the propagule bank, *i.e.* the collection of eggs, spores, seeds and other propagules in the soil (*cf.* Ng *in* Sutton *et al.*, 1983). As in all biological information stores, the vast majority of carriers die. This has been established for seeds like for gametes. The information which is left has to be mobilized by a key-event, fertilization in case of gametes, some opening impact in the case of an ecosystem. The *mechanisms* of ecosystems and organisms have nothing in common. The reproductive *function* within these living systems is an analogue.

The attribute basic for all others is the existence of a border layer. If this is lacking, the analysis of species composition, information transfer or production processes refers to an arbitrary volume. This indeed is the assumption in all ecological production models, from the polygons examined by Odum and Pigeon (1970) to the circular gaps used in the models described by Shugart (1984). This latter author implicitly assumes "forest patches" to be separated from each other by some border layer, because he explicitly considers them as islands, being liable to analysis in terms of island biogeography. All forest models developed by Shugart and colleagues deserve the attention of tropical biologists, because they originate from species-rich forests of the warm temperate zone in the Appalachians, USA and have been applied successfully in the even more species-rich rain forests of Queensland, Australia.

Gap models have been developed for the tropical rain forest during more than fifteen years (*e.g.* Oldeman, 1978). The main controversy in these models has become the question of their floristic identity. If this had been resolved, many calculations and correlations used by Ashton, or Gentry, or Bruenig and Huang (this volume) would have been superfluous. But, in 1982 Orians denied that tree-falls influence species richness in the tropics. Brokaw and other authors (*in* Leigh *et al.*, 1982) found no clear-cut relationships between gap characteristics and floristics including the effects of plant-animal relations, and in 1988 authors like Popma and Bongers or Alexandre (unpubl.) doubted the existence of clear links between gap characteristics and tree temperaments. It seemed that all evidence pointed to the necessity of a return to the classical mixed-stand models. Only one ecological group of trees with the light-demanding "pioneer" temperament would remain clear, the others being just forest inhabitants. The time seemed ripe to return to the results of the Kandy Symposium (UNESCO, 1958).

Architecture and dynamics of silvatic mosaics: ecological interference

The way out of the dilemma is to consider the eco-units not as islands, but as interacting subsystems of the silvatic mosaic. This mosaic shows comparable gap dynamics under comparable site conditions (Bruenig and Huang, this volume). However complex the impacts that cause the opening of new eco-units may be (*e.g.*, Bruenig, 1987), their result is a mosaic with predictable dynamics. Given that the impacting factors often depend on climate or distant biotic events such as migration of pests, the prediction of mosaic dynamics is by nature stochastic. The *probabilities* involved then concern the reaction of the forest system, which is not stochastic because it follows a growth programme, its interaction with the soil resource which also is structured and only shows slow change, and the composition of the propagule bank as long as the forest is not greatly disturbed.

On the contrary, the impacts themselves are most stochastic. Still, some rules can be formulated which apply to such impacts. Let us distinguish two extreme cases: (i) *the sweeping event* which is rare, affects large surfaces and involves the application of large quantities of raw energy like cyclones, earthquakes, large fires, volcanic eruptions or armies of bull-dozers; and (ii) *the buckshot event* which is frequent, affects small forest surfaces of one or a few tree crowns and involves the application of reduced gusts of concentrated energy, like gusts of wind, lightning strokes, whirlwinds, local mud pockets favouring uprooting, or selective wood harvest (*e.g.* Hartshorn, this volume).

In reality, the impacts may be placed most often between these extremes, whereas different factors work out differently, like cyclones and large fires leaving quite different situations behind (Pickett and White, 1985).

The simplest event that can be thought of is an uneven-aged mortality in tree populations. This is used in figure 2, in which two phenomena are visible that determine the architecture of a forest mosaic at any one moment. These are: (i) *fragmentation*, the breaking up of a large eco-unit into smaller ones, for whatever cause there may be; and (ii) *fusion of eco-units*, when the youngest catches up with the oldest one and the whole mosaic gives the illusion of being one larger eco-unit.

Figure 2A shows *early fragmentation*, occurring shortly after the establishment of pioneer herbs mixed with woody seedlings, in which the role of lianas in creating fragmentation (central eco-unit) illustrates one of the reasons, which is heterogeneity of the propagule bank. Site micro-heterogeneity is another cause of early fragmentation. Figure 2B shows *late fragmentation* of a silvatic mosaic resembling one large eco-unit, because of earlier fusion. This can be seen from the different crown depth of the trees.

It is to be emphasized that strata in forests are of two different kinds. In one eco-unit they are determined by the trees of the present (Fig. 2C). This is sometimes indicated as "horizontal canopy closure." In silvatic mosaic they are on the contrary determined by eco-units of different height and architecture that are adjacent and convey to the mosaic a "stepped" architecture (Fig. 2D). This is sometimes called "vertical canopy closure." Both are the result of forest dynamics, the first one belonging to the biostatic phase of an eco-unit (Fig. 1B) and the second one to a certain

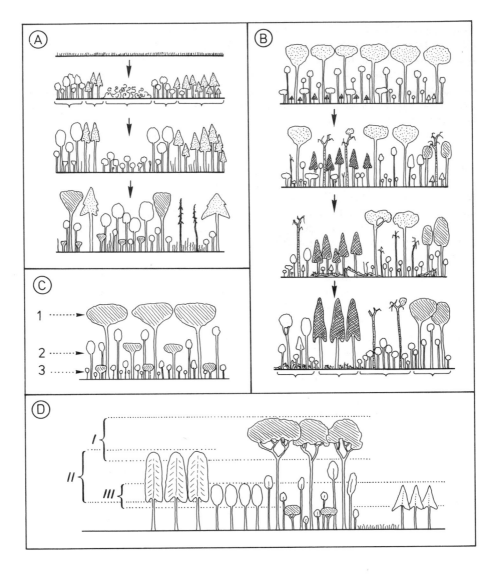

Figure 2. Fragmentation in forest mosaics and strata at different hierarchical levels. **A** - early fragmentation of a herbaceous, large-sized, pioneer eco-unit, note the liana-induced small eco-unit in the middle and the resulting mosaic state with composition of small eco-units. **B** - late fragmentation of a eco unit-like mosaic in which fusion of eco-units has taken place earlier (see difference in crown depth). Results comparable with A. Large-scale, sweeping impacts may establish one large-sized eco-unit again, which will fragment later. **C** - strata in one eco-unit are horizontal layers of crowns of the present (1, 2, 3), horizontal closure. **D** - strata in a forest mosaic are collections of eco-unit canopies at the same height level (I, II, III), the German *Stufung* or vertical canopy closure.

stage of a sufficiently ripe silvatic mosaic (Fig. 1A). These differences are nearly never mentioned by authors "placing" non-arborescent organisms at some level or stratum in tropical rain forests.

In this image of the forest, which is the same for all forests including its variants in tropical rain forests, the structural and the stochastical processes can be well balanced. Forest architecture yields a diagnosis of the eco-unit pattern in a mosaic and the kind of strata to be found predominantly. Variables like size and shape of the eco-units and the state of observed crowns show the structure of the coherent and structured habitat patterns for non-arborescent forest components.

Figure 3 shows that this corresponds to reality. It represents a rain forest in Queensland, Australia, where eco-units in the aggradation phase can be seen to be distributed in a network among the islands of biostatic eco-units with their huge and very reiterated crowns. Such documents already may serve to pinpoint the most probable habitats of animals living in the canopy and eating fruits.

Figure 4 shows the same situation as seen from below one of the openings in the canopy, *i.e.* from the bottom of a very young eco-unit, illustrated by a fish-eye photograph made in French Guyana. This image demonstrates that the border zones of eco-units are far from being impermeable to light for instance , and that the pattern of light intercepted or transmitted by the neighbouring eco-units is a living, structured reality. The light spots towards the border of the photograph show such light sources from elsewhere in the forest.

As a first approximation, these interactions based on differential light interception could be stylized as the transmission of light through horizontal and vertical grids. The interference patterns resulting from light waves passing through grids are classical in physics. The image becomes less regularly elegant below the rain forest canopy, because the grids has irregularly spaced holes of different sizes, the vertical lattice has still other properties and the light source is traveling from east to west and nearly never is vertically above the forest.

The sun does not only radiate light, but also heat. The interference patterns therefore should also have some bearing on the atmospheric humidity. Wind patterns are determined by canopy roughness. And finally, fallen trees interfere with each other by creating patterns of mineral soil, at places of uprooting mostly, and surfaces rich in organic matter, *e.g.* where trunks or crowns have come down. Interference then may be expressed as local organic matter content, being the sum of the local imported and exported quantities.

Ecological interference patterns, because of this addition-and-subtraction principle, can create niches, in the n-dimensional sense, for species in places where one would not a priori expect such species. Koop (1981) established this for herbaceous plants in North-Western Germany

Figure 3. Real mosaic as seen from the air in Queensland rain forest, with "islands" of biostatic eco-units (see large reiterated crowns of the present) separated by a network of innovation and aggradation phases of eco-units.

(Urwälder Neuenburg and Hasbruch), which were locally found in such mini-biotopes within forest communities where they "ought not" to grow according to classical vegetation science. Interference here was linked to treefalls and organic matter turnover and also to a certain limit of available light.

Current research in Fontainebleau Forest Reserve (France) shows certain tree growth phenomena that would be difficult to explain except in terms of ecological interference (De Kort, Koop, pers. comm.). The places where Florence (1981) found pioneer trees in and around the *chablis* that he investigated, with reference to isophotic lines that he mapped in percentages of macroclimatic light, point in the same direction for a very dynamic tropical rain forest in Gabon.

Figure 5 gives an impression of the hypothetical interference line for added and substracted light, and the derived places where light-demanding pioneers and shade-tolerant seedlings would occur.

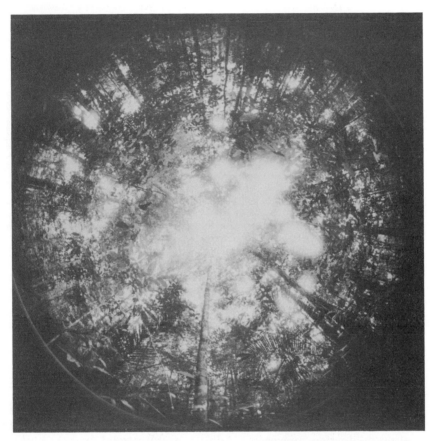

Figure 4. Same kind of mosaic as in figure 3, seen from below in French Guyana with fish-eye taken at 1.20 m height. Note the bright spots of light sources, where light inputs in interference patterns have reached high sums of intensity at the place of the camera.

Conclusion

The proposed hypothesis of ecological interference, linked to the architecture of the silvatic mosaic, the build-up and dynamics of its eco-units, and the distribution pattern of habitats is the way out of the dilemma that there is an undoubted relation between "gap dynamics" and species richness and diversity (*e.g.* Gentry, Ashton or Bruenig and Huang, this volume),

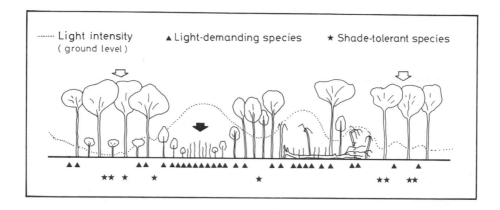

Figure 5. Schematical transect on which the resultant light intensities in an interference situation like figure 4 have been indicated. Light-demanding species (▲) not only grow in the apparent open places, but also in spots of light sums. Shade-tolerant species (★) show the inverse pattern. Open arrows: habitats for frugivores. Black arrow: high leaf production, hence habitat for herbivores. This model unites architectural criteria (mosaic build-up) with production criteria (*e.g.*, light intensity/photosynthesis) and floristic criteria (species counts and distributions). In this way, most existing theories and hypotheses can be linked together (*e.g.* Ashton, Gentry, this volume).

whereas there is certainly no floristic identity per eco-unit in the tropics. This is not so in the temperate forests (Barkman, 1973), where *microcoena* were defined, so that the floristic identity of certain eco-units seems indeed to exist outside the tropics.

This hypothesis permits to the mutual adjustment of the chapters of the present book mentioned above, and which have mostly chosen the population/production approach. Adding architectural criteria yields a map of the forest mosaic (Bruenig and Huang, this volume; Fig. 5) which can be used to draw conclusions about species richness and diversity. But on the other hand, the ecological interference hypothesis allows the linking of architectural pattern solidly to production factors and their distribution in the forest mosaic, composed by vividly interacting eco-units.

In this context, apparent contradictions can be resolved and explained as different aspects of the same truth, as suggested by Gentry (this volume) for evolutionary models implying either refugia or hydrological dynamics in the Amazonian models. Equilibrium and non-equilibrium approaches of the tropical rain forest (Ashton, this volume) can be seen as emphasizing short-term aspects of long-term processes and the inverse, respectively. Scaling in time can never be seen separately from spatial scales, and the long-term is rather linked to silvatic mosaics whereas the short term belongs at smaller spatial realms, *e.g.* the eco-unit and the component organisms.

When considering the *patria* of a species and hence the sympatric, parapatric and other aspects in speciation (Gentry, this symposium), it is certain that the concept of immediate environment, which is linked to the systems hierarchy proposed in the present chapter, can serve to increase precision in the image of *patriae*.

Though imperfectly verified, the ecological interference hypothesis shows that there are intellectual instruments which explain earlier and more modern theories as well as seemingly opposite views simultaneously. We may be constrained to believe less in concepts that we worked with during long years and to see them in another context. In our view, this is necessary if a proper theoretical framework is to be erected that is operational enough to design methods for saving the living treasure-houses which are the tropical rain forests.

Literature cited

Alexandre, D. Y. (1977). "Régénération naturelle d'un arbre caractéristique de la forêt équatoriale de Côte d'Ivoire: Turraeanthus africana. " *Pellegr. Oecologia Plantarum* **12(3)**, 241-262.

Alexandre, D. Y. (1978). "Observations sur l'écologie de Trema guineensis en basse de Côte d'Ivoire." *Cah. ORSTOM, ser Biol.* **13(3)**, 261-266.

Alexandre, D. Y. (1986). "Croissance et démographie des semis naturels en forêt de Tai (Côte d'Ivoire)." *Mém. Mus. Nat. Hist. Nat., sér. A (Zool.)* **132**, 193-200.

Barkman, J. J. (1973). "Synusial approaches to classification." pp. 437-491 *In* Whittaker, R. (ed.), "Handbook of vegetation science, V." *Junk Publ. The Hague.*

Bongers, F. and Popma, J. (1988). "Trees and gaps in a Mexican tropical rain forest." *Dr. thesis, Rijksuniv. Utrecht*

Boyce, S. G. (1978). "Theory for new directions in forest management." *USDA, For. Serv. Res. Pap.* **SE-193** 19 p. *Asheville, North Carolina.*

Bruenig, E. F. (1987). "The forest ecosystem: tropical and boreal." *Ambio* **16(2/3)**, 68-79.

Florence, J. (1981). "Chablis et silvigénèse dans une forêt dense humide sempervirente du Gabon." *Thèse de Spécialité, Univ. Strasbourg .*

Graaf, N. R. de, (1986). "A silvicultural system for natural regeneration of tropical rain forest in Suriname." *PUDOC, Wageningen.*

Grene, M. (1987). "Hierarchies in biology." *Amer. Sci.* **75,** 504-510.

Hommel, P. W. F. M. (1987). "Landscape ecology of Ujung Kulon (West Java, Indonesia)." *Publ. by the author, P.O. Box 98, 6700 AB, Wageningen.*

Jacobs, M. (1988). "The tropical rain forest. A first encounter." *Springer, Heidelberg.*

Jonkers, W. B. J. (1987). "Vegetation structure, logging damage and silviculture in a tropical rain forest in Suriname." *PUDOC, Wageningen.*

Koop, H. (1981). "Vegetatiestructuur en dynamiek van twee natuurlijke bossen: het Neuenburger en Hasbrucher Urwald." *PUDOC, Wageningen.*

Leigh, E. G., Rand, A. S. and Windsor, D. M. (eds.), (1982). "The ecology of a tropical forest: seasonal rythms and long-term changes." *Oxford Univ. Press, Oxford.*

Odum, H. T. and Pigeon, R. F. (eds.), (1970). "A tropical rain forest." *Off. Inform. Serv. U. S. Atomic Energy Commission, Oak Ridge*

Oldeman, R. A. A. (1974a, 2nd ed.). "L'architecture de la forêt guyanaise." *Mémoires ORSTOM,* **73.**

Oldeman, R. A. A. (1974b). "Ecotopes des arbres et gradients écologiques verticaux en forêt guyanaise." *La Terre et la Vie,* **28(4),** 487-520.

Oldeman, R. A. A. (1978). "Architecture and energy exchange of dicotyledonous trees in the forest." pp. 535-560 *In* Tomlinson, P. B. and Zimmerman, M. H. (eds.), "Tropical trees as living systems." *Cambridge Univ. Press, London, New York.*

Oldeman, R. A. A. (1980). "Grondslagen van de bosteelt." *AUW-Silviculture, Wageningen.*

Oldeman, R. A. A. (1983a). "On rural silvicultural systems." 3 poster papers. *Intern. Symp. Let there be forest,Wiersum.*

Oldeman, R. A. A. (1983b). "The design of ecologically sound agroforests." Chapter 14. *In* Huxley, P. (ed.), "Plant research and agroforestry." *ICRAF, Nairobi.*

Oldeman, R. A. A. and Dijk, J. van (1988). "Tree characteristics, silvigenesis and architectural diversity." *Biology International, spec. issue* **18,** 18-22.

Orians, G. H. (1982). 'The influence of tree-falls in tropical forests on tree species richness." *Trop. Ecol.* **23(2),** 255-279.

Pickett, S. T. A. and White, P. S. (eds.), (1985). "The ecology of natural disturbance and patch dynamics." *Academic Press, Orlando.*

Poels, R. L. H. (1987). "Soils, water and nutrients in a forest ecosystem in Suriname." *Dr. thesis, AUW, Wageningen,*

Shugart, H. H. (1984). "A theory of forest dynamics: the ecological implications of forest succession models." *Springer, Heidelberg.*

Sutton, S. L., Whitmore, T. C. and Chadwick, A. C. (eds.), (1983). "Tropical rain forest: ecology and management." *Blackwell, Oxford.*

UNESCO (1958). "Study of tropical vegetation: proceedings of the Kandy symposium." *UNESCO, Paris.*

Whitmore, T. C. (1988). "Forest dynamics and questions of scale." *Biology International* **18,** 13-17.

Quaternary geological history of the Amazon lowlands

G. IRION

Forschungsinstitut Senckenberg, Wilhelmshaven, FRG

Occurrence and distribution of plants and animals is explained in many parts of the world by climatological changes and by changing sea-levels during the Pleistocene. Large parts of the world's continental surface were reshaped during that time. Northern Eurasia and North America are characterized by landscapes resulting from glacial activity during the Pleistocene. Sediments along more than 10 000 km of the world's coastal and island areas were deposited during the last sea-level rise.

In North America and northern Europe, pollen analyses of peat bogs, lake sediments, together with ^{14}C-dating and radio isotope analyses contribute important information for understanding the geological and climatological history of the Quaternary. Details of plant migrations are known and paleotemperature curves have been drawn. Considering the excellent knowledge of temperate Quaternary geology, surprisingly little is known about the tropical lowlands during this period of the earth's history. The climatological history of the Amazon basin, which is the largest tropical lowland area of the world, is disputed. Some workers assume an arid to semiarid climate during the Pleistocene cold periods (*e.g.* Damuth and Fairbridge, 1970), while others do not find any proof of climatic changes (Irion, 1982). The main difficulty is that suitable deposits for studying Pleistocene history do not occur or are difficult to find in the Amazon lowlands. This is in contrast to the well developed sediments found at higher latitudes, in particular glacial deposits, lake sediments, or organic deposits dating back to Pleistocene cold periods.

Geological framework

Northern South America is characterized by the contrast between the very old Precambrian shields south and north of the lower Amazon and the relatively newly formed Andean arc. The Andean arc is a high mountain range emerging through the collision of the Nazca and South American plates since the Miocene. Most of the western parts of the Amazon lowlands are covered by sediments eroded from the Andes. These sediments

TROPICAL FORESTS
ISBN 0–12–353550–6

reach thicknesses of more than 1000 meters. In the basin area between the Precambrian shields, sediments have accumulated since the Paleozoic. The surface of this area is formed by Cretaceous sediments, and outcrops of Paleozoic sediments occur only at the edges of the shield.

According to Grabert (1983), the pre- and mid-Tertiary Amazon River was separated in two branches of which one entered the Pacific and the other the Atlantic. The formation of the modern Amazon drainage system is assumed to have taken place in the late Tertiary or early Pleistocene.

Erosion rates in the Andes are extremely high, reaching values of almost 1000 tons per square kilometer per year (NEDECO, 1973). Most of the material eroded from the eastern slope of the Andes is deposited in the sub-Andean sedimentary basin where it forms rapidly changing river systems (Salo *et al.*, 1986). Each year the Amazon and Madeira rivers carry 1.2 billion tons of river suspended and bedload material into the Atlantic ocean (Meade *et al.*, 1985). The amount of sediments delivered to the lowlands may not have changed fundamentally since middle to late Quaternary, but the amount reaching the delta is controlled by sea-level changes and varies to some extent. Damuth and Flood (1984) estimated that about 700 kubic kilometers of sediments have been deposited in the Amazon deep-sea fan since the Miocene.

Floodplain-formation during Pleistocene high sea-level

To understand the Quaternary geology in the Amazon basin, the study of its morphology is one of the most important prerequisites. At Iquitos, 3600 km from the Atlantic ocean, the level of the river is about 80 meters above recent mean sea-level, and at the confluence with Rio Negro 1500 km from the sea, it is 23 meters (Fig. 1). Large parts of the upper and middle Amazon basin do not exceed elevations of more than 100 meters above recent mean sea-level. West of Santarém, which is 600 km from the Amazon River discharge into the Atlantic Ocean, the 100 meters contour line encloses an area of 1.3 million square kilometers with its largest proportion between 60° and 70° western longitude. In the lower Amazon the relief is higher, in some places reaching 300 meters elevation.

During high sea-levels, the ocean dammed up the waters of the Amazon drainage system, affecting the water tables far away from the coast. The sea-levels during the Pleistocene are subject to controversy, but it may be assumed that maxima of at least some tenth of meters above recent mean sea-level occurred. Klammer (1971, 1984) described, from his own observation and from literature going back to Katzer (1903), river terraces from various areas within the lower Amazon region. Assuming

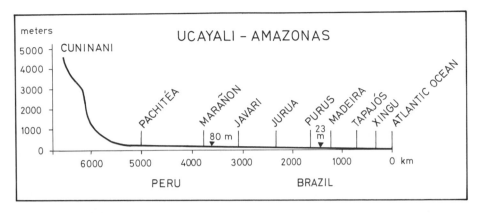

Figure 1. Watertable for the main body of the Amazon river. The low inclination reflects the low altitudes of the Amazon basin (After Soares, 1959).

that they were of Pleistocene age, he correlated these terraces with those formed during high sea-level stages on the southeastern coast of the United States and in the Thames basin.

If sea-level reached a height of 50 meters above recent mean sea-level in one of the earlier Pleistocene warm periods, the Amazon valley would have been drowned at least up to 1500 km distance (by air) from the mouth. Large fresh water lakes may have formed, but after sufficient sediments were deposited new water ways developed with their characteristic ridges, swales, and levées as evidence of their presence (Irion *et al.*, 1983). Due to great differences between the yearly high and low waters of the Amazon, which reach 10 to 15 meters in the middle Amazon area, the ridges and swales formed are very distinct. Due to the dense cover of forest, systems of Pleistocene ridges and swales are hard to identify, but from radar maps showing the surfaces without vegetation cover, the former floodplains are visible (Fig. 2). Analyses of radar maps for the upper Amazon shows extensive areas where the Pleistocene floodplains occur (Irion, 1978). The floodplains are of different heights. We recovered sediment cores nearly 15 meters long from some selected floodplains (Irion, 1976a, 1976b, 1987). In the lower part, the mineral association in the unweathered sediments is the same as that of the modern river sediment load of the area, and hence it can be assumed that the sediment load during former Pleistocene warm periods was similar to the recent ones. In the uppermost decimeters of the sediment cores, the mineral association changes due to the intensive tropical weathering. In higher, and assumingly older floodplains, weathering reaches more advanced stages than in the younger floodplains which have formed during the last Pleistocene sea-level rise.

The formation of these Pleistocene floodplains is not understood in

G. Irion

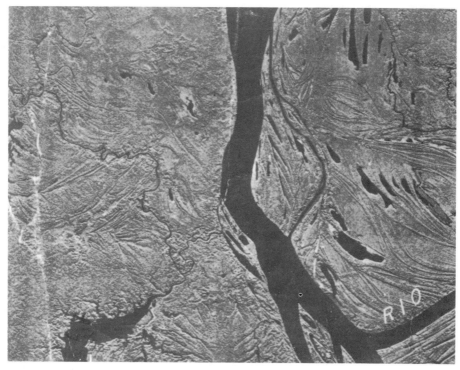

Figure 2. Rio Amazonas 200 km above the confluence with Rio Negro. In the upper left Pleistocene floodplains are present (Radar map, 1971).

detail and the extent of the floodplains is known only from the published radar maps. The geochemical and mineralogical studies are incomplete, particularily when comparing the extent of the floodplains with the areas studied. Nevertheless, it seems possible from our sedimentological and geochemical/mineralogical studies to establish the basic principles for the genesis of Pleistocene sedimentary deposition in the Amazonian Quaternary floodplains (Figs. 3 and 4).

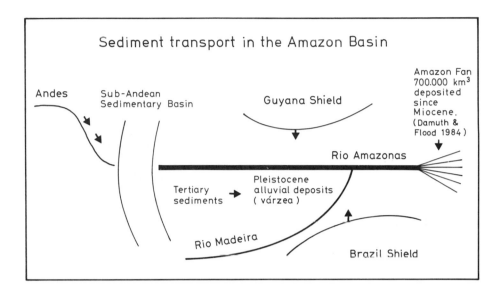

Figure 3. Sediment transport and sediment deposition in the drainage area of the Amazon River.

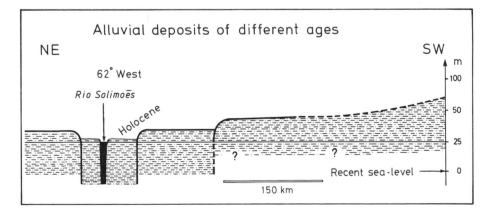

Figure 4. Cross-section through the upper Amazon lowlands showing floodplains of different Quaternary ages.

Pleistocene low sea-level

During the maximum of the last Pleistocene cold period, the sea-level dropped more than 100 meters below the recent mean sea-level. During that time there was a low erosional base level and the river beds of the Amazon and its tributaries were subjected to intensive erosion. The sediments of the Pleistocene river valleys may have been predominantely soft, and could be easily eroded away. The river beds cut down as far as 100 meters below present levels. In the lower parts of recent Rio Negro, the river bed which was eroded during sea-level depression 18 000 years ago is preserved because of the low sediment transport in this river. In front of Manaus, the water depth is today close to 100 meters (Fig. 5). The water depth of Rio Negro during the last sea-level depression may not have exceeded 40 meters, hence its water table may have been situated about 70 meters below the recent one. Therefore, the whole Amazon water table below this location was lowered. Lowering of the water table of the rivers may have reached as far west as to the confluence of the Rio Amazonas with the Río Napo.

With increasing erosion of the deeper horizons, erosion in un-weathered bedrocks and sediments must have taken place. This explains the high amount of arkosic (feldspar rich) Amazonian sands encountered in sediment cores from the Amazon deep-sea fan (Damuth and Fairbridge, 1970).

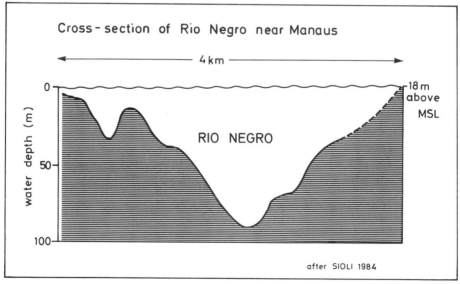

Figure 5. Cross-section of Rio Negro near Manaus. The valley formed during the last sea-level depression is well preserved.

It does not indicate a dry climate in the drainage area of Amazon River, as supposed by some authors.

After 15 000 B. P., when sea-level rose about two cm per year, the main stem of the Amazon valley was drowned because sedimentation rates in the river beds were not high enough to balance the rising water table. A large fresh water lake may have developed with a length of about 1500 km, extending from the mouth of the Amazon to 65° W, and a width not exceeding 100 km (Irion, 1976c). The maximum extent of the lake was reached when the velocity of sea-level rise decreased around 6000 years B.P.

A hint for the supposed lake stage could be the fine grained sediment deposits in the Amazon valley. Echo soundings show fine grained bottom sediments in the middle and upper Brazilian Amazon section (Irion, 1976a). They cannot have been deposited by the present river because its water velocity is too high for the deposition of fine grained material. Dredged samples from these river sections show a clay to silt composition which corresponds to the recent Amazon suspension load not only in grain size distribution but also in geochemistry and mineralogy. It seems reasonable to assume that these sediments have been deposited during the assumed lake stage.

The lower reaches of lower- and middle Amazon tributaries were drowned, as mentioned above, during the sea-level rise. There are many of these drowned river sections along the Amazon and its tributaries (Sioli, 1957).

In the middle Amazon area, sediment cores from lakes formed in drowned rivers were studied (Irion, 1982, 1984a). The ^{14}C records show a good agreement between sediment deposition and rising lake watertables. It can be shown that the rising of the lake watertable is in accordance with the world-wide rising sea-level.

During this time of the lake stage, the sediment delivery to the Atlantic Ocean decreased. In the sediment cores recovered from the Amazon deep-sea fan, Damuth and Fairbridge (1970) observed "an abrupt change to pelagic sediments at about 10 000 to 11 000 years B.P. in all 39 cores demonstrating that with the beginning of the Holocene, large quantities of continental detritus no longer reached the continental rise and abyssal plains of the Guyana basin." This observation coincides with the existence of an Amazon lake which must have been an effective sediment trap.

Quaternary terra-firme

By far the largest part of the five million square kilometers of Amazon lowlands has not been affected by the Pleistocene high inland watertables.

This area may be named the *Quaternary Terra-firme.*

In the lower Amazon, one of the most obvious features is the appearance of high plains at elevations of 100 to 300 meters above sea-level. Sombroek (1966) assumed that a clay which forms the uppermost horizons of these plains was deposited from a lake. This lake should have existed during the Calabrium at a sea-level height of 180 meters above recent mean sea-level. Based on the *locus typicus* of these clays, Sombroek (1966) named them *Belterra clays.* In the lower beds of the *Belterra clays,* a horizon of iron-oxide and -hydroxide concretions often appear with thickness ranging from a few centimeters to decimeters.

Journeaux (1975), Brown and Ab'Saber (1979), and the authors of Volume 18 of *Projeto Radambrasil* (1978) interpreted this horizon as a *stone line,* a feature which is believed to have its origin in areas of dry climate . The formation of a *stone line* is thought to take place in areas without or with very little vegetation. Strong winds promote blowing out of clay, silt and sand, whereas coarse-grained material remains in place and, therefore, accumulates on the surface. Under suitable conditions this process leads to the formation of a stone pavement. When this stone pavement later is covered by other material, it appears in a vertical profile (*e.g.* along a road cut) as a band which may be named *stone line.* Journeaux, (1975) attributes the *stone line* in the lower beds of the *Belterra clays* to a period of dry climate before the deposition of the clays. Like Sombroek (1966), Journeaux (1975) also places *Belterra clays* into the Calabrium. Hence, it follows that the *stone line* should be older than 2.5 million years. Brown and Ab'Saber (1979) do not deal with the *Belterra clays* but believe the *stone line* horizon to be of much later origin. They place its formation during the maximum of the Würmian glaciation, for which reason they postulate a dry climate throughout the Amazon lowlands.

Our mineralogical, geochemical, and sedimentological surveys, which were carried out along the cuts of newly constructed Amazonian roads and in other areas with the help of small soundings, do not coincide with the stratigraphical interpretation of the genesis of the "*stone line*" and *Belterra clay.* On the contrary, it was shown (Irion, 1976b, 1984b, 1987) that *Belterra clays, stone lines,* and the horizon below - which is a first-stage weathered bed rock - belong to the same weathering unit. Their formation dates back into the Tertiary and is still continuing.

In our survey, we rarely reached the unweathered bed rock. In general, the studied sequences started in horizons where the structure of the bed rock was still preserved, but feldspars have disappeared totally and kaolinite has been formed. Above this zone, in a transition horizon of alternating stages of humidity and changing Eh-pH conditions, the formation of haematite rich pisolithes takes place through the dissolution of iron and precipitation of iron hydroxides. Here, very fine-grained kaolinites were also formed and the kaolinites of the lower horizon dissapear. Simultaneously, the quartz content drops, generally to less than 20% and clay accumulates as residue, mainly due to dissolution of quartz. In a million years of weathering, the upper horizon may reach a thickness up to 10 meters or sometimes more. The kaolinites of the top layer are very resistant to further weathering.

The thickness of the uppermost clay horizon varies on the shields and the Cretaceous sediments, and is greater than those on the Paleozoic sediments. Weathering of the shield and of the Cretaceous sediments results in the development of good vertical drainages, whereas the Paleozoic sediments, mainly consisting of slates of Devonian age, are badly drained, and therefore weathering is comparatively restricted. In the clays above the slates, geochemical and mineralogical features derived from their original composition are admixed. Considering this and the different thicknesses, it seems unlikely that the clays are lake sediments. If that was the case, the thickness and mineral admixture would not depend on bed rock prerequisites. Additionally, the *Belterra clay* does not show any sedimentological structures indicating that it was deposited in a lake, and its grain-size distribution, finer than one micron and coarser than 63 microns, is very atypical for lake sediments.

From the difference between the quartz content in the bed rock and in the uppermost layers, it is possible, when assuming that the sequence is weathered *in situ*, to calculate a minimum age of the sediments. The calculated age of 10 million years (Irion, 1984a) fits well with the hypothesis that the formation of the weathered units, the *Quaternary terra firma* of eastern Amazonia, started in Middle Tertiary or even earlier.

Due to low sediment load of most rivers of the eastern lowlands, it can be assumed that surface erosion throughout the Quaternary was slow. For example Rio Uatumá, like many other rivers draining the Guyana shield and Cretaceous sedimentary deposits, does not carry more than three mg/l of suspended load. Together with the observations described above, it seems likely that in most surfaces of the shields and the Paleozoic and Cretaceous sedimentary deposits, deep weathered horizons have developed over millions of years, probably starting in mid Tertiary or earlier. Only in the southwestern lowlands, in the upper reaches of Rio Purus, Rio Juruá, and in the sub-Andean arc are the surfaces of younger age.

Conclusion

The Quaternary history of the Amazon lowlands is characterized by deposition of sediments of Andean provenance and by the influences of changing sea-levels. Most Andean sediments where deposited in the sub-Andean region, but on the average one billion tons per year may have reached the sea during Quaternary history. During periods of high sea-levels which dammed the Amazon drainage system, deposition of sediments in the Amazon lowlands was favored. The high sea-level stages affected areas as far away as 3000 km upstream the Amazon river. Floodplains corresponding to the different Pleistocene sea-level heights were formed. During low sea-level, erosion in the drainage areas increased and the water levels of the central Amazon river systems were lowered.

Due to the dammings, valleys drowned and lakes formed in the lower reaches of rivers and creeks. These lakes remained in those valleys with rivers having a low sediment load. Results from the ^{14}C-dating of sediment cores recovered from these lakes correspond closely with the Holocene sea-level curve, after taking into account the inclination of the water table between the inland lakes and the sea.

Areas well above the present water tables were not reached by Pleistocene high water stages. These areas have been intensively weathered since the Tertiary, forming mighty lateritic weathering horizons. In their upper part, the laterites consist mainly of fine-grained kaolinite together with underlying horizons of pisolithes. The pisolites are generally *in situ,* and were simultaneously with the kaolinites formed during humid tropical climates.

The results of our studies do not show any climatic change or change in the vegetation cover in the Amazon lowlands during Pleistocene. But they reveal that during high sea-level extensive areas of the western Amazon lowlands were dominated by a fresh water aquatic system, and during low sea-level former existing floodplains where inactive and eroded near the main channels. During the latter periods the extent of water bodies may have decreased.

Sediment distribution, weathering horizons, and surface relief of the Amazon lowlands are best explained by relatively constant humid tropical climate throughout the Quaternary. Vertical changes in the composition of sediments in the Amazon Deep-Sea Fan may be to a large extent related to sea-level changes.

Acknowledgments

I thank the Max-Planck-Institute in Plön and INPA in Manaus for their collaboration, and Deutsche Forschungsgemeinschaft for financial support.

Literature cited

Brown, K. and Ab' Saber A. (1979). "Ice-age forest refuges and evolution in the Neotropics." *Univ. Sao Paulo, Inst. Geograph.*

Damuth, J. E. and Fairbridge R. W. (1970). "Equatorial Atlantic deep-sea arcosic sands and ice-age aridity in tropical South America." *Geol. Soc. Am. Bull.* **81**, 189-206.

Damuth, J. E. and Flood R. D. (1984). "Morphology, sedimentation processes, and growth pattern of Amazon deep-sea fan." *Geo-Marine Letters* **3**, 109-117.

Grabert, H. (1983). "Der Amazonas - Geschichte eines Stromes zwischen Pazifik und Atlantik." *Nat. u. Mus.* **113**, 61-71.

Irion, G. (1976a). "Quarternary sediments of the upper Amazon lowlands of Brasil." *Biogeographica* **7**, 163-167.

Irion, G. (1976b). "Mineralogisch-geochemische Untersuchungen von der pelitischen Fraktion amazonischer Oberböden und Sedimente." *Biogeographica* **7**, 7-25.

Irion, G. (1976c). "Die Entwicklung des zentral- und oberamazonischen Tieflandes im Spätpleistozän und im Holozän." *Amazoniana* **4**, 67-69.

Irion, G. (1978). "Soil infertility in the Amazonian rain forest." *Naturwissenschaften* **65**, 515-519.

Irion, G. (1982). "Mineralogical and geochemical contribution to climatic history in central Amazonia during Quaternary time." *Trop. Ecol.* **23**, 76-85.

Irion, G. (1984a). "Sedimentation and sediments of Amazon rivers and evolution of the Amazon landscape since Pliocene times." pp. 201-214 *In* Sioli (1984).

Irion, G. (1984b). "Clay minerals of Amazon soils." pp. 537-579 *In* Sioli (1984)

Irion, G. (1987). "Die Tonmineralvergesellschaftung in Fluss sedimenten der Feuchten Tropen (Amazonas-Becken, West Papua Neuguinea) als Ausdruck der Verwitterung im Einzugsgebiet." *Habilitationsschrift, Univ. Heidelberg*.

Irion, G., Adis, J., Junk, W. J., and Wunderlich, F. (1983). "Sedimentological studies of the Ilha de Marchantaria in the Solimoes/Amazon River near Manaus." *Amazoniana* **8**, 1-18.

Journeaux, M.A. (1975). "Geomorphologie des Bordures de l'amazonie Brailienne: Le Modele des Versandts; Essai d'evolution Paleoclimatique." *Bull. Ass. Geogr. Fr.*, 422-423.

Katzer, F. (1903). "Grundzüge der Geologie des unteren Amazonasgebietes." *Leipzig*.

Klammer, G. (1971). "Über plio-pleistozäne Terrassen und ihre Sedimente im unteren Amazonasgebiet." *Z. Geomorph. N. F.* **15**, 62-106.

Klammer, G. (1984). "The relief of the extra-Andean Amazon Basin." pp. 47-84 *In* Sioli (1984).

Meade, R. H., Dunne, T., Richey, J. E., de M. Santos, U. and Salati, E. (1985). "Storage and remobilization of suspended sediment in the lower Amazon River of Brazil." *Science* **228,** 488-490.

NEDECO (1973). "Rio Magdalena and Canal del Dique Survey Project." *Netherl. Engin.Consult., The Hague.*

Projeto Radambrasil (1978). "Folha SA.20 Manaus, Geologia, Geomorfologie, Pedologia, Vegetacao, uso potencial da terra. Ministerio das Minas e Energia Departemento Nacional da Producao Mineral." *Volume* **18,** *Rio de Janeiro.*

Salo, J., Kalliola, R., Häkkinen, I., Mäkinen, Y., Niemela, P., Puhakka, M. and Coley, P. D. (1986). "River dynamics and diversity of the Amazon lowland forest." *Nature* **322,** 254-258.

Sioli, H. (1957). "Sedimentation im Amazonasgebiet." *Geologische Rundschau* **45,** 608-633.

Sioli, H. (1984). "The Amazon and its main affluents: Hydrography, morphology of the river courses, and river types." pp. 127-166 *In* Sioli (1984).

Sioli, H. (ed.), (1984) "The Amazon - Limnology and landscape ecology of a mighty tropical river and its basin." *Dr. Junk, Hague-Boston-Lancaster.*

Soares, L. de Castro (1959). "Hydrografia." pp. 128-194 *In* Guerra, A. T. (ed.), "Geografia do Brasil", Vol. 1, *Grande Regiao Norte. Rio de Janeiro, IBGE. Cons. Nac. de Geografia.*

Sombroek, W.G. (1966). "Amazon soils."*Center for Agric. Publ. Document, Wageningen.*

Hierarchy of landscape patterns in western Amazon

J. SALO AND M. RÄSÄNEN

University of Turku, Finland

The current debate on the factors which have contributed to the present species richness of the Amazon has been strongly affected by the climatic Pleistocene refuge theory (see papers in Whitmore and Prance, 1987). The refuge theory has its major domain in explaining mechanisms of the biological differentiation - allopatric speciation in isolated forest patches during the supposedly arid Pleistocene glacial maxima - and it has had its strongest support in the documented patchy distribution patterns of selected modern forest biota.

The Pleistocene refuge dynamics still need further geoscientific documentation in order to be generally accepted as a feasible model for biological differentation in the Amazon lowlands. However, there is a clear lack of surveys of the modern edaphic factors and landscape structures which would contribute to the observed patterns of endemism, contact zones and species richness. The need for these analyses is obvious because geologic surveys of the western Amazon basin have revealed dynamic patterns which contradict the general view of a stable Quarternary geologic history, which was based on information from the central Brazilian Amazon.

Major parts of the western and central Amazon lowland rain forests are located on fluvially deposited plains (Khobzi *et al.*, 1980; Putzer, 1984; Räsänen *et al.*, 1987; Salo *et al.*, 1986; Salo, 1988). The deposition history of these plains is related to the Sub-Andean foreland dynamics which have contributed to the relief history of the area since the Tertiary (Räsänen *et al.*, 1987). These findings show that also the dissected lowland *terra firme* relief is of apparent fluvial origin, originally deposited by ag-grading river systems. This contrasts the widely held view of the lacus-trine origin of Amazon *terra firme* (Sombroek, 1966).

The predominantly fluvial origin of the uppermost sedimentary beds in the western Amazon calls for landscape ecology analyses (see Pickett and White, 1985; Forman and Godron, 1986) which would document the levels and hierarchies of landscape patches which have their origins in the fluvial processes. The part of the Amazon basin which is fluvially deposited, inherently bears various levels of mosaicism and age-heterogeneity, starting from recent aggradation processes along the fluvial corridors and ending in the large dissected *terra firme* reliefs which are composed of smaller former deposition units. These patch dynamics, still in their active phases in the western Amazon, may have a crucial role in the formation of distinct forest types characterizing the area (Encarnación, 1985), as well as differences in forest regeneration mechanisms (Salo and Kalliola, 1989) and faunistic and floristic assemblages.

Floodplain processes

The upper Amazon whitewater rivers (Fig. 1), which in most cases have Andean origin, are characterized by their high sediment load (Gibbs, 1967). Suspended sediment transport in the Solimões-Amazon River has been surveyed on the basis of measurements of sediment transport during 1982 to 1984 (Meade, 1985; Meade *et al.*, 1985; Mertes, 1985; Sternberg, 1987).

At Obidos in Brazil, the average Amazon River discharge is 157 000 cubic meters per second (Oltman, 1968) and the total suspended sediment discharge averages 1.1-1.3 billion metric tons per year (Meade *et al.*, 1985). The deposition of whitewater river sediments takes place along the floodplains, which in the case of the Brazilian Amazon-Solimões River cover more than 60 000 square kilometers (Melack, 1983). The recent total floodplain area of the Peruvian lowland Amazonia, which is loosely comparable to the Brazilian *várzea*, has been estimated to cover 62 100 square kilometer (Salo *et al.*, 1986).

Suspension sediments with a grainsize greater than 0.25 mm comprises more than 95% of the total sediment carried in the Amazon River (Meade, 1985). This fraction of the sediment load is deposited along the channel systems mainly through overbank flow and through crevasses. The coarser material, however, which is mainly carried either in intermittent suspension or as bedload, predominantly stays within the main channel or chutes. This fraction is the main component in forming the point bar and mid-channel deposits which are the sites for the riverine forest primary succession (Kalliola *et al.*, 1987, 1988).

Mertes (1985) has demonstrated great variation in lateral migration rate of the Solimões-Amazon River between São Paulo de Olivenca and Obidos. These results, together with the observations made in the upper Amazon basin (da Cunha, 1906; Salo *et al.*, 1986) have shown that the earlier suggestion of Sternberg (1960) that there is virtually no lateral migration of the Amazon River cannot be widely generalized. Along the study reach of Mertes, three distinct types of channel change were observed: (i) lateral migration of main channel bends; (ii) migration of chutes (*paranas*); (iii) change in island shape resulting from channel migration.

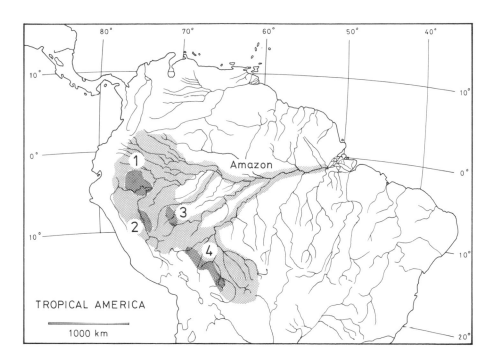

Figure 1. A schematic presentation of the western Amazon, characterized by the present landscape processes classified in Fig. 2 (shaded area, following the concept of the Western Periphery of the Amazon Basin by Fittkau, 1971). The central areas of the four structural basins are depicted and numbered: 1 Pastaza-Marañon; 2 Ucayali; 3 Acre; 4 Madre de Dios-Beni

Sternberg (1960) and Mertes (1985) have shown that chute type of side-channel migration is the predominant agent in floodplain modification.

Within the Amazon basin proper, the rate of lateral migration is by far fastest at the upper reaches of the Amazon River and its tributaries Ucayali, Marañon, and upper Solimões. Recent analyses, based on comparison of two Landsat MSS images (from 1979 and 1983) show that the lateral migration may in some cases exceed 250 meters per year along the lower meandering Ucayali reach and the anastomozing reach of the Solimões near Iquitos, Peru (Salo *et al.*, in prep.).

The activity of channel and floodplain processes provide constant unstability and reorganization of the floodplain biotic communities. They repeatedly create a heterogeneous setting of open habitats like point and mid-channel bars, abandoned chutes and channel cut-offs for primary and secondary succession. These habitats are often short-lived and they form an uniquely dense-packed habitat mosaic.

Foreland dynamics

Although the fluvial processes described in the previous chapter contribute to site-turnover dynamics of the aggrading whitewater floodplains, their effect on large-scale forest dynamics in the area would remain relatively minor unless they formed a part of a larger process which is the Andean foreland evolution. These dynamics are driven by the active foreland tectonics which is part of the Andean evolution (Räsänen *et al.*, 1987) active also during the Quaternary. These dynamics are characterized by formation of infilling basinal areas, within which the aggrading floodplain dynamics are most active. These basins are separated by higher lying denudated *terra firme* areas, arranged along the western Amazon arch system.

The Quaternary Amazon deposition cycle was first initiated by widespread syntectonic fluvial aggradation, induced by Andean uplift (Räsänen *et al.*, 1987). This sequence was followed by further foreland fractioning which triggered erosional and depositional processes that characterize the modern rainforest area. Related geodynamic processes have concurrently functioned in a wide foreland area along the Cordillera Oriental from Venezuela to Bolivia (Pflafker, 1964; Khobzi *et al.*, 1980; Lowrie *et al.*, 1981).

In the Andean-Amazonian transition area, there is a sequential network of foreland basins with the oldest completely deformed and partially eroded Jurassic basins in the Cordillera Oriental. In the modern lowland area there are four major structural basins: (i) Pastaza-Marañon basin (the Ucamar depression, Villarejo, 1988); (ii) Ucayali basin; (iii) Acre basin; and (iv) Madre de Dios-Beni basin. The central part of the Pastaza-Marañon basin forms the largest known ponded inland foreland basin with modern fluvial aggradation. It is even larger than the corresponding area in the Beni basin (Pflafker, 1964). The aggradation is due to elevation of base level, either by active rise of the flanks of the basin in the east, or greater relative subsidence in the basin.

The processes described here range from the megaform change induced by plate tectonics to the microform sites of contemporary plant colonization. Together they have caused the vastly complex shifting mosaic of the Amazon basin (Fig. 2). It is probable that the influence on the biota of these landscape processes equally range from biological differentiation and extinction to local colonization dynamics of the patches. It remains to be studied whether the observed contact zones of certain well studied animal groups like birds (Haffer, 1978, 1985, 1987a,b;Haffer and Fitzpatrick, 1985) and Nymphalid butterflies (Brown, 1987) follow some of the ecotone zones created by the fluvial dynamics and relief evolution.

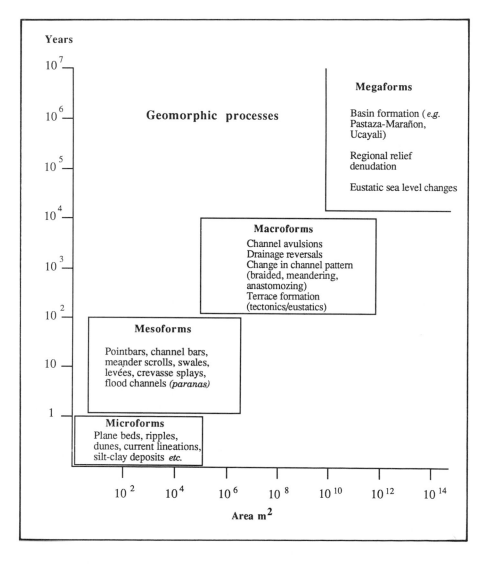

Figure 2. Geomorphological landscape processes in western Amazon, arranged according to their spatiotemporal continuum (see Lewin, 1978; Urban *et al.*, 1987; di Castri and Hadley, 1988 and Fig. 3).

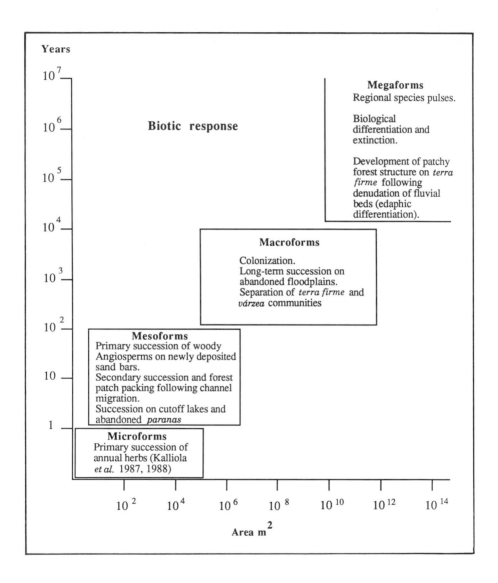

Figure 3. Biological correlatives of geomorphological processes in western Amazon, arranged according to their spatiotemporal continuum (see Urban *et al.*, 1987, di Castri and Hadley, 1988 and Fig. 2).

Species richness and landscape patches

In addition to the Pleistocene refuge theory, a number of new models have been put forward to explain the observed Amazon patterns in endemism and species richness. Among these are the edaphic differentiation concept (Gentry, 1982, 1988), flashflood massive disturbance and large-scale Holocene colonization scheme of the western Amazon (Campbell and Frailey, 1984), unpredictable floods and wet periods as disturbance factors (Colinvaux *et al.*, 1985; Colinvaux, 1987), and long-term fluvial dynamics and landscape mosaicism (Räsänen *et al.*, 1987; Salo *et al.*, 1986). The unifying concept for all these diversity considerations is that there is a clear neoendemism character suggesting the Pleistocene differentiation events to be of primary interest.

The landscape classification, presented in figure 2, gives some tentative explanations of the observed differences in species richness and floristic and faunistic compositions. The necessity to postulate the existence of Pleistocene broad-leaved forest refuges has largely risen from the intuitive view of a homogeneously structured present forest. However, the forest structure in the western Amazon is obviously not homogenous. On the contrary, fluvial dynamics have created a multi-level mosaic of forests. Due to this, many of the restricted species distribution patterns may reflect true edaphic differences in the forest bed.

As long as there are no specific data on the depositional history and age of the rainforest bed (evidenced by absolute dating techniques) on the sites characterized by high levels of neoendemism, there are only few possibilities to determine whether the observed species distribution patterns represent *in situ* biological differentiation or long-term species dynamics regulated by the emergence of various landscape patches.

To sum up the landscape ecology setting of the western Amazon lowlands, the following levels of processes can be distinguished:
River dynamics as a disturbance factor. In the western Amazon, channel migration affects floodplain biota by destroying the old forest and by initiating primary succession on deposited river sediments. Doing so, the river erosion prevents competitive exclusion within various forest types. However, as these river dynamics have characterized the area since the Tertiary, and because of the relatively predictable nature of channel processes within the present floodplain, perturbance by lateral erosion is not a truly unpredictable disturbance factor in terms of the intermediate disturbance hypothesis (Connell, 1978).
Within-floodplain site-turnover and forest patch packing. The mosaic forest structure characterizing the western Amazon *várzea* forests is created by channel diversions within the contemporary flood basins. The packing of differentially aged forests diversifies the overall forest structure because more habitats are opened than would be available in the

case of an even-aged forest bed. As the channel diversion processes are more unpredictable than those in lateral migration of the channel, the community structure of mosaic floodplain forests is highly complex.

Floodplain diversions. The aggradational floodplains in the modern Pastaza-Marañon and Ucayali basins show a high degree of contemporary floodplain abandonment and diversion. These activities leave behind a network of abandoned floodplains and they have also been present during the evolution of the Acre and Madre de Dios basins.

The foreland structural basins and their marginal denudated areas. The current scheme of western Amazon is composed of four major molasse basins separated by slightly higher areas exhibiting surface erosion (denudation). It is obvious that even this level of mosaicism, with the largest components, represents a dynamic Pleistocene relief evolution due to a tectonic or downwarp-induced shift between erosional and depositional surfaces.

These levels of landscape processes and the resulting mosaic represent a genuine chronological continuum. Biologically the continuum may also represent a continuum from an ecological to an evolutionary theater.

Literature cited

Brown, K. S. Jr. (1987). "Biogeography and evolution of Neotropical butterflies." pp. 66-104 *In* Whitmore, T. C. and Prance, G. T. (eds.), (1987).

Campbell, K. E. Jr. and Frailey, C. D. (1984). "Holocene flooding and species diversity in southwestern Amazonia." *Quat. Res.* **21**, 369-375.

Colinvaux, P. A., Miller, M. C., Liu, K-b., Steinitz-Kannan, M. and Frost, I. (1985). "Discovery of permanent Amazon lakes and hydraulic disturbance in the upper Amazon basin." *Nature* **313**, 42-45.

Colinvaux, P. A. (1987). "Amazon diversity in light of the paleoecological record." *Quat. Sci. Rev.* **6**, 93-114.

Connell, J. H. (1978). "Diversity in tropical rain forests and coral reefs." *Science* **199**, 1302-1310.

da Cunha, E. (1906). "Relatório da comissão mista brasileiro-peruana de reconhecimento do Alto Purus; notas complementares do comissário brasileiro; 1904-1905." *Imprensa Nacional, Rio de Janeiro*.

di Castri, F. and Hadley, M. (1988). "Enhancing the credibility of ecology: Interacting along and across hierarchial scales." *Geo Journal* **17**, 5-35.

Encarnación, F. (1985). "Introducción a la flora y vegetación de la Amazonia peruana: estado actual de los estudios, medio natural y ensayo de una clave de determinación de las formaciones vegetales en la llanura amazónica." *Candollea* **40**, 237-252.

Fittkau, E. J. (1971). "Ökologische Gliederung des Amazonas-Gebietes auf geochemischer Grundlage." *Münster Forsch. Geol. Paläont.* **20/21**, 35-50.

Forman, R. T. T. and Godron, M. (1986). "Landscape Ecology." *John Wiley and Sons, New York.*

Gentry, A. (1982). "Patterns of Neotropical plant species diversity." pp. 1-84 *In* Hecht, M. K., Wallace, B. and Prance, G. T. (eds.), "Evolutionary Biology" *Plenum Press, New York.*

Gentry, A. (1988). "Tree species richness of upper Amazonian forests." *Proc. Natl. Acad. Sci. USA* **85**, 156-159.

Gibbs, R. J. (1967). "The geochemistry of the Amazon River system: Part I. The factors that control the salinity and the composition and concentration of the suspended solids." *Geol. Soc. Amer. Bull.* **78**, 1203-1232.

Haffer, J. (1978). "Distribution of Amazon forest birds." *Bonn. Zool. Beitr. Heft* **1-3**, 38-78.

Haffer, J. (1985). "Avian zoogeography of the neotropical lowlands." *Ornith. Monogr.* **36**, 113-146.

Haffer, J. (1987a). "Quaternary history of Tropical America." pp. 1-18 *In* Whitmore, T. C. and Prance, G. T. (eds.), (1987).

Haffer, J. (1987b). "Biogeography of Neotropical birds." pp. 105-150 *In* Whitmore, T. C. and Prance, G. T. (eds.), (1987).

Haffer, J. & Fitzpatrick, J. W. (1985). "Geographic variation in some Amazonian forest birds." *Ornith. Monogr.* **36**, 147-168.

Kalliola, R., Salo, J. and Mäkinen, Y. (1987). "Regeneración natural de selvas en la Amazonia peruana 1: Dinamica fluvial y sucesion ribereña." *Memorias 18B, Univ. Nac. Mayor de San Marcos, Lima.*

Kalliola, R., Mäkinen, Y. and Salo, J (1988). "Regenracion natural de selvas en la Amazonia peruana 2: Autecologia de algunas especies sucesionales." *Memorias 19, Univ. Nac. Mayor de San Marcos, Lima.*

Khobzi, J., Kroonenberg, S., Faivre, P. and Weeda, A. (1980). "Aspectos geomorphologicos de la Amazonia y orinoquia Colombianas." *Revista CIAF* **5**, 97-126.

Lewin, J. (1978). "Floodplain geomorphology." *Progr. Phys. Geogr.* **2**, 408-437.

Lowrie, A., Cureau, S. A. and Sarria, A. (1981). "Basement faults and uplift in the Colombian Llanos." *Z. Geomorph. Suppl. Bd.* **40**, 1-11.

Meade, R. H. (1985). "Suspended sediment in the Amazon River and its tributaries in Brazil during 1982-1984." *U.S. Geol. Survey Open-*

File Report 85-492.
Meade, R. H., Dunne, T., Richey, J. E., Santos, U. de M. and Salati, E. (1985). "Storage and remobilization of suspended sediment in the lower Amazon River of Brazil." *Science* **228**, 488-490.
Melack, J. M. (1983). "Amazon floodplain lakes: shape, fetch, and stratification." *Int. Assoc. Theor. Appl. Limnol. Proc.* **22**, 1278-1282.
Mertes, L. A. K. (1985). "Floodplain development and sediment transport in the Solimoes-Amazon River, Brazil." Unpubl. *M. Sc. thesis, Univ. of Washington*.
Oltman, R. E. (1968). "Reconnaissance investigations of the discharge and water quality of the Amazon River." *U. S. Geol. Survey Circular 552*.
Pflafker, G. (1964). "Oriented lakes and lineaments of northeastern Bolivia." *Geol. Soc. Am. Bull.* **75**, 503-522.
Pickett, S. T. A. and White, P. S., Eds. (1985). "The Ecology of Natural Disturbance and Patch Dynamics." *Academic Press, Orlando*.
Putzer, H. (1984). "The geological evolution of the Amazon basin and its mineral resources." pp. 16-46 *In* Sioli, H. (ed.), "The Amazon. Limnology and Landscape ecology of a mighty tropical river and its basin." *Dr. W. Junk Publ., Dordrecht/Boston/Lancaster*.
Räsänen, M. E., Salo, J. S. and Kalliola, R. J. (1987). "Fluvial perturbance in the western Amazon basin: Regulation by long-term Sub-Andean tectonics." *Science* **238**, 1398-1401.
Salo, J. (1988). "Rainforest diversification in the western Amazon basin: The role of river dynamics." *Rep. Dept. Biol. Univ. Turku* **16**.
Salo, J., Kalliola, R., Häkkinen, I., Mäkinen, Y., Niemelä, P., Puhakka, M. & Coley, P. D. (1986). "River dynamics and the diversity of Amazon lowland forest." *Nature* **322**, 254-258.
Salo, J. S. and Kalliola, R. J. (1989). "River dynamics and natural forest regeneration in Peruvian Amazonia." *In* Jeffers, J. (ed.), "Rainforest Regeneration and Management." *MAB(UNESCO) Book Series, UNESCO and Cambridge Univ. Press, Paris and Cambridge, in press*.
Sombroek, W. G. (1966). "Amazon Soils. A Reconnaissance of the Soils of the Brazilian Amazon Region." *Center for Agric. Publ. Document, Wageningen*.
Sternberg, H. O'R. (1960). "Radiocarbon dating as applied to a problem of Amazonian morphology." *Comptes rendus 18eme Congres International de Geographie, International Geographical Union, Rio de Janeiro*, **2**, 399-424.
Sternberg, H. O'R. (1987). "Aggravation of floods in the Amazon as a consequence of deforestation?" *Geogr. Ann.* **69A**, 201-219.
Urban, D. I., O'Neill, R. V. and Shugart, H. H. Jr. (1987). "Landscape

ecology." *Bio Science* **37,** 119-127.

Villarejo, A. (1988). "Así es la Selva." *CETA, Iquitos.*

Whitmore, T. C. and Prance, G. T. (1987). "Biogeography and Quarternary History in Tropical America." *Clarendon Press, Oxford.*

Flood tolerance and tree distribution in central Amazonian floodplains

W. J. JUNK

Max-Planck-Institute for Limnology, Plön, FRG, and INPA, Manaus, Brazil.

The Amazon River basin covers seven million square kilometers and it is the largest catchment area in the world. Annual precipitation in its central region ranges from 2000 to 3000 mm, but on the east Andean slopes it exceeds 5000 mm and towards the north and the south of the basin it decreases to about 1500 mm. In most of the catchment area precipitation is not equally distributed over the year (Salati and Marques, 1984). High precipitation feeds a dense network of streams and rivers which transport rainwater surplus to the Atlantic Ocean. While the water level of small streams fluctuates according to local rainfall which is unpredictable, that of the Amazon River and its large tributaries is monomodal and predictable and follows the seasonality of rains over the whole catchment area. Near the coast tidal influence creates short and predictable flood pulses of fresh or brackish water. Large and poorly drained areas are flooded every year during the rainy season.

Our studies in central Amazonia were concentrated on two major types of floodplain forests: (i) the forest of the *várzea* floodplain of white water rivers; (ii) the forest of the *igapó* floodplain of blackwater- and clearwater rivers. White water rivers, such as the Amazon, are rich in suspended and dissolved solids, have neutral *pH*, and are fertile. Black water rivers, such as Rio Negro, are poor in suspended and dissolved solids, they are acidic, brownish because of high concentrations of humic and fulvic acids, and have low fertility. Clear water rivers, for example Tocantins and Tapajos, are not hydrochemically uniform. They have low to neutral pH, are transparent with little dissolved and suspended inorganic matter and have intermediate fertility (Sioli, 1965).

The large Amazonian rivers and their floodplains cover about 300 000 square kilometers of which 30% are covered by flood tolerant forests. Duration and amplitude of the flood pulse influence tree distribution but few data exist on these pulses in the Amazon floodplain forests. All plots studied so far cover a large part of the flood level gradient and only Keel and Prance (1979) refer to the specific position of tree species on the gradient. The present paper evaluates resistance to flooding of some common trees in central Amazonian *várzea* forest and interprets existing phenological, morphological and physiological information about Amazonian floodplain tree species.

The floodplain of the Amazon River is a mosaic of habitats which include lakes, periodically dried out lakes, levées, swampy depressions, sand or mud-flats, sedimentation or erosion-areas, *etc.* (Junk, 1984). Every place in the mosaic occurs on a flood level gradient from permanently aquatic to permanently terrestrial habitats. Because the yearly flooding amplitude varies and trees are longlived organisms, we have examined hydrological data for a 80 years period to show average, maximum and minimum numbers of dry and flood days (Fig. 1A,B).

To evaluate maximum flood tolerance of some of the most common tree species in the *várzea* near Manaus, the positions of adult individuals growing at low water levels were examined. The Manaus harbor hydrograph was used to calculate the level and the period of flooding, defined here as the period when the river level is above the soil surface. This gives conservative estimates because flood stress starts when the water reaches the roots. Flood tolerance of adult trees may be greater than suggested by their position on the flood level gradient, because seedling establishment may be more important for tree distribution than flood tolerance of adult trees. Data were collected at Lago do Rei and Careiro Island 60 km downstream from Manaus, at Marchantaria Island and in the Janauacá area 30 and 60 km upstream of Manaus. Hydrological data collected by Schmidt (1973) in the Janauacá area show only few centimeters of difference compared to the Manaus hydrograph.

Plant distribution

Low-lying shrub community: A shrub community which is submersed for several months grows at elevations of about 20.5 meters AASL (above average sea-level), and undergo on the average 270 days of flooding and, at most, three years of permanent flooding over a period of 80 years (Table 1). In Lago do Rei and Lago Janauacá we observed large stands of dead individuals which had grown in areas about one meter below present live stands, where there is an average flooding of 300 days per year and, at most, five years of permanent flooding. They were alive in 1968 but had died in 1974. The waterlevel data indicate that from 1970 to 1973 low water levels did not fall below 20 to 21 meters (Fig. 2). This is the most extreme situation at Manaus since 1902. Shrubs growing near the 19.5 meters level were unable to tolerate this long period of flooding, whereas they did tolerate it at the 20.5 meters level. Until today, no extreme dry period has occurred which could have encouraged the successful re-establishment of

Table 1. Plant communities along the flood level gradient on the Amazon River.

Low-lying shrub community	Mid-level tree community	High-level tree community
Coccoloba ovata	*Acosmium nitens*	*Aspidosperma carapanauba*
Eugenia inundata	*Bowdichia virgilioides*	*Calycophyllum spruceanum*
Myrcia sp	*Buchenavia macrophylla*	*Ceiba pentandra*
Ruprechtia ternifolia	*Casearia grandiflora*	*Couroupita guianensis*
Symmeria paniculata	*Cassia leiandra*	*Genipa americana*
	Cecropia latiloba	*Guazuma ulmifolia*
	Crataeva benthamii	*Hevea brasiliensis*
	Ficus anthelmintica	*Lecointea amazonia*
	Holopyxidium jarana	*Licania sp.*
	Licania kunthiana	*Malouetia furfuracea*
	Machaerium leiophyllum	*Manilkara amazonica*
	Macrolobium angustifolium	*Ocotea cymbarum*
	Nectandra amazonum	*Rheedia brasiliensis*
	Piranhea trifoliata	*Spondias lutea*
	Pseudobombax munguba	*Sterculia elata*
	Tabebuia barbata	
	Unonopsis guatterioides	

the shrub community at the 19.5 meters level. Species occuring in this extremely flood tolerant shrub community are restricted to the lowest-lying areas, probably because of the disadvantages they face when competing for light with taller trees in higher-lying areas.

Mid-level tree community: The lake basins forest communities appear above 22 meters AASL, where flooding occurs on an average of 230 days per year (Table 1). On steep and stable channel borders the trees of this community can be seen at levels as low as 21 meters, but on the shores of the Amazon River they commonly appear at 23 meters and continue upwards. In a few cases adult *Salix humboldtiana* and *Alchornea castaneifolia* were found at 22 meters. Many tree species which colonize low-lying areas are also normally found higher up. Pioneer species such as species of *Cecropia, Salix humboldtiana,* and *Alchornea castaneifolia* are substituted by species of greater competitive strength when environmental conditions become suitable.

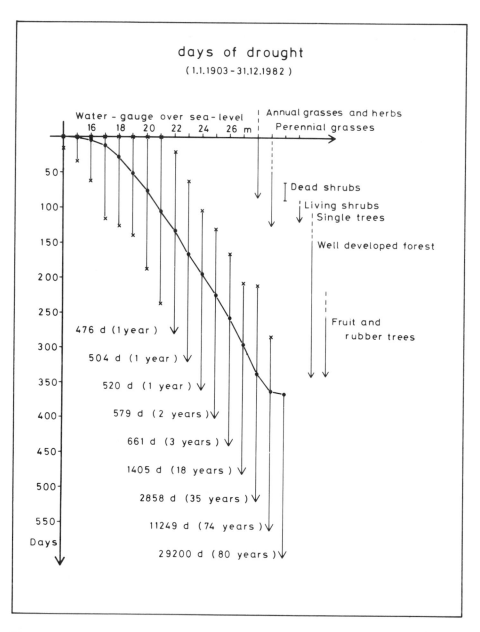

Figure 1A, B: Distribution of major terrestrial plant communities of the Amazon River floodplain near Manaus on a level gradient in relation to 80 years average, maximum and minimum numbers of dry and flood days (hydrological data from Manaus harbor). Total number of years without dry respectively flood periods are indicated in (). Values are calculated according to the hydrological year.

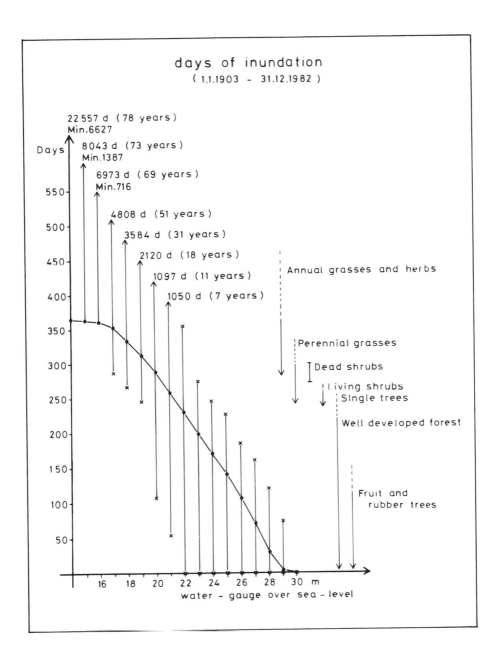

days of inundation
(1.1.1903 – 31.12.1982)

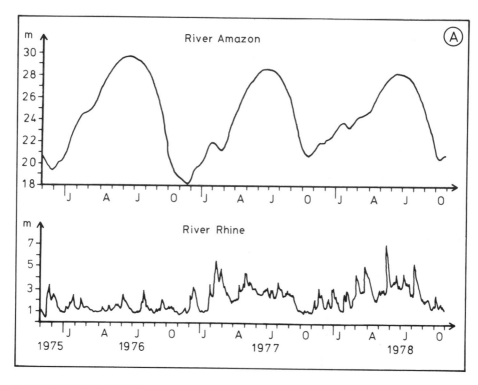

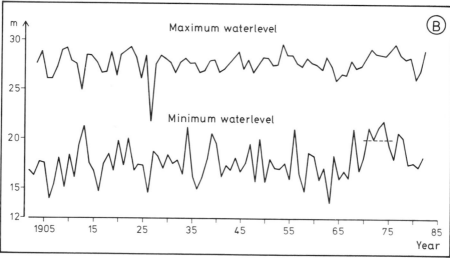

Figure 2. A - Maximum and minimum water levels of the Amazon River near Manaus (hydrological data from Manaus harbor). B - Flood regime of the Amazon River near Manaus (hydrological data from Manaus harbor) and the Rhine river near Worms (according Dister, 1980).

High-level tree community: Many species colonize higher levels. These form a community which grows mainly from the 25 meter level and upwards, where the flooding lasts for an average of 140 days per year (Table 1). Some species, for example the giant *Ceiba pentandra,* are even found in *terra firme* (Kubitzki, in press.). As seen in figure 1, the deviation in numbers of flood and dry days from average values is often large. During the last 80 years the 25 meter level was dry for as long as two consecutive years, whereas the 27 meter level was dry for about four years without interruption. At the same time, the 25 meter and 27 meter levels suffered flood periods of 230 and 160 days respectively.

These unusually long periods of wetness and dryness probably have much greater influence on tree distribution than average values do. The humid tropical climate does not provide a period for plants to rest. In temperate zones flooding during winter or early spring has little effect because trees then operate at a low level of physiological activity. The flood pattern of most large temperate rivers is spiky and with low amplitude (Fig. 2). The annual period of flooding is made up of several short periods of flooding distributed throughout the year (Dister, 1980, 1983). In the Amazon, on the contrary, high flooding is concentrated in one long period which exerts a greater stress on trees than would many short periods of flooding of the same total length .

The forest communities on the flood level gradient represent seral stages of a succession from low-lying pioneer communities with few species to a high-lying floodplain forest with many species. Considering the complex edaphic conditions in the floodplain, the interaction of allogenic and autogenic factors and the actual stage of knowledge at the community level, such a generalized interpretation are simplified.

Sedimentation and erosion

Water movements, erosion and sediment deposition strongly influence community structure because they affect habitat stability (Salo *et al.,* 1986). On the middle Amazon River, low-lying areas near the main channel are often subject to as much as one meter sediment deposition per year. In such areas roots of trees quickly become buried under thick layers of fine sediments and suffer from hypoxia even during the dry period. Therefore high sedimentation rates result in low tree species diversity. Considering the long lifespan of many tree species, deposition of only a few centimeters of fine sediments every year may be critical for their growth.

The pioneer trees *Salix humboldtiana* and *Alchornea castanei-folia,* which colonize sedimentation areas, overcome the heavy sediment deposition by producing new lateral roots near the soil surface. At a level of 24 meters we excavated several three year old specimens, 80 cm in height, with an underground part of about 1.5 meters and three lateral root systems at 1, 0.5, and 0.1 meters below the surface. Seedlings of *Pseudobombax munguba* growing nearby could not establish themselves despite adequate flood tolerance.

Eugenia inundata and other species of the extremely flood tolerant shrub community do not tolerate high sedimentation rates in the *várzea* of the Amazon River. Yet, in the floodplains of the Tocantins and Araguaia rivers *Eugenia* species occur partially buried under as much as two meters of coarse sand. In the coarse sediments of clearwater rivers water exchange and oxygen supply may therefore be sufficient to avoid asphyxia of the root system. Better oxygen supply may also be the reason for the more frequent occurrence of *Astrocaryum jauari* in low-lying areas of the floodplain of the Rio Negro compared to the *várzea* of the Amazon River.

Erosion of the levées along the main channels destroy forest communities. The velocity of meandering is more pronounced in the upper reaches of the Amazon river than in the lower ones. Near Manaus, only small changes have been recorded during the last centuries. On the island of Careiro 2000 year old Indian potsherds were found which indicates a long period of low disturbance (Sternberg, 1960). This would allow the establishment of advanced seral stages of floodplain forest communities. The oldest tree dated on Careiro Island was a large *Piranhea trifoliata* which was more than 400 years old (Worbes, pers. comm.).

Diversity

There is little information on total numbers of tree species in the floodplains of the Amazon rivers. For the *igapó* Revilla (1981), Worbes (1983) and Keel and Prance (1979) reported between 54 and 111 species in plots ranging from 0.18 to 1.1 hectares. For *várzea* forest Klinge (pers. comm.), Worbes (1983), Rankin de Merona and de Merona (1988) and Revilla (1988) reported between 33 and 220 species in plots ranging from 0.2 to 10 hectares. Most of the plots are small, and may not represent the full range of species in the area. Many forested areas which still exist near Manaus are disturbed or in an early stage of succession. Species diversity is low in low-lying areas and in areas strongly disturbed by the river. Revilla´s plots with greatest diversity have suffered less human influence than the other areas investigated. Both areas are situated at rather high elevations where the impact of flooding is not extreme and they are far away from the main channel, and sedimentation and erosion are of little importance.

With 70 to 130 species per hectar, diversity in well developed and stable *várzea* forest is relatively high (Revilla, 1988), but lower than the nonflooded *terra firme* forest, where maximum values of up to 500 species per hectar have been reported (Hueck, 1966; Klinge, 1973; Kubitzki, 1977; Keel and Prance, 1979). It has been suggested by Worbes (1983) that the *igapó* forest is more diverse than the *várzea* forest. Prance (1979a), however suspects that there is lower diversity in the *igapó* because of drought stress in dry periods and low nutrient levels. Kubitzki (in press) suggests a higher local species diversity in the *várzea* forest but a greater regional differentiation of the *igapó* forest.

The total number of flood resistant tree species in the floodplains of the Amazon can not be determined yet. In the *várzea* it exceeds 400 species and may reach 500. A similar or larger number may occur in floodplains of the large blackwater and clearwater tributaries, which have distinct sets of species (Kubitzki, in press; Prance, 1979a; Revilla, 1981; Worbes, 1983). If a total number of 4000 to 5000 tree species is assumed for the Amazon lowlands (Rodrigues, pers. comm.) probably about 20% would tolerate a periodical flooding of several weeks to many months.

The diversity of trees in Amazonian floodplain forests is very high compared to that of temperate zones. The total number of tree species in bottomland forests of the United States is about 100 including those with a flood tolerance of a few days only (Clark and Benforado, 1981).

Dispersal and establishment

Seeds may be dispersed by water or animals or a combination of these. Gottsberger (1978), Goulding (1980) and Rankin de Merona and de Merona (1988) point to the great importance of fish in seed dispersal. Probably all seeds and fruits are consumed at least occasionally or in part by one or more of the numerous Amazonian fish species, but it is not known to what extent consumption and dispersal can be correlated. Birds, bats and monkeys disperse seeds as well. Their importance for floodplain species is still unknown. Wind dispersal has only been observed in some species which may as well be water or fish dispersed. Only some fruits and seeds can float (*e.g. Macrolobium acaciaefolium, Vatairea guia-nensis, Aldina latifolia* and *Lecointea amazonica*). Many seeds are too heavy and sink to the bottom (*e.g. Crataeva benthamii, Hevea spp., Astrocaryum jauari* and other palms). These heavy seeds may be dispersed by fish or by river currents along the bottom (Piedade, 1985).

The main fruiting period depends on seed dispersal strategies, and therefore mostly occurs during the flood period. Shrubby species which colonize the lowest areas fruit when the water rises in January through February, and the seeds are dispersed by water and fish. Fruit production depends on the amount of time during which the branches are out of the water (Peters and Vasquez, 1987). *Cecropia latiloba* colonizes a large spectrum of disturbed habitats over a wide range of the flood level gradient and fruits from February to August (Rankin de Merona and de Merona, 1988). Wind dispersed species fruit when the waterlevel has already passed its maximum and the floodplain begins to dry. This is the case in *Pseudobombax munguba, Ceiba pentandra, Salix humboldtiana* and *Triplaris surinamensis.*

Resistance of trees to flooding increases with age and size (Gill, 1970). Therefore species distribution in periodically flooded habitats is not only related to flood tolerance of adult individuals, but also on the successful establishment of seedlings (Demaree, 1932; Hosner, 1957, 1960; Dister, 1980).

There are no studies on flood tolerance and seedling establishment of trees in Amazonian floodplains. Our observations indicate that during the dry period, many seedlings start colonizing low-lying areas but then become eliminated during the following flood period. Seedlings of *Cecropia* species show stem tip necrosis when totally inundated. Necrosis increases with longer periods of submergence and, eventually, kills the seedling. The critical period is over when the seedling is greater than the flood peak. A succesfull establishment of seedlings at the lowest level of their habitats is only possible in prolonged low-water periods or in periods of several consecutive dry years, which allow seedlings to reach a size which enables them to survive years with normal and later with extreme hydroperiods. The same holds true for catastrophic events, which eliminate species from the extreme ends of their distributional range. Therefore, current species distribution can only be understood, when taking into consideration extreme flood events occurring over the last decades or even centuries.

Most shrubs which colonize the lowest lying habitats further propagate by sprouting from roots or branches in contact with the sediment. This may compensate for the harsh conditions under which seedlings must establish themselves.

The main growth period usually starts with the beginning of the unflooded phase, but leaf sprouting, leaf fall, flowering and fruiting may occur throughout the year depending on the species. The behavior of a species may vary according to its position on the flood level gradient (Revilla, 1981). Species like *Pseudobombax munguba, Triplaris surinamensis, Crataeva benthamii, Vitex cymosa* and *Crescentia amazonica* loose their leaves completely at high water, whereas *Bonafusia tetrastachya, Elaeoluma glabrescens, Symmeria paniculata* and the palms maintain their leaves throughout the year.

Morphological and anatomical adaptations

The main problem for flooded trees is that of obtaining oxygen for submerged organs. Oxygen conditions are more critical in tropical waters and waterlogged soils than in comparable temperate habitats; high temperatures reduce solubility of oxygen in water and increase oxygen demand by accelerating physiological activities and chemical processes such as decomposition of organic material. According to Junk *et al.* (1983); Melack and Fisher (1983), and Schmidt (1973) water in *várzea* lakes is chronically deficient in oxygen, often with considerable amounts of H_2S near the bottom.

Oxygen transport occurs in aerenchymatic tissue in connection with lenticels in species of *Alnus* and *Nyssa* and others from temperate and subtropical zones or by pneumatophors in the mangroves and the neotropical palm *Mauritia flexuosa*. The efficiency of such systems depends on the distance to be overcome. Several meters of high flooding during several months require large gas transport systems. This strategy is only little developed in trees in the floodplains of the large Amazon rivers. Worbes (1986) did not find aerenchymatic tissue in the wood of about 90 tree species from the Amazon River and the Rio Negro floodplains, although palms may be an exception (Schlüter, pers. comm.). Prance (1979a) interprets large root bunches on the stems of some trees in the floodplain of blackwater and clearwater rivers as adaptations to flooding. Adventitious roots can also be found at high water on the submersed parts of the stems of *Salix* in the *várzea*. They may substitute the function of the basal root at high water.

Persistence of submersed leaves during flood periods, may be observed in many Amazonian tree species. Leaves of *Symmeria paniculata* are maintained functionally intact under eight meters of water for more than eight months. Soon after emergence they begin photosynthesis as shown by measurements with a portable infrared gas analyzer.

According to Furch (1984) the sclerophylous leaves do not show external adaptations to flooding but thylacoids are well developed. The ratio of chlorophyll a to b is much lower than in leaves of typical low light plants. Light measurements in the water of Rio Negro and in the floodplain lakes indicate that, even at 6-8 meters depths, there is a very low photon flux (Furch *et al.*, 1985). We believe that these small amounts of energy may be sufficient to maintain the photosynthetic pigments intact. According to Schlüter (pers. comm.) leaves are photosynthetically active under water.

Physiological adaptations

Under anoxic conditions many plants switch from aerobic to anaerobic metabolism which is combined with low energy output and the accumulation or excretion of highly energetic metabolic products such as malathe and sometimes even toxic products such as ethanol (Crawford, 1982).

There is little information available on the respiratory metabolism of Amazon floodplain trees.   Schlüter (pers. comm.) studied root respiration of *Astrocaryum jauari, Macrolobium acaciaefolium* and *Salix humboldtiana* during terrestrial and aquatic phases.   During prolonged flooding *Salix* and *Macrolobium* satisfy energy requirements anaerobically, producing and accumulating malate and ethanol, which are metabolised at the beginning of the terrestrial phase.   When submersed in shallow water, young *Astrocaryum jauari* have mixed anaerobic and aerobic respiration which is possible because the roots contain aerenchymatic tissue with large lacunae. It cannot be said to what extent the aerobic pathway is effective in adult and deeply submersed individuals, since only young plants covered by one meter of water could be investigated.   *Astrocaryum* is a slow growing species, with a permanently low metabolic rate which may allow permanent growth even when submersed.

Roots of many species accumulate large amounts of starch in different parts of the tissue to fuel anaerobic respiration during submergence. Considering the high number of flood tolerant tree species, further studies on the respiratory physiology should be very promising.

Formation of tree rings

Morphological and physiological adaptations of tree species growing under prolonged submergence obviously do not guarantee similar growth rates during dry and flood periods. The main growth period of trees occurs during the non-flooded period (Worbes, 1986).   When flooding starts, cambial activity of many species is reduced and finally stopped.   This produces a ring in the wood, the width of which varies according to the length of the dry period.

The boundary layers of the rings are marked by one or more of the following characteristics: (i) bands of alternating widths of parenchyma and fibre-cells in species such as *Nectandra amazonum, Rourea sp.* and *Vitex cymosa*; (ii) marginal parenchyma bands often containing metabolic products in species such as *Parkia auriculata* and *Macrolobium acaciaefolium;* (iii) decreasing size and thicker walls of fibre cells in most of the species investigated except Sapotaceae. The structure of the boundary layer varies within families but it is similar within a genus because of similar wood structure. The rings are inconspicuous and cannot be followed throughout the whole stem disc in some species such as *Crescentia amazonica, Swartzia argentea, Caraipa grandiflora* and *Poecilanthe amazonica*. Other species such as *Crataeva benthamii* and *Macrolobium acaciaefolium* have clear rings which are often discontinuous or divided. *Vitex cymosa* forms visible growth rings, whose borders cannot be determined when analyzed under the microscope. Other species have a well developed ring structure in the center, but hardly distinguishable rings near the cambium. This is the case in *Pseudobombax munguba, Laetia suaveolens* and *Labatia glomerata*. Some species show a clear ring structure throughout the whole stem disc, *e.g. Nectandra amazonum, Parkia auriculata, Pseudoxandra polyphleba* and *Sorocea duckei*.

Positive correlation between length of dry period and width of the respective ring proved that rings were formed on an annual basis. For the first time, in the humid tropics, it permitted determination of age and growth rate of native trees by ring counting. Additional studies on the ^{14}C bomb carbon content of anatomically predetermined rings of trees from the Amazon floodplain confirmed the results which are now being used to determine the age of tropical trees from other areas (Worbes and Junk, in press).

Conclusions

In the floodplains of large Amazonian rivers there is a large number of highly floodtolerant tree species. Morphological, anatomical, physiological and phenological adaptations to flooding are diverse but still little understood. The great species diversity observed in Amazonian floodplain habitats may be related to three factors: (i) the great diversity of habitats, caused and maintained by river dynamics, which allow the coexistence of a great number of species; (ii) the existence of floodplains under tropical climate during geologically long periods, which allowed speciation to occur; (iii) the predictability of flooding, which favored the development of adaptations.

The great diversity of habitats in any floodplain is related to river dynamics, which create patchy arrangements of soil types with different nutrient status, age and stability, and which are subjected to varying hydrological conditions. Communities of different ages and seral stages occur side by side in a highly dynamic system, subject to permanent modification (Kalliola *et al.*, 1987; Salo *et al.*, 1986).

The periodicity of precipitation over large catchment areas creates a high amplitude and a pronounced monomodal, or in a few cases bimodal, flood pattern in all large tropical rivers. It may be postulated that large rivers under tropical climate have always been subjected to similar annual pulses although with different amplitudes. With the uplifting of the Andes in the Tertiary, central Amazonia became a large sedimentation basin where freshwater sediments of riverine and lacustrine origin were deposited. This makes the existence of large floodplains and swamps over long periods very probable (Irion and Adis, 1979; Kubitzki, in press).

Some authors postulate major climatic changes in the Amazon basin during the last 150 000 years with drier and cooler climate during the glacial periods and a warmer climate with higher precipitation during interglacial times (Prance, 1979b). Such changes however would influence only the total discharge, amplitude and duration of floods but affect slightly the monomodal flood pattern of the large Amazon rivers. The huge eustatic changes of sea-level by as much as 130 meters during the last 150 000 years have had a strong impact on the position and the extent of the floodplains (Adis, 1984; Irion, 1982, 1984; Milliman and Emery, 1968). However they did not affect their existence. Accordingly, trees have had sufficient time and habitats available in the Amazon basin to adapt to flooding at least since the Tertiary. This is especially important when considering the evolution of structural elements such as those, which keep seeds afloat and complex coevolutionary aspects, such as those between fish and floodplain forests. The development of physiological adaptations to flooding may have required much shorter periods as indicated by the fact that some Amazonian species which occur in non floodable habitats have developed flood tolerant ecotypes in *várzea* and *igapó* (Kubitzki, in press).

In floodplains, the regular change from a pronounced terrestrial to a pronounced aquatic habitat represents a very strong stress to all organisms colonizing these areas and forces such organisms to develop adaptations. The predictability of the flood pulse of large rivers favors this tendency (Junk *et al.*, in press). This is clearly shown by the great number of adaptations recently discovered in nearly all groups of plants as well as in animals which colonize the floodplains of the Amazon River and Rio Negro (Adis, 1981, 1984; Goulding, 1980; Irmler, 1981; Junk, 1984).

Beside its impact on the local fauna and flora, the monomodal floodpulse of large tropical rivers may have had a general importance for evolution. In the humid tropics it represents a very potent abiotic timer, provoking a pronounced annual seasonality in an otherwise aseasonal environment. In the floodplains of large tropical rivers such as the Amazon River and its large tributaries, annual floods cause the development of annual biorhythms and may play a role similar to the one performed by the light temperature regime in higher latitudes, or by the change from the rainy to a dry season in the semiarid tropics and subtropics.

Literature cited

Adis, J. (1981). "Comparative ecological studies of the terrestrial arthropod fauna in Central Amazonian inundation forest." *Amazoniana* **8**, 87-173.

Adis, J. (1984). "Seasonal Igapó - forest of Central Amazonian blackwater rivers and their terrestrial arthropod fauna." pp. 246-268 *In* Sioli, H. (ed.), "The Amazon: limnology and landscape ecoly of a mighty tropical river and its basin." *Dr. Junk Publ. Dordrecht, Boston, Lancaster.*

Clark, J. R. and Benforado, J. (1981). "Wetland of bottomland hardwood forests." *Dev. Agric. Managed-Forest Ecol.* **11,** *Elsevier Scient. Publ. Comp., Amsterdam, Oxford, New York.*

Crawford, R. M. M. (1982). "Physiological responses to flooding." pp. 453-477 *In* Pirson, A. and Zimmermann, M. H. (eds.), *Enc. plant physiol. NS* **12B**, *Berlin,Heidelberg,New York.*

Demaree, D. (1932). "Submerging experiments with Taxodium." *Ecology* **13,** 258-262.

Dister, E. (1980). "Geobotanische Untersuchungen in der hessischen Rheinaue als Grundlage für die Naturschutzarbeit." *Diss. Univ. Göttingen.*

Dister, E. (1983). "Zur Hochwassertoleranz von Auenwaldbäumen an lehmigen Standorten." *Verh. Ges. Ökol.* **10**, 325-336.

Furch, B. (1984). "Untersuchungen zur Überschwemmungstoleranz von Bäumen der Várzea und des Igapó. Blattpigmente." *Biogeographica* **19**, 77-83.

Furch, B., Correa, A. F. F., Mello, J. A. S. N. de and Otto, K. (1985). "Lichtklimadaten in drei aquatischen Ökosystemen verschiedener physikalisch-chemischer Beschaffenheit. Abschwächung, Rückstreuung und Vergleich zwischen Einstrahlung, Rückstrahlung und sphärisch gemessener Quantenstromdichte (PAR)." *Amazoniana* **9**, 411-430.

Gill, C. J. (1970). "The flooding tolerance of woody species - a review." *Forestry Abstracts Vol.* **31**, No. 4. *Commonwealth Agric. Bureau, Farnam Royal, England.*

Gottsberger, G. (1978). "Seed disperal by fish in the inundated regions of Humaitá, Amazonia." *Biotropica* **10**, 170-183.

Goulding, M. (1980). "The fishes and the forest. Explorations in Amazonian natural history." *Univ. California Press, Berkeley,Los Angeles,London.*

Hosner, J. F. (1957). "Effects of water upon the seed germination of bottomland trees." *Forest Sci.* **3,** 67-70.

Hosner, J. F. (1960). "Relative tolerance to complete inundation of fourteen bottomland tree species." *Forest Sci.* **6,** 246-251.

Hueck, K. (1966). "Die Wälder Südamerikas." *Stuttgart.*

Irion, G. (1982). "Mineralogical and geochemical contribution to climatic history in Central Amazonia during quarternary time." *Trop. Ecol.* **23,** 76-85.

Irion, F. (1984). "Sedimentation and sediments of Amazonian rivers and evolution of the Amazonian landscape since Pliocene times." pp. 201-204 *In,* Sioli, H. (1984).

Irion, G. and Adis, J. (1979). "Evolução de florestas amazônicas inundadas, de igapó- um exemplo do rio Tarumá mirim." *Acta Amazonica* **9,** 299-303.

Irmler, U. (1981). "Überlebensstrategien von Tieren im saisonal überfluteten Überschwemmungswald." *Zool. Anz. Jena* **206,** (1/2), 26-38.

Junk, W. J. (1984). "Ecology of the várzea, floodplain of amazonian whitewater rivers." pp. 216-243 *In* Sioli, H. (1984).

Junk, W. J., Soares, M. G. M. and Carvalho, F. M. (1983). "Distribution of fish species in a lake of the Amazon River floodplain near Manaus (Lago Camaleao), with special reference to extreme oxygen conditions." *Amazoniana* **7,** 397-431.

Junk, W. J., Bayley, P. B. and Sparks, R. E. (in press). "The flood pulse concept in river-floodplain systems." *Can. J. Fish. Aquat. Sci.*

Kalliola, R., Salo, J. and Mainen, Y. (1987). "Regeneración natural de selvas en la Amazonia Peruana. 1: Dinamica fluvial y sucesion ribereña." *Mem. Mus. Hist. Nat. Javier Prado, 18, Univ. Nat. Mayor de San Marcos, Lima.*

Keel, S. H. and Prance, G. T. (1979). "Studies on the vegetation of a whitesand black-water igapó (Rio Negro, Brazil)." *Acta Amazonica* **9,** 645-655.

Klinge, H. (1973). "Struktur und Artenreichtum des zentral-amazonischen Regenwälders." *Amazoniana* **4,** 283-292.

Kubitzki, K. (1977). "The problem of rare and of frequent species.: The monographer's view." pp. 331-336 *In* Prance, G. T. and Elias, T. D. (eds.), "Extinction is forever." *New York Botanical Garden, Bronx.*

Kubitzki, K. (in press). "The ecogeographical differentiation of Amazonian inundation forests." *Pl. Syst. Evol.*

Melack, J. M. and Fisher, T. R. (1983). "Diel oxygen variations and their ecological implications in Amazon floodplain lakes." *Arch. Hydrobiol.* **98,** 422-442.

Milliman, J. D. and Emery, K. O. (1968). "Sea-levels during the past

35 000 years." *Science* **162**, 1121-1123.
Peters, C. M. and Vasquez, A. (1987). "Estudios ecologicos de Camu-Camu (Myricaria dubia). I. Produccion de frutos en poblaciones naturales." *Acta Amazonica* **16/17**, 161-173.
Piedade, M. T. F. (1985). "Ecologia e biologia reprodutiva de Astrocaryum jauari Mart. (Palmae) como exemplo de população adaptada as áreas inundáveis do Rio Negro (igapós)." *M.Sc. thesis INPA, FUA, Manaus.*
Prance, G. T. (1979a). "Notes on the vegetation of Amazonia III. The terminology of Amazonian forest types subject to inundation." *Brittonia* **31**, 26-38.
Prance, G. T. (1979b). "Distribution patterns of lowland neotropical species with relation to history, dispersal and ecology with special reference to Chrysobalanaceae and Caryocaraceae." pp. 59-88 *In* Larsen, K. and Holm-Nielsen, L. B. (eds.), "Tropical Botany." *Academic Press, London*
Rankin de Merona, J. and de Merona, B. (1988). "Les relations poissons - foret. " pp. 202-228 *In*"Conditions écologiques et économiques de la production d'une île de várzea: 1, île du Careiro." *Rapport terminal, ORSTOM, INPA, CEE.*
Revilla, J. D. (1981). "Aspectos florísticos e fitosociológicos da floresta inundavel (Igapó) Praia Grande, Rio Negro, Amazonas, Brasil." *M. Sc. thesis, INPA Manaus.*
Revilla, J. D. (1988). "Aspectos florísticos e ecológicos da floresta inundável (Várzea) do Rio Solimões, Amazonas, Brasil." *Ph. D. thesis INPA / FUA. Manaus.*
Salati, E. and Marques, J. (1984). "Climatology of the Amazon region." pp. 85-126 *In* Sioli, H. (ed.). "The Amazon: limnology and landscape ecology of a mighty tropical river and its basin." *Dr. Junk Publ., Dordrecht, Boston, Lancaster.*
Salo, J., Kalliola, R., Hakkinen, J., Mäkinen, Y., Niemela, P., Puhakka, M. and Coley, P. D. (1986). "River dynamics and the diversity of Amazon lowland forest." *Nature* **322**, 254-258.
Schmidt, G. W. (1973). "Primary production of phytoplankton in the three types of Amazonian waters. II. The Limnology of a tropical floodplain lake in central Amazonia (Lago do Castanho)." *Amazoniana* **4**, 139-203.
Sioli, H. (1965). "Bemerkungen zur Typologie Amazonischer Flüsse." *Amazoniana* **1**, 267-277.
Sternberg, H. O. R. (1960). "Radiocarbon dating as applied to a problem of Amazonian morphology." *Comptes Rendues du 18. Congr. Intern. de Geogr., Rio de Janeiro, 1956.*
Worbes, M. (1983). "Vegetationskundliche Untersuchungen zweier Überschwemmungswälder in Zentralamazonien - vorläufige

Ergebnisse." *Amazoniana* **8,** 47-65.

Worbes, M. (1986).   "Lebensbedingungen und Holzwachstum in
    zentralamazonischen Überschwemmungswäldern."   *Scripta Geo-
    botanica* **17**.

Worbes, M. and Junk, W. J. (in press). "Dating tropical trees by means of
    bomb carbon analysis." *Ecology*.

Gap-Phase dynamics and tropical tree species richness

G. S. HARTSHORN

World Wildlife Fund, Washington, USA

Over the past 25 years, modern studies of tropical forest dynamics have revolutionized our general concepts about the dynamics of forest structure, the role of canopy openings in natural regeneration, species packing in forest communities, internal heterogeneity of forest gaps, *inter alia*. The pioneering studies of gap-phase dynamics in different tropical forests (Hartshorn, 1978; Oldeman, 1978; Whitmore, 1978) called attention to the high frequency of natural tree falls and canopy openings. A calculated turnover time of 118±27 years for the La Selva primary forests in Costa Rica (Hartshorn, 1978) is representative of several lowland forests in tropical America (Brokaw, 1985; Hartshorn, 1989a).

Transient canopy openings, or gaps as they are popularly known, stimulate natural regeneration of many shade-intolerant plant species. Hartshorn (1978) estimated that 50% of the La Selva tree species are dependent on gaps for successful regeneration. Of the La Selva tree species that attain the canopy, 63% are estimated to be gap-dependent. Many of the valuable timber trees of tropical America are gap species, *e.g.*, *Cedrela odorata*, *Swietenia macrophylla*, and *Cedrelinga catenaeformis*. Furthermore, most light hardwoods (*sensu* Swaine and Whitmore, 1988) are classic gap species with fast growth rates, such as canopy trees of the Myristicaceae and Vochysiaceae.

The great richness of tree species (Gentry, 1988) as well as other plant species in many tropical forests (Gentry and Dodson, 1987) has focused interest on species-packing (=α-diversity) in tropical forests. Not surprisingly, the role of gap-phase dynamics in the maintenance of high species diversity has attracted much attention. The key role of disturbance in promoting species diversity in such disparate ecosystems as tropical forests and the marine rocky intertidal generated the intermediate disturbance theory. The importance of disturbance in maintaining high species diversity in non-static communities (Hubbell and Foster, 1986; Ashton, this volume; Swaine, this volume).

The high proportion and dependence of tropical tree species on gaps has stimulated research on how so many species can co-exist. The stochastic nature of gap occurence in time and in space and the internal

TROPICAL FORESTS
ISBN 0–12–353550–6

heterogeneity of gaps appear to play key roles in the differential availability and suitability of gaps for colonization by a large number of tree species (Hartshorn, 1980; Denslow, 1980; Orians, 1982; Brandani *et al.*, 1988).

Tree composition in natural gaps

The natural regeneration of trees was inventoried in 51 natural gaps in the La Selva primary forests (Hartshorn, 1978; Brandani *et al.*, 1988; Hartshorn and Hammel, 1989). Maximum age of regeneration, *i.e.*, from the date of the tree fall, ranged from 13 months to 16.4 years. Of the 51 study gaps, 90% were created by tree falls during the six wettest months. The 7672 individuals inventoried represented 273 species, which was 87% of the tree species reported for La Selva in 1978.

Hartshorn (1978, 1980) and Orians (1982) proposed that within-gap heterogeneity, specifically the occurrence of root, bole and crown zones within a gap, as well as temporal differences in zone availability for seedling colonization, should influence the species composition of natural regeneration in gaps. The gap inventory of trees recorded the following: (i) each individual as a survivor of the tree fall or colonizer; (ii) presence in root, bole or crown zone; and (iii) the area of each zone. Severity of disturbance from the tree fall was usually root>crown>bole zones, although some gaps did not have a root zone. The bole zone had the highest proportion (20%) of survivors, which also had the greatest diversity of tree species. Overall, 84% of the regenerating trees representing 253 species were colonizers, indicating that the primary determinants of tree composition in a gap occur after the gap is formed.

Because of the analytical problems with rare species, the analysis of species distributions in gap zones was done on the 30 most abundant species, each with at least 51 individuals (Brandani *et al.*, 1988). The expected numbers were calculated according to the actual sizes of the three zones in each gap. The results show that actual distributions are higher than expected from a random distribution for each of the three zones as well as for the combinations of root+bole and root+crown zones, whereas the bole+crown and root+bole+crown combinations had actual distributions less than expected. These results confirm our hypothesis of zone-specific mortality in gaps and that differential mortality probably occurs within the first few years after gap formation.

Most of the species included in the analysis of gap-zone composition are over- or underrepresented in one or more zones. Far more species (20 of 30) are over-represented in the root zone, whereas nearly half (14 of 30) are not under-represented in any zone. The strong over-representation of tree species in the root zone, which averages much smaller than the other two zones, confirms the importance of exposed mineral soil as a prime regeneration niche for several shade-intolerant tree species. These root-zone gap specialists range from small treelets (*e.g., Ocotea atirrensis*) to large canopy trees (*e.g., Laetia procera*) (*cf.* Lieberman *et al.*, 1985).

Multiple Discriminant Analysis (MDA) was used to test for similarity of species composition among gap zones, using 104 species with 10 or more individuals (Brandani *et al.*, 1988). A total of 137 gap zones was classified into three groups based on the abundances of the 104 species in each zone of each gap. The three groups (root, bole and crown zones) are clearly separated by the first two discriminant scores, with very little overlap. These results are particularly interesting because of the statistically significant association of species abundances with a specific gap zone, despite great overlap in the size of gap zones and the creation of all three zones in a gap at the same time. In other words, root zones of different aged gaps have greater similarity in composition and abundances of species than does a root zone with either the bole or crown zone of the same gap.

We used canonical correlation to analyze which factors (gap age, gap size and tree species that fell) influence the composition of tree species in regenerating gaps (Brandani *et al.*, 1988). Although we identified nine species as causal trees in 32 gaps, 21 were *Pentaclethra macroloba* , the dominant tree species in the La Selva forests (Hartshorn 1983). Hence, we treated each causal tree as *Pentaclethra* or non-*Pentaclethra*. Results of the canonical correlation show that the independent canonical variates are highly correlated with different factors in each gap zone: (i) age of gap with the root zone; (ii) size of gap with the bole zone; and (iii) causal tree species with the crown zone. Furthermore, 24 gap-colonizing species show a direct correlation with zone size in at least one of the gap zones. Non-*Pentaclethra* gaps have more species over-represented in root and crown zones than in *Pentaclethra*-caused gaps. About one-third of the tree species occur patchily in relatively few gaps, even though many are fairly common components of the La Selva forest. The dominant *P. macroloba* had 881 individuals in the study gaps, however, it was absent from two of the gaps.

Tree composition in strip clear-cuts

In 1984, USAID contracted the Tropical Science Center (TSC) of San José, Costa Rica to provide technical assistance for the forestry and land-use components of the Central Selva Resources Management Project (CSRM) in the Peruvian Amazon. The USAID project focused on the Palcazu valley, a small 20x70 km watershed transitional from high jungle to low in Peru's Central Selva. The Palcazu valley is exceptionally wet with three years of rainfall data averageing 6700 mm, with acid, nutrient-poor soils that limit sustainable land-use primarily to production forestry on the rolling hills of the lower valley (Hartshorn, 1981; Tosi, 1981; Bolaños and Watson, 1981).

TSC designed a vertically integreated forest management and production system based on local processing of forest products under the control of the forest owners, who manage their production forests on sustained yield principles (Tosi, 1982). USAID and TSC assisted with the creation of the Yanesha Forestry Cooperative (YFC), comprised exclusively of Amuesha native communities and individuals - the first Indian forestry cooperative in Amazonia (Moore, 1988; Stocks and Hartshorn, 1989). The YFC owns the local processing center, consisting of a portable sawmill, a bank of 44 PresCaps for preserving roundwood, and a portable charcoal kiln (Krones, 1987). YFC personnel (Amuesha Indians) not only run the processing equipment, they also have been trained to classify land-use capability, identify production forests and prepare operational plans for harvesting and managing their production forests (*e.g.*, Sanchoma *et al.*, 1986).

The key feature of the Palcazu forest management system is the rotation of long, narrow (30-40 m) strips through a production stand. All the timber on a strip is harvested in order to promote natural regeneration of shade-intolerant tree species (Hartshorn 1981, 1989b). In effect, each narrow strip is clear-cut, which simulates an elongated gap and promotes excellent colonization of the harvested strip by native trees. The long (100-300 m), narrow strips are located and scheduled for harvesting over a 30-40 years rotation in such a way to maintain the structural matrix of the primary forest and facilitate seed dispersal into recently harvested strips. For example, in the first production unit at Shiringamazu the 3-5 strips scheduled for cutting in any year are no closer than 150 meters to each other. Furthermore, the total stand area allocated to strips over the entire harvesting rotation is only 46% of the 493 hectare production forest (Sanchoma *et al.*, 1986).

Two demonstration strips were harvested in 1985 in the research forest near the project headquarters at Iscozacin. Because of uncertainty about optimum width of a strip, the first strip (F1) cut in May was 20x75 meters (=0.15 hectare) and the second strip (F2) cut in October-December was 50x100 meters (=0.5 hectare). Harvesting proceeded from small to large trees with each log removed from the strip after felling and the crown chopped up. Small poles were moved manually or bunched for skidding. All extraction of logs was by handwinch or by animal traction (oxen), hence skidding damage to soil was negligible.

Post-harvest slash was not burned, nor did any planting of agricultural crops occur on the two demonstration strips. The considerable quantity of organic debris (20-40 cm thick) covering the soil caused a 5-9 month lag in the aggressive colonization of the strips by seedlings. However, vigorous stump sprouting in the first months after harvesting created much spatial heterogeneity on the young strip. Nevertheless, by the end of the first year the rapid height growth of seedling colonists of gap species, including pioneers (*Cecropia montana, C. sciadophylla, Pourouma minor, P. spp., Vismia spp.*) had surpassed the stump sprouts and formed a more homogeneous canopy about two meters tall.

Inventories of the natural regeneration of trees have been done more or less annually on the first strip, but only once on the second strip. Complete inventories of all trees more than 50 cm tall were done at 30 months for F1 and 24 months for F2. Because of the above-mentioned lag due to decomposing slash and that seedling establishment continues well into the second year, the age of the regenerating stand should be considered as a maximum. More detailed information on procedures, analyses and results is in a technical report by Hartshorn (1988).

Tree species richness totaled 209 for the first demonstration strip (at 30 months) and 285 for the second strip (24 months). The higher total for the younger second strip is not surprising because of its much greater area. In fact, the perimeter tier of 10x10 meters subplots in the second strip has the same average number of tree species (50/subplot) as does the entire first strip. Average tree species richness on the inner subplots of the second strip is significantly lower than for the outer tier, indicating the importance of the forest edge in enhancing the natural regeneration of tree species.

Vegetative regeneration from cut stumps contributes to the richness of tree species regenerating on the two demonstration strips. On each strip, 13% of the species richness is due to species represented only by stump sprouts. Many tree species are represented by both seedlings and stump sprouts on the demonstration strips; 67% of the species on the second strip and 45% of those on the first strip have both types of regeneration. The greater number of tree species with both types of regeneration on the second strip appears to be a simple consequence of the much greater number of individuals due to the larger area.

There appears to be a fairly clear pattern of regeneration mode related to abundance and wood density. Rare or uncommon heavy hardwoods (*e.g., Hymenolobium elegans, Mezilaurus palcazucensis, Tabebuia obscura, Trichilia pleeana*) are present in the strip only as a few stump sprouts. More common heavy woods, such as *Diplotropis purpurea, Myrocarpus sp., Vatairea sp., Calocarpum sp., Micropholis spp.* and *Pouteria spp.* are well-represented in the natural regeneration on the two strips, mostly by stump sprouts. All these heavy hardwoods appear to be shade-tolerant species with good sprouting capability.

Medium hardwoods appear to regenerate largely by seedlings on the strips, which agrees with the gap requirement of many of these fast-growing species. Many examples of these medium hardwoods occur in the well-represented groups Burseraceae, Lauraceae, Myristicaceae and *Inga*. The last group consists of the shade-intolerant pioneer species that regenerate exclusively by seedlings. These heliophilic species aggressively colonize large gaps, including the strip clear-cuts, *e.g., Cecropia spp., Pourouma* spp., *Jacaranda copaia, Miconia spp.*, and *Vismia spp.*

Despite the great richness of species and the abundance of trees on the two demonstration strips, dominance is concentrated in a small number of species. On the first strip it takes nine species to account for 50% of the importance value, whereas it takes only five species to tally 50% of the importance value on the second strip. The greater concentration of dominance on the second strip is due to *Cecropia sciadophylla*, whose importance value is three times greater on the second than on the first strip. Similarly, the second through fourth most important species on the second strip all have greater importance values than the correspondingly ranked species on the first strip. The greater concentration of dominance by the classic pioneer species on the second strip is probably related to a more favorable light regime, which is a function of strip aspect and width. Interestingly, 15 of the top 20 species are the same on the two strips. In addition to the characteristic pioneer tree species of the Palcazu valley, the top 15 also includes some of the common gap species of primary forests, such as *Laetia procera, Mollia stellaris* , and several *Inga* species.

Many of the tree species present on the demonstration strips are represented by few individuals. For typically rare species the strip clear-cut may offer favorable regeneration sites, *e.g.*, *Minquartia guianensis* has two juveniles on the first strip and seven on the second, yet I have been unable to find a parent tree in the surrounding forest. These rare juveniles of valuable heavy hardwoods can be favored and protected during silvicultural treatments of the growing stand. Not only does it appear that strip clear-cuts can be used to promote excellent natural regeneration of hundreds of native tree species, natural forest management using strip clear-cuts may be able to maintain or even enhance tree species diversity.

Conclusions

Tree falls in tropical forests create gaps in the canopy that become prime regeneration sites for hundreds of native tree species. In the dynamic La Selva forests juveniles that survive the tree fall contribute less than 20% to the number of individuals and species comprising the natural regeneration established in gaps. The distribution of colonizers is clumped, both among as well as within gaps, probably due to non-random input of seeds into a gap and differential mortality of seedlings within a gap.

In natural gaps, specific zones (root, bole and crown) associated with the fallen trees are more similar to one another in species composition than they are to other zones of the same gap. Although total species richness is least in the small root zones, more tree species are over-represented in the root zone than in the other two zones. Species abundances in gap zones are significant variables for clearly discriminating these units, suggesting that the environmental features of a particular zone quickly override the influence of spatial and temporal contiguity within a gap. For those species with more than ten individuals, almost half show patchiness in their distribution among gaps.

The establishment of narrow strip clear-cuts in primary forests of the Peruvian Amazon promotes outstanding natural regeneration of hundreds of native tree species. Width of the strip clear-cut affects the richness of tree species regeneration, with a significantly greater number of tree species occuring within 10 meters of the forest edge. Although more than half of the tree species are represented by both seedlings and stump sprouts, only 13% of the tree species are represented only by vegetative regeneration. Major regeneration modes appear to be related to wood density: heavy hardwoods are shade-tolerant and occur almost exclusively as stump sprouts on the strips; medium hardwoods tend to be shade-intolerant and regenerate primarily by seedlings; and the heliophilic pioneers establish only from seed.

Strip clear-cuts appear to effectively simulate natural gaps as excellent regeneration sites for a high proportion of the native tree species in some tropical forests. A well-designed and implemented rotation of strips in a production forest should maintain the forest structure, while producing timber on a sustained yield basis. Natural forest management using strip clear-cuts will permit the economic use of tropical production forests without destroying their biological habitats and ecological services. Sound forest management based on the rapidly expanding knowledge about tropical forest dynamics and gap-phase regeneration may not only help save tropical forests, but it could also help maintain and possibly enhance the conservation of biological diversity.

Literature cited

Bolaños M., R. A. and Watson C., V. (1981). "Report on the ecological map of the Palcazu valley." *In* "Central Selva natural resources management project: USAID project No. 527-0240," appendix C. JRB Assoc., 2 vols *McLean, Virginia,*

Brandani, A., Hartshorn, G. S. and Orians, G. H. (1988). "Internal heterogeneity of gaps and species richness in Costa Rican tropical wet forest." *J. Trop. Ecol.* **4**, 99-119.

Brokaw, N. V. L. (1985). "Treefalls, regrowth, and community structure in tropical forests." pp. 53-69 *In* S. T. A. Pickett and P. S. White (eds.),"The ecology of natural disturbance and patch dynamics." *Acad. Press, New York.*

Denslow, J. S. (1980). "Gap partitioning among tropical rain forest trees."

Biotrop. **12,** (suppl.), 47-55.

Gentry, A. H. (1988). "Tree species richness of upper Amazonian forests." *Proc. Nat. Acad. Sci.* **85,** 156-159.

Gentry, A. H. and Dodson, C. (1987). "Contribution of nontrees to species richness of a tropical rain forest." *Biotrop.* **19,** 149-156.

Hartshorn, G. S. (1978). "Tree falls and tropical forest dynamics." pp. 617-638 *In* Tomlinson P. B. and Zimmermann M. H. (eds.), "Tropical trees as living systems." *Cambridge Univ. Press, Cambridge.*

Hartshorn, G. S. (1980). "Neotropical forest dynamics." *Biotrop.* **12,** (suppl.), 16-23.

Hartshorn, G. S. (1981). "Forestry potential in the Palcazu valley." *In* "Central Selva natural resources management project: USAID project No. 527-0240." appendix *G. JRB Assoc., McLean, Virginia.*

Hartshorn, G. S. (1983). "Plants." pp. 118-157 *In* Janzen D. H. (ed.), "Costa Rican natural history." *Univ. Chicago Press, Chicago.*

Hartshorn, G. S. (1988). "Natural regeneration of trees on the Palcazu demonstration strips." *Tech. Report* **88-2-E,** *58 p., Trop. Sci. Center, San José, Costa Rica.*

Hartshorn, G. S. (1989a). "An overview of neotropical forest dynamics." *In* Gentry A. H. (ed.), "Four neotropical forests" [in press] *Yale Univ. Press, New Haven.*

Hartshorn, G. S. (1989b). "Application of gap theory to tropical forest management: natural regeneration on strip clear-cuts in the Peruvian Amazon." *Ecol.* [in press]

Hartshorn, G. S. and Hammel, B. E. (1989). "Introduction to the flora and vegetation of La Selva." *In* McDade, L., Bawa, K., Hespenheide H.,and Hartshorn G. (eds.), "La Selva biological station: ecology and natural history." [in press] *Univ. Chicago Press, Chicago*

Hubbell, S. P. and Foster, R. B. (1986). "Biology, chance, and history and the structure of tropical rain forest tree communities." pp. 314-329 *In* Diamond, J. and Case, T. (eds.), "Community ecology." *Harper and Row, New York.*

Krones, M. (1987). "Informe final sobre las actividades desarrolladas en la implementacion y puesta en marcha del primer nucleo de transformacion en la cooperativa forestal "Yanesha." *Tech. Report* **114-C,** *131 p. Trop. Sci. Ctr., San José, Costa Rica.*

Lieberman, D., Lieberman, M., Hartshorn, G. and Peralta, R. (1985). "Growth rates and age-size relationships of tropical wet forest trees in Costa Rica." *J. Trop. Ecol.* **1,** 97-109.

Moore, T. (1988). "La cooperativa forestal Yanesha: una alternativa autogestionaria de desarrollo indigena." *Amazonia Indigena* **13,** 18-27.

Oldeman, R. A. A. (1978). "Architecture and energy exchange of dicotyledonous trees in the forest." pp.535-560 *In* Tomlinson, P. B.

and Zimmermann, M. H. (eds.), "Tropical trees as living systems" *Cambridge Univ. Press, Cambridge.*

Orians, G. H. (1982). "The influence of tree-falls in tropical forests in tree species richness." *Trop. Ecol.* **23,** 255-279.

Sanchoma R. E., Simeone G. R., Velis, M. and Vílchez, B. H. (1986). "Plan de manejo forestal: bosque de produccion de la Comunidad Nativa Shiringamazu, 1987-1989." *Tech. Report* **105-C,** *37 p. Trop. Sci. Center, San José, Costa Rica.*

Stocks, A. and Hartshorn, G. (in press). "The Palcazu project: forest management and native Amuesha communities." *In* Hecht, S. and Nations, J. (eds.), "The social dynamics of deforestation in Latin America: processes and alternatives." *Univ. California Press, Berkeley.*

Swaine, M. D. and Whitmore, T. C. (1988). "On the definition of ecological species groups in tropical rain forests." *Vegetatio* **75,** 81-86.

Tosi, J. A., Jr. (1981). "Land use capability and recommended land use for the Palcazu valley." *In* "Central Selva natural resources management project: USAID project No. **527-0240**" appendix *N. JRB Assoc., McLean, Virginia.*

Tosi, J. A., Jr. (1982). "Sustained yield management of natural forests: forestry subproject, Central Selva resources management project, Palcazu valley, Peru." *Tech. Report, 68 p., Trop. Sci. Ctr., San José, Costa Rica.*

Whitmore, T. C. (1978). "Gaps in the forest canopy." pp. 639-655 *In* Tomlinson, P. B. and Zimmermann, M. H. (eds.), "Tropical trees as living systems" *Cambridge Univ. Press, Cambridge.*

Patterns of tree species diversity and canopy structure and dynamics in humid tropical evergreen forests on Borneo and in China

E. F. BRUENIG AND HUANG Y-W.

Hamburg University, FRG

Diversity has become a fashionable and diversely applied term. In the context of this paper, we confine the term to two conditions. First, the diversity or evenness (Haeupler, 1982) is a measure of the pattern of plant species composition in a plant community. Secondly, the diversity is a measure of difference of floristic or morphological characteristics between plant communities. This conforms basically to the alpha (species) and beta (community) diversities of Whittaker (1965, 1970), but includes morphological features.

In this paper we compare diversity of structure in the *kerangas* (heath forests) and peatswamp forests of Sarawak and Brunei and rain forests in China. Canopy gap dynamics will be illustrated by an example of *Shorea albida* forest in Sarawak. A further comparison of patterns in 1963 and 1988 in the 20 hectare sample plot in Sabal Forest Reserve, Sarawak (Newbery, 1985; Newbery *et al.*, 1986) and a comparison with the new ecosystem research areas in Bawang Ling have been initiated but will not be completed before 1990. Further extension to a physiognomic-structural comparison with Australian rainforests (Webb, 1968) is planned.

The data base for this paper are the sample plots of the ecological study of the kerangas forests in Sarawak and Brunei, studied from 1954 to 1963, by the Sarawak Forest Department and the senior author, and data of an ecological survey of evergreen rainforests in Hainan and Guangdong collected in 1983 by the junior author and collaborators. The evaluation and supplementary data are part of a special project of comparative structural analysis. Additional data on the regeneration of *Shorea albida* in the area defoliated in 1947 by a hairy caterpillar *ulat bulu* were supplied by Ngui (1986).

TROPICAL FORESTS
ISBN 0–12–353550–6

Species richness, evenness of mixture and site

Analyses of floristic and morphologic forest community structure in the Bornean *kerangas* and in the the Amazon *caatinga*, showed surprisingly consistent patterns (Bruenig *et al.*, 1979; Bruenig and Klinge, 1977; Kurz, 1983; Weischke, 1982). The same happened in the comparison between Mixed Dipterocarp and *kerangas* forest in Borneo and lowland dipterocarp-bearing forest, *kerangas* and montane forests in China. The conclusions of these comparisons with respect to diversity were that tree species richness in pristine forest is primarily controlled by site factors, especially by the physical and chemical soil properties. The evenness of species mixture, or the values of any of the diversity indices, are more closely related to the dynamics of the regenerative cycle or to successional changes of vegetation and soil, or to the phases of recovery after disturbance, than to site.

The values of the McIntosh index (Greig-Smith, 1983), which was used for the comparison, vary much within a community and the differences between communities are so small that the index is not as useful for analysis on small scales as it is for global and broad comparisons. The first exploratory test of this index in a comparative study of rainforest structure in Borneo and Hainan proved the usefulness of the index for such a regional analysis. The relative positions of the stands in Borneo and China (Fig. 1) corroborate the hypothesis that the tropical evergreen rain forest stands in China show structural features which in Borneo occur on sites with more unfavourable growing conditions such as on podzols, peatbogs or exposed, higher altitude sites. This opens the question for the underlying causal factors, which is being pursued by the on-going comparative ecosystem research in Bawang Ling (Bruenig and Huang, 1987). This also opens the question of what taxa really mean in terms of ecology. The similarity of species richness on equally poor soils in equatorial Borneo and Amazonia, and on better soils at a higher latitude in South China indicates significance. The close association of physiognomic and structural features to soil and site type units and their inter-regional similarity justify the hypothesis that these features are adaptations. If this were so, then the question about the ecological role of the various physiognomically similar species and of the species richness and diversity arises.

Physiognomic-structural diversity and site

If the hypothesis is right that morphologic and in particular physiognomic features of vegetation communities are adaptions which express the growth

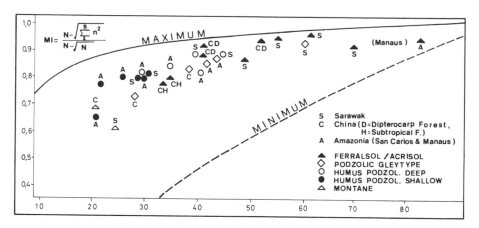

Figure 1. Mean species richness and mean diversity of predominantly evergreen lowland and montane forests on a variety of soils in *kerangas* in Sarawak and Brunei (S), Hainan (C) and Guangdong (CH) and Amazonia (A). Species richness is in Borneo and Amazonia similar on similar soils but lower in the forests at higher latitudes in China. The diversity index declines somewhat with lower species richness, but is distinctly lower only in stand with such high dominance of a single species that no index would be required to detect and demonstrate it. MI : McIntosh Index, N = 100 individuals, n = number at trees of each species in the sample of 100 trees.

potentials and survival risks of a site (Bruenig, 1970, 1971), then an ordination of vegetation stands with different structural features should produce patterns of scatter which can be ecologically interpreted. One of the authors (Huang) ordinated by various parameters 48 forest stands in Borneo (41 plots) and China (7 plots) in one set. The ordination of sample plots, representing vegetation communities such as a stand of trees, in principle orders the sample plots by their similarity or dissimilarity respectively, with respect to a set of chosen parameters, such as leaf sizes or any suitable other structural feature, or species composition. In the present study, one series, or ordination run, applied the Decorana ordination method (Hill, 1979) to arrange the plots in a multidimensional parameter space by their leaf size spectra. This was done for the canopy as a whole, and separately in four layers delimitated by stem dbh in cm (2.4-8.9, 9.0-13.7, 13.8-19.4, > 19.4 cm). One of the resulting ordinations is, for two dimensions only, reproduced in Figure 2, which shows:
- the lower left corner is occupied by predominantly meso-notophyll stands on (a) medium humus podzol and (b) peatswamp (*Shorea albida*-bearing);
- sample plot 16 (*Dryobalanops beccari - Shorea flavescens*-bearing forest in Bako N. P.) appears from its edaphic position deplaced into the

range of noto-microphyll forest stands on a variety of ferralsols, bleached clays (ultisols) and deep humus podsols (*Agathis borneensis*-bearing); - the nano-leptophyll segment are forests on shallow humus podzols (sample plots 15, 39), shallow gley-clay (33, 53), upland peats (54, 40), medium-deep humus podzol at higher altitude (52, 56) and secondary *kerangas* (17); - the forest stands in China generally grow on better soils than their leaf-size spectra would suggest, and are accordingly displaced into the opposite direction as plot 16 in Bako N. P.The tentative interpretation is that: (i) leaf-size spectra of trees of a natural tree community reflect soil and site conditions, especially soil-controlled rooting depth and altitude; (ii) noto-phyll-microphyll forests occur on deeper and deeper rooted soils which are more consistently well water supplied, and; these forests are taller than nano-leptophyll forests; (iii) the exceptionally tall nano-leptophyll forest in sample plot 17 is old (approx. 100 years), secondary forest on a secon-dary podzol developed after clearing Mixed Dipterocarp forest and gam-bier growing on a sandy ultisol (Bruenig, 1974); (iv) the aberrant positions of forest stand 16 can be explained by additional exogenous inputs of

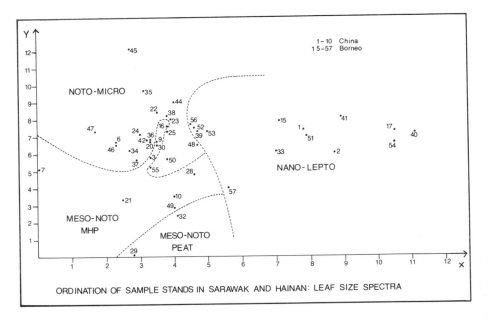

Figure 2. Result of the ordination by leaf-size spectra of 37 forest stands in Sarawak and Brunei and seven forest stands in China, calculated with Decorana (Hill, 1979).

water and nutrients into the site from up-slope and/or root-spreading into adjacent better soils down-slope (Bruenig, 1971).

In Borneo both the physiognomic and floristic ordinations reflect again the close relationship between montane (sample plots 54, 57) and lowland *Shorea albida* bearing stands (28, 29, 32, 49) which was demonstrated in an earlier ordination (Bruenig, 1970).

The tentative conclusions are: (i) the trend in physiognomic differences between the forest stands is related to soil/site conditions in a very similar fashion as the trends expressed in the floristic ordination, but differences due to exogenous supplies of growth factors (sample plot 16) or succession (17) are more explicitly demonstrated by the physiognomic ordination; (ii) the trends in physiognomic diversity are similar along the 3 gradients from oligotrophic, droughty soils to more mesic sites, lowland to montane and equatorial to tropical.

Diversity of gap formation, canopy topography and forest type.

The single-species dominated *Shorea albida* peatswamp forests of Sarawak provide an unique opportunity to study the pattern of gap formation along an ecological gradient of peat-soils and associated forest communities. These forest communities form a phasic sequence from the edge of the peatswamp to its central raised dome (Anderson, 1961a). The phasic communities are distinguished by their stature and structure (Fig. 3). *Shorea albida* forms the canopy in forest types 36 (*alan* forest) and 37 (*alan bunga* forest) and 38 (*padang alan*). In 36 and 37 and less in 38, lightning and wind-throw cause gaps of very different sizes. An unidentified hairy caterpillar (*ulat bulu*) caused an at least exceptionally rare and episodic, if not unique catastrophic canopy die-back over large areas in 1948, of which there has been no previous records and no indications of such occurrence in pollen analysis of peatswamp profiles and which has not recurred since (Anderson, 1961b and 1964).

The peculiarly uniform canopy of *Shorea albida* forests makes canopy gaps clearly recognizable on aerial photographs. A time-series of aerial photographs of the forests between Sungai Nyabor and Sungai Karap in northern Sarawak had been interpreted for gap frequencies and sizes in 1947, 1963 and 1968 (Bruenig, 1964 and 1973). The conclusions then were that damage patterns were closely related to canopy surface topography and varied markedly along the sequence of the phasic communities from 36 to 38. While damage from *ulat bulu* has remained a singular episodic event, windbreak, windthrow and lightning strikes, and subsequent gap formation are chronic events, but also these peaked markedly between 1963 and 1968. It was concluded that generally, but particularly in forest type 37, the chances of a tree to be damaged by lightning or wind during its lifetime and the area proportion of gaps were high enough particularly in the *alan* and *alan bunga* forest (Tab. 1) to be ecologically and economically important factors of disturbance.

New aerial photographs of 1981 became available for interpretation recently, bringing the whole time series to almost 34 years. The comparison of the situations in 1968 and 1981 gave the following results: (i) the number and area proportion of lightning gaps increased again in the *alan bunga* consociation (Tab. 1, 2, Fig. 4, 5 and the map in Fig. 6) indicating that the rate of gap formation is not equal in different years and periods; (ii) the edges of old gaps remained unchanged or gradually closed by crown extension along the margins; (iii) new wind-throw occurred at the edges of existing gaps; (iv) while the area of the old 1968 gaps changed only little, the total wind-throw area increased slightly between 1968 and 1981 as a result of new break and throw; (v) there is no regeneration of *Shorea albida* visible in large old gaps; (vi) in the *ulat bulu* area scattered small whitish crowns, characteristic for *Shorea albida* (Bruenig, 1969b), appear on the 1981 air-photos but lack of regeneration of

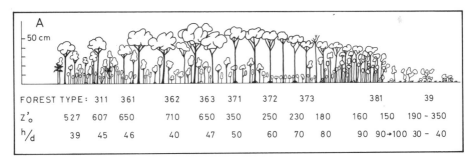

Figure 3. Profile from edge to center of a Borneo peatswamp. The diagram shows mature natural deltaic peatswamp forest types in everwet (p 2000 - 4000 m), nonseasonal to weakly seasonal equatorial lowland climate in Sarawak. The forest types are described in Bruenig (1969b) and correspond to the phasic communities of Anderson (1961), 311 - Mixed peatswamp forest (*Gonystylus bancanus*), 361-363 - Alan (*Shorea albida* Sym.) association, 371-373 - *Alan bunga* (*S. albida*) consociation, 381 - *Alan* (*S. albida*) *padang* forest, 39, Open *padang* woodland (mixed species, common *Combretocarpus rotundatus*). zo: estimator of the dimensionless parameter expressing aerodynamic roughness of canopy, also indicator of the capacity and intensity of intake of energy and turnover of energy, water, nutrients and organic matter, and of species richness and diversity. h/d: mean height/diameter ratio of upper-canopy trees calculated as mean values of sample plots; this index indicates vigour and mechanical stability (resistence against wind) of the trees.

Table 1　Area percentage occupied by lightning and windthrow gaps in 1451 hectare of *Shorea albida* forest in 1968, 363 = *alan* association, 37 = *alan bunga* consociation, 38 = padang.　Forest types according to Bruenig, 1969b; compare Fig. 3.

Forest type	362	363	37	38
lightning gaps (%)	0.7	3.0	1.0	0.2
windthrow (%)	0	1.4	1.5	< 0.1

362:　transitional forest in the ecotone between *Shorea albida-* forest and peripheral mixed swamp forest with very open upper canopy in which gaps do not show well.

363:　dominated by tall and big-crowned *Shorea albida* Sym., very rough and irregular canopy.

37:　dominated by tall but slender *S. albida*, moderately rough to rather smooth, uniform canopy.

38:　mixed forest of lower, slender and small-crowned trees, including *S. albida*, transitional to the central padang woodland with *Combretocarpus rotundatus* (Miq.) Dans., very smooth, uniform canopy.

Shorea albida in the *ulat bulu* area and its replacementby *Litsea grassifolia (medang padang)* has been reported from ground surveys (Ngui, 1986).

Conclusion

The kind of scatter of an ordination of forest communities or tree stands, represented by small sample plots in Borneo, China and Amazonia, according to species richness and evenness (diversity) of species mixture confirms earlier contentions that mature forests of tropical moist evergreen forests within a climatic zone have the same floristic structure on the same soil/site units along the edaphic gradients from mesic ("climatic climax") to xeric or alternatingly dry and water-logged and oligotrophic soils.　Climatic gradients related toaltitude and /or latitude have

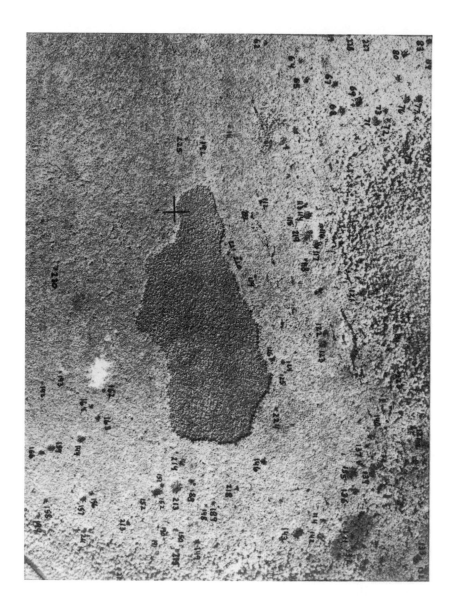

Figure 4. Aerial view of parts of the gap observation area. 1963: Lightning gaps are marked, the *ulat bulu*-gap of 110 hectare size shows the original understory of *alan bunga* forest, type 372 in Fig. 3. (Courtesy Sarawak Government)

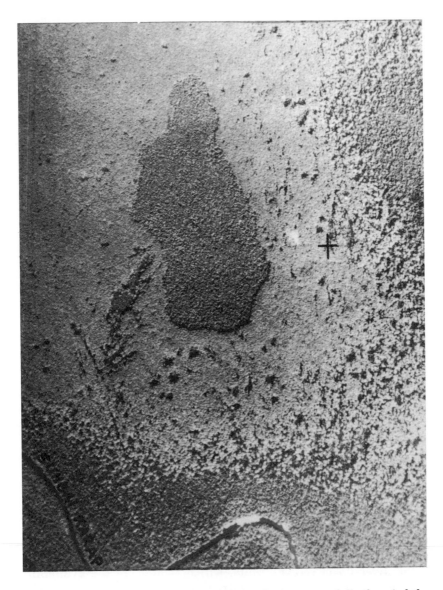

Figure 5. Same area as Fig. 4 in 1981: note the changes, especially the wind-throws caused by episodic violent micro-bursts of descending cold air-streams and line-squalls in thunderstorms and by occasional tornadoes. (Courtesy Sarawak Government)

Table 2 Forest type areas and distribution and size of new (1968-1981) lightning and windthrow gaps in the 1186 hectare reassessed area of *Shorea albida* forest. 36 = *alan* association, 37 = *alan bunga* consociation, 37D = *ulat bulu* area, 38 - *padang*. Forest types according to Bruenig, 1969b. For scientific names see Table 1. The caterpillar *ulat bulu* has not yet been authentically identified and named.

Forest type	36	37	37D	38
area (ha)	226	491	110	359
%	19.1	41.4	9.3	30.3
gaps				
number	19	74	9	14
area (ha)	3.98	7.71	0.50	0.86
%	1.76	1.57	0.45	0.24
mean area (ha)	0.20	0.10	0.05	0.06
size range (ha)	0.02-0.71	0.02-0.58	0.02-0.12	0.02-0.16
windthrow				
number	12	23	0	0
area (ha)	3.84	11.38	0	0
%	1.70	2.32	0	0
mean area (ha)	0.32	0.50	0	0
size range (ha)	0.14-0.28	0.12-1.30	0	0

the same effect with respect to species richness and to some extent to diversity as the edaphic gradients related to nutrient and water availability and rooting depth.

The ordination of the same stands in Borneo and China by physiognomic parameters produced a scatter of forest stands which showed that diversity of leaf-size spectra and the related degree of sclerophylly differs in the same manner as species diversity and that it also coincides with the diversity of soil/site units, which relates to the availability of growth factors such as water and nutrients.

The continued observation of the formation and development of lightning gaps and the pattern of scatter among the phasic communities of peatswamp forests corroborates the previous hypothesis that the diversity of forest stand structure (*i.e.* canopy tallness and topography) are reflected

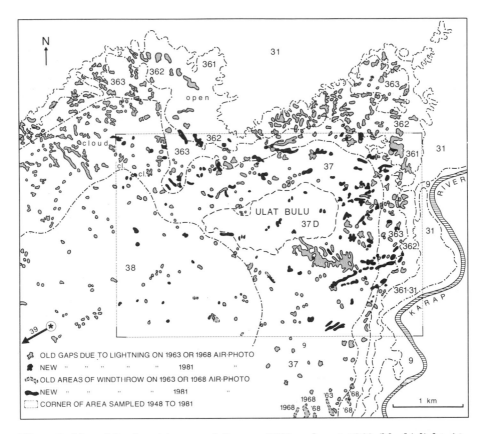

Figur 6. Map of the forest types and the pre-1968 and post-1968 (black) lightning holes and windthrows. The boundary of the surveyed part of the area on 1981 aerial photography is marked by dotted line. 39 * = the arrow points along the gradient from *alan* forest (38) toward open *padang* woodland (39), the epicenter of the peat dome lies approx. 1.5 km in direction of the arrow. The previous states 1948 and 1963 are shown in Bruenig, 1973.

by the pattern of successful (killing) lightning strikes. Similarly, windthrow is most abundant in *alan bunga* forest where aerodynamic roughness of the canopy is still high but trees are more slender. The resulting higher h/d ratio makes the trees more susceptible to swaying in wind and consequently to break and throw. The role of the canopy roughness estimator Zo also as an ecological indicator, especially for climatic, meteorological and growth processes, is well known. More recently, air-photo interpreted Zo in broadleaf and coniferous forests in northwest Germany has been shown to be closely and significantly correlated with the richness in bird species (Boettcher, 1987; Thieme, 1988).

The means and ranges of gap sizes in the *Shorea albida* forest correspond to those reported from structurally similar forests in the Amazon *caatinga* in San Carlos de Rio Negro and from successionally dynamic ecotones in Bornean *kerangas, e.g.* Gunong Pueh National Park and the Merurong plateau in Sarawak (Bruenig, 1974).

The pattern of tree species distribution in Sabal Forest Reserve, Sarawak, suggests that gap formation in connection with mobilization of the seed bank is the driving force in maintaining tree species diversity. Their function in *Shorea albida* forest - maintaining regeneration cycles or successional phasic developments - is not yet sufficiently understood and requires further long-term observation and basic ecosystem and eco-physiological research.

Regeneration of *Shorea albida* seems to be erratic and on a whole slow to appear and develop. Regeneration is not noticeable on the aerial photographs in gaps, but whitish crowns appear scattered in the *ulat bulu* area, while ground-surveys record absence of regeneration. Gaps close from the sides and this is most effective in small gaps in the *Shorea albida* forest type 37, *alan* forest.

The failure of *ulat bulu* to reoccur during the period 1948-88 points to the episodic uniqueness of the *ulat bulu* damage. The very slow recovery of the vegetation in the *ulat bulu* gap and the reported failure of *Shorea albida* to regenerate in this and other areas suggests that such events are extremely rare. Also the periodicity of the increases of gap frequency and gap area emphasizes that sporadic episodic events and periodicism may be contributing factors to the dynamics of the forests and must be considered in assessing dynamic processes in the vegetation and soil and must be also considered in silviculture and management planning (Bruenig, 1969a; Bruenig and Schmidt-Lorenz, 1985).

Acknowledgments

We thank the German Research Foundation (DFG, projects Br 316-6 and 316-11), the Volkswagenwerk Foundation (project II/63 132 1) and the Chinese-German Cooperative Ecological Research Programme (CERP) funded by the Federal Ministry for Research and Technology and administered by Unesco, Division of Ecological Sciences. Special thanks go to Mr. Lee Hua Seng, Deputy Director for Research, Forest Department Sarawak and Mr. S. K. Ngui, National Parks Office, Forest Department, Sarawak.

Literature cited

Anderson, J. A. R. (1961a). "The ecological types of the peatswamp forests of Sarawak and Brunei in relation to their silviculture." *Ph.D. Thesis, Univ. Edinburgh.*

Anderson, J. A. R. (1961b). "The destruction of Shorea albida forest by an identified insect." *Empire Forest Rev.* **40,** 103, 19-28.

Anderson, J. A. R. (1964). "Observations on climatic damage in peatswamp forest in Sarawak." *Commonwealth Forestry Rev.,* **43,** 116, 145-158.

Boettcher, J. H. (1987). "Zusammenhänge zwischen der Vegetationsstruktur und dem Artenreichtum der Vogelwelt (Relationships between vegetation structure and species richness of the avifauna)." *Diploma Thesis, Biol. Dept. Univ. of Hamburg.*

Bruenig, E. F. (1964). "A study of damage attributed to lightning in two areas of Shorea albida forest in Sarawak." *Commonwealth Forestry Rev.* **43,** 134-144.

Bruenig, E. F. (1969a). "On the seasonality of droughts in the lowlands of Sarawak (Borneo)." *Erdkunde* **23,** 127-133.

Bruenig, E. F. (1969b). "Forest classification in Sarawak." *Malay. Forest.* **32,** 143-179.

Bruenig, E. F. (1970). "Stand structure, physiognomy and environmental factors in some lowland forests in Sarawak." *Trop. Ecol.* **11,** 26-43.

Bruenig, E. F. (1971). "On the ecological significance of drought in the equatorial wet evergreen (rain) forest of Sarawak (Borneo). Transact. First Aberdenn-Hull Symposium on Malesian Ecology." *Univ. Hull, Dept. Geogr., Misc. Ser.*No. **11,** 66-97. Hull, U.K.

Bruenig, E. F. (1973). "Some further evidence on the amount of damage attributed to lightning and wind-throw in Shorea albida-forest in Sarawak." *Commonw. For. Rev.* **52,** 3, 260-265.

Bruenig, E. F. (1974). "Ecological studies in the kerangas forests of Sarawak and Brunei." *Borneo Literature Bureau. Govt. Printer, Kuching.*

Bruenig, E. F. and Klinge, H. (1977). "Comparison of the phytomass structure of tropical rainforest stands in Central Amazonas and in Sarawak, Borneo." pp. 81-101 *In*: Essays presented to E. J. H. Corner for his 70th birthday. *Garden's Bull., Singapore,* 1977.

Bruenig, E. F., Alder, D. and Smith, J. P. (1979). "The international MAB Amazon rainforest ecosystem pilot project at San Carlos de Rio Negro: vegetation classification and structure." pp. IV x 295 *In* Adisoemarto, S. and Bruenig, E.F. (eds.), "Transaction of the second international MAB-IUFRO workshop on tropical rainforest ecosystems research". *Chair of World Forestry, Hamburg-*

Reinbek, Special Report No. 2.

Bruenig, E. F. and Schmidt-Lorenz, R. (1985). "Some observations on the humic matter in kerangas and *caatinga* soils with respect to their role as sink and source of carbon in the face of sporadic episodic events." *SCOPE/UNEP Special Issue* **58**, 107-122.

Bruenig, E. F. and Huang, Y.-w. (1987). "Bawang Ling Nature Reserve: A potential international research and demonstration site and MAB Biosphere Reserve." *Tuebingen, Plant Research and Development, 26,* 19-35.

Greig-Smith, P. (1983). "Quantitative plant ecology." 3rd ed., *Blackwell Sci Publ., Oxford London.*

Haeupler, H. (1982). "Evenness als Ausdruck der Vielfalt in der Vegetation (eveness as expression of the diversity in the vegetation)." *Dissertationes Botanicae,* **65**, 1-267 Cramer, Vaduz.

Hill, M. O. (1979): "DECORANA: A FORTRAN Program for detrended correspondence analysis and reciprocal averaging." *Ecology and Systematics, Cornell Univ., Ithaca, New York.*

Kurz, W. A. (1983). "Biomasse eines Amazonischen immergrünen Feuchtwaldes: Entwicklung von Biomasseregressionen." *Mitt. Bundesforsch.anst. Forst- Holzwirtsch.* No. 139.

Newbery, D. McC. (1985). "Analysis of spatial pattern at San Carlos de Rio Negro." *Unpubl. research report.*

Newbery, D. McC., Renshaw, E., and Bruenig, E. F. (1986). "Spatial pattern of trees in Kerangas forest, Sarawak." *Vegetatio* **65**, 77-89.

Ngui, S. K. (1986). "Regeneration survey of "Ulat bulu" areas in the Baram peatswamp forests." *Unpubl. report, Forest Dept. Sarawak.*

Thieme, F. (1988). "Kronendachrauhigkeit und Vogelartenreichtum in nordwestdeutschen Nadelwäldern." *Thesis, Chair of World Forestry, Univ. Hamburg.*

Webb, L. J. (1968). "Environmental relationships of the structural types of Australian rainforest vegetation." *Ecology* **49**, 296-311.

Weischke, A. (1982). "Struktur und Funktionen in Waldkosystemen: Strukturvergleich zwischen Kerangas und *Caatinga*." *Thesis, Chair of World Forestry, Univ. Hamburg.*

Whittaker, R. H. (1965). "Dominance and diversity inplant communities." *Science,* **147**, 250-260.

Whittaker, R. H. (1970). "Communities and ecosystems. Curent Concepts in Biology." *The Macmillan Co. London.*

Architectural concepts for tropical trees

D. BARTHÉLÉMY, C. EDELIN AND F. HALLÉ

Institut Botanique, Montpellier, France

Following Corner's ideas (1949, 1953, 1954) on the growth, form, and evolution of plants, the study of plant architecture became a discipline in its own right. The first syntheses were published by Hallé and Oldeman (1970), Oldeman (1974), and Hallé *et al.* (1978). Architectural studies started in tropical regions and were concerned with the aerial vegetative structure of trees. Architectural concepts, however, provided a powerful tool for studying plant form, and investigations quickly spread to temperate species, mostly herbs and lianas, and root systems were also investigated.

The architecture of a plant depends on the nature and relative arrangement of each of its parts; it is, at any given time, the expression of an equilibrium between endogenous growth processes and exogenous constraints exerted by the environment. The aim of architectural analysis is to identify these endogenous processes by means of observation (Edelin, 1984). Considering the plant as a whole, from germination to its death, architectural analysis is a global and dynamic analysis of plant development. For each species and each stage of development, observations are made on varying numbers of individuals, depending on the complexity of the architecture. The results are summarized in diagrams which symbolize successive growth stages. The validity of the diagrams is then checked by comparing them with reality, and they must apply to the architecture of any individual of the same species for the analysis to be considered complete. Architectural analysis depends on three major concepts: architectural model, architectural unit, and reiteration.

The architectural model

For a tree, the growth pattern which determines the successive architectural phases is called its architectural model, (Hallé *et al.*, 1978).

The architectural model (Hallé and Oldeman, 1970) is an inherent growth strategy which defines both the manner in which the plant elaborates its form and the resulting architecture. It expresses the nature and the sequence of activity in the endogenous morphogenetic processes of the organism and corresponds to the fundamental growth program on which the entire architecture is established. The identification of the architectural model of any given plant is based on the observation of four major groups of simple morphological features which are well documented (Hallé and Oldeman, 1970; Hallé *et al.*, 1978): (i) *the type of growth* (rhythmic or continuous growth); (ii) *the branching pattern* (presence or absence of vegetative branching, terminal or lateral branching, monopodial or sympodial branching, rhythmic, continuous, or diffuse branching); (iii) *the morphological differentiation of axes* (orthotropy or plagiotropy) and; (iv) *the position of sexuality* (terminal or lateral).

Each architectural model is defined by a particular combination of these morphological features and named after a well-known botanist. Although the theoretical number of combinations is high, there are apparently only 23 architectural models in nature. These models apply to arborescent or herbaceous plants from tropical or temperate regions of closely related or distant taxa.

In figure 1 some of these models are demonstrated. Each of them are represented by hundreds if not thousands of species. *Corner's model* (1a) concerns unbranched plants with lateral inflorescences, such as the commonly cultivated *Carica papaya*. *Leeuwenberg's model* (1b) consists of a sympodial succession of equivalent units called "modules" (Prevost, 1967, 1978; Hallé, 1986), each of which is orthotropic and determinate in its growth by virtue of its terminal inflorescence. Branching is three dimensional. Examples of this model are *Manihot esculenta* and *Ricinus communis*. *Rauh's model* (1c) is represented by numerous woody plants from both tropical (*e.g. Hevea brasiliensis*) and temperate areas (*e.g. Pinus spp.*). Growth and branching are rhythmic on all monopodial axes and sexuality is lateral. *Aubreville's model* (1d) is much less frequent than the previous ones and represented by *Terminalia catappa*. The trunk is monopodial and grows rhythmically bearing whorled branch tiers. Branches grow rhythmically but are modular, each of them being plagiotropic by apposition. *Massart's model* (1e) differs from the previous one only in that the branches are plagiotropic either by leaf arrangement or symmetry but never by apposition. Numerous trees exhibit this growth strategy in both temperate (many gymnosperms, *e.g. Abies, Araucaria*) and tropical areas (*e.g.* most species of Myristicaceae or Bombacaceae).

Figure 1. Seven of the 23 known architectural models: a - Corner's model, b - Leeuwenberg's model, c - Rauh's model, d - Aubreville's model, e - Massart's model, f - Roux's model, g - Troll's model. (From Hallé and Edelin, 1987; with permission).

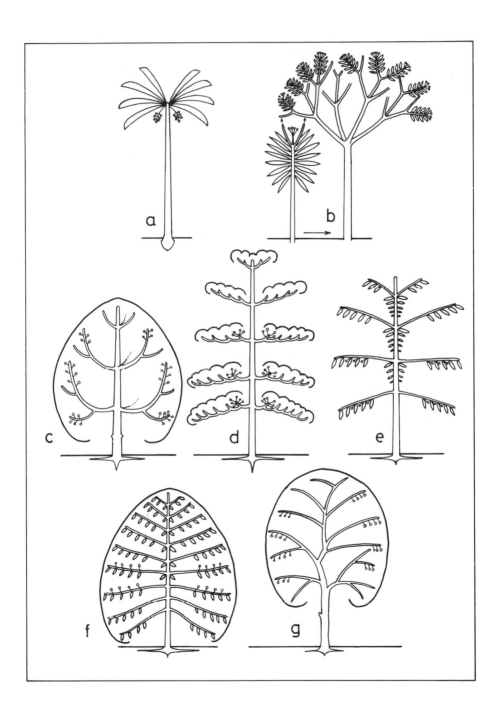

Roux's model (1f) is very close to Massart's model from which it differs only in the continuous or diffuse, not rhythmic, growth and branching of the trunk. This model is mainly represented by tropical species such as *Coffea arabica.* Finally, *Troll's model* (1g) which seems to be the most frequent in both tropical and temperate woody species. The axes are plagiotropic and the architecture is built by their continual superposition. Many examples of this model may be found in the Leguminosae.

Growth patterns defined by architectural models are genetically determined. Only under extreme ecological conditions is their expression affected by the environment (Temple, 1975; Hallé, 1978; Barthélémy, 1986a). Different models can be represented by plants belonging to closely related species. Architectural analysis also shows that some plants frequently exhibit morphological features which are apparently related to two or three models (Hallé and Ng, 1981; Edelin, 1977, 1984). These intermediate forms prove that there is no real disjunction between the models. On the contrary, it must be considered that all architectures are theoretically possible and that there could be a gradual transition from one to the others. In this "architectural continuum," the models themselves represent the forms that are the most stable and the most frequent, that is to say the biologically most probable ones.

The architectural unit

As we have seen, the architectural model represents the basic growth strategy of a plant. Nevertheless, the characters used in its identification are much too general to describe the complete and precise architecture of a plant. So, as illustrated in figure 2, in the case of Rauh's model, within the context of the same model, each species expresses its own architecture. For any given plant, the specific expression of its model has been called its architectural unit (Edelin, 1977).

Figure 2. Specific expression of Rauh's model (a) in four Gymnospermous species: b - *Pinus sylvestris*, c - *Cunninghamia lanceolata,* d - *Araucaria rulei*, e - *Metasequoia glyptostroboides* (From Edelin, 1977, with permission).

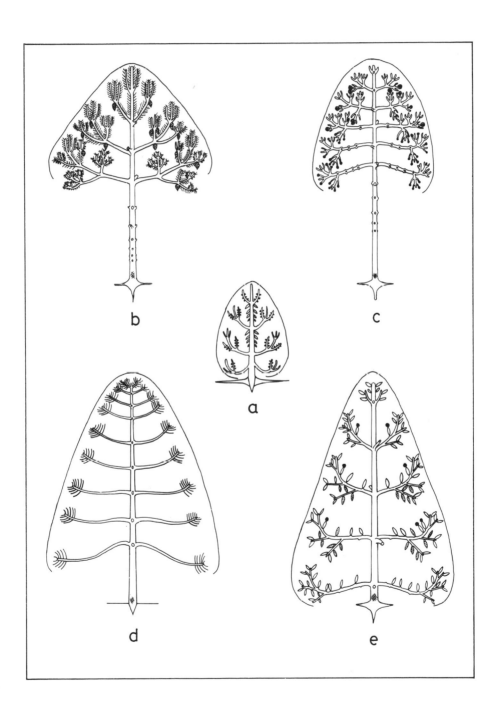

The architecture of a plant can be seen as a hierarchical branched system in which the axes can be grouped into categories. The structure and function of each category is characteristic of its rank, and for each species, the number of axes categories is finite. The identification of the architectural unit is achieved by a complete diagnosis of the functional and morphological features of all its axes categories. For each of them the observation of all the architectural characteristics previously described is necessary, but the observations have to be as exhaustive as possible and may concern any kind of morphological feature including growth direction, phyllotaxis, syllepsis or prolepsis, form and size of foliar organs, and presence or absence of sexuality. The results may be summed up in a table and with the help of a diagram, they describe and define the specific elementary architecture of each plant, *i.e.* its architectural unit. Within the context of a general organization, the differences between architectural units are thus represented by the number of categories of axes, their functional and morphological features, and their relative positions.

This indicates that the architecture of a fully established branched system, whatever its complexity, can be summarized in terms of a very simple sequence of axes which represents its fundamental organization. In this sequence, leading from axis one to the ultimate axis following the specific branching pattern, each branch is the expression of a particular state of meristematic activity and the branch series as a whole can be considered to be tracking the overall activity. In this sense, the architectural unit represents the fundamental architectural and functional elementary unit of any given species.

Reiteration

As defined by Oldeman (1974), reiteration is a morphogenetic process through which the organism duplicates totally or partially its own elementary architecture, *i.e.* its architectural unit. The result of this process is called a "reiterated complex". Two situations illustrate this phenomenon: (i) the traumas undergone by a plant throughout its whole life damage the vegetative structure more or less seriously. Generally, the destruction of an axis, involving the disappearance of its terminal meristem, allows the development of some previously dormant or suppressed meristems subjacent to the injury. This gives rise to branched systems, reiterated complexes, which develop an architecture identical to that of the bearing axis; (ii) within the crown of an old tree it is common to observe small branched systems that look like the juvenile one and seem to be naturally grafted on the bearing plant. Depending on whether the development of the reiterated complex is due to a trauma or not, one speaks respectively of "traumatic reiteration" or "adaptive reiteration".

As already noted, the development of a plant conforming to its model implies the notion of a sequence in the activity of the whole set of meristems; one speaks of a differentiation sequence. The occurrence of reiterated complexes in this sequence seems to be a move backwards within this sequence, a real dedifferentiation. A supernumerary trunk, resulting from the transformation of a growing branch (sylleptic reiteration) or from the development of a dormant meristem (proleptic reiteration), implies that the plant expresses all over again the juvenile growth pattern. This reversion in the growth pattern can be complete and thus involve again the total expression of the architectural unit leading from axis one to the ultimate branching order (complete reiteration), or it can be partial and duplicate only a part of the architecture of the plant (partial reiteration). This is well illustrated in cases of regeneration; when a trunk is cut, it produces sprouts identical to the bearing trees, whereas reiterated complexes that develop after a branch has been damaged have the same architecture as this branch.

In fact, reiteration encompasses several aspects including sprouts and root-suckers which have been known to botanists for a long time, but the fundamental interest of this concept is to regroup all these phenomena into a coherent whole to bring out a common morphogenetic event. Reiteration at first was considered as an opportunistic process. However, recent investigations (Edelin, 1984) have demonstrated that of adaptive reiteration is an automatic event that occurs after a definite threshold of differentiation during the normal development of a tree. This will be illustrated, in the context of complete reiteration by the following two examples.

Figure 3 shows the architectural sequence of *Shorea stenoptera* from South East Asia which conforms to Roux's model. Up to a height of one to two meters (3a) the trunk is not branched. Then (3b), the orthotropic trunk, a continuously branching monopodium, bears plagiotropic distichous branches. Each of these consists of a single second axis, rarely branched, which dies and self-prunes rapidly. Proleptic reiterated complexes grow out of some dormant buds borne on the trunk (3c), they branch rapidly and their plagiotropic branches are identical to those described above. This type of reiteration occurs although the tree is still in the understory shade most of the time. These reiterated complexes do not exceed 4-5 m before precociously dying and falling. When the tree reaches the forest canopy the same process goes on, but this time the reiterated complexes do not die (3d). On the contrary, they produce strong limbs (R). Within the crown of the adult tree (3e), new proleptic reiterated complexes (r) appear on the established infrastructure. There, the longer ones are formed on the upper side of the limbs. The others grow out from

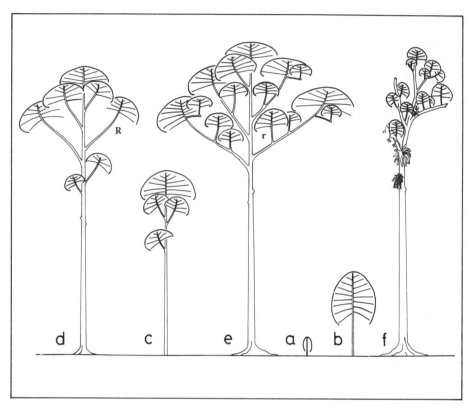

Figure 3. Diagrammatic representation of the developmental sequence of *Shorea stenoptera*. (From Edelin, 1984, with permission).

the branches and appear perpendicularly to them. The development and maintenance of the crown throughout the years is ensured by the constant renewal of these short-lived reiterated complexes. This process continues on the old tree which is marked by the death and gradual dislocation of the crown (3f).

    Dipterocarpus costulatus (Fig. 4), another Dipterocarpaceae from South East Asia, conforms to Massart's model. It has an orthotropic monopodial trunk with spiral phyllotaxis and rhythmic growth. When juvenile, (4a) its branches are plagiotropic, distichous, and arranged in regular tiers. As it develops (4b), the trunk produces larger and more ramified branches. These features of development become more pronounced as the tree grows (4c). The appearance of a new branching order is followed by an architecturally stable period of variable duration that precedes and prepares the following stage of the development. As the tree grows, this process goes on and becomes more pronounced. The general production of 5th order branches (Fig. 4d-e) coincides with formation of sylleptic reiterated complexes, which represent the major limbs of the adult tree and on which sexuality occurs laterally among the ultimate axes. So, the development of *Dipterocarpus costulatus* is marked by a progressive transformation of its branches. From plagiotropic and poorly branched at the beginning, they become more and more orthotropic, and the adaptive reiteration represents a true "architectural metamorphosis" (Edelin, 1984).

Whether the limbs of the old tree are established by proleptic reiteration (*Shorea stenoptera*) or by sylleptic reiteration (*Dipterocarpus costulatus*), the crown of the old tree is a true colony of individuals of various sizes (Fig. 3e-f; Fig. 4e).

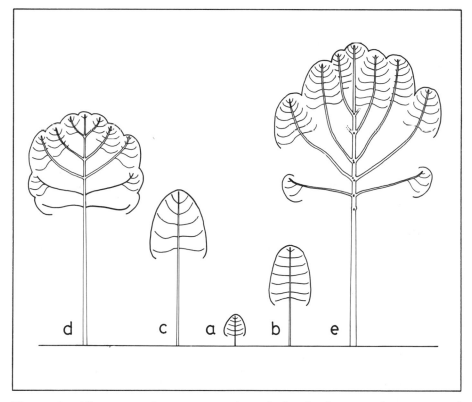

Figure 4. Diagrammatic representation of the developmental sequence of *Dipterocarpus costulatus*. Ultimate axes are not represented. (From Edelin, 1984, with permission).

The reiterated complexes, naturally "grafted" together as it were, are relatively independent and exert a strong selective competition on each other (Torquebiau, 1979). Following a progressive and precise sequence of events, each species expresses its own "reiterative strategy" (Edelin, 1986). However, these events are incorporated into a more general cycle which leads the "colonial tree" from germination to death.

The life of a forest tree is thus marked by three major architectural stages (Oldeman, 1974): (i) the tree of the future, corresponds to growth in the understory, and possesses an architecture that strictly conforms to the architectural unit. The tree expresses its architectural unit from axis one to the ultimate axis which is the threshold of reiteration; (ii) the tree of the present has reached the forest canopy and has developed a large crown by means of adaptive and traumatic reiteration. It is composed of the juxtaposition of increasingly smaller reiterated complexes, which are termed arborescent, frutescent, and herbaceous; (iii) the tree of the past is marked by death and gradual dislocation of the crown. The limbs start breaking, leaving stumps on the trunk on which populations of epiphytes become established. The crown progressively breaks up heralding the death and the fall of the tree.

Conclusion

The discovery of the concepts of architectural model, architectural unit, and reiteration has provided botanists with a powerful tool for studying plant form and structure. More than 150 families have undergone architectural analysis, and these observations concern herbaceous as well as woody plants from both temperate and tropical regions.

Architectural analysis has proved to be probably one of the most efficient means that we presently possess for the study of the organization of arborescent plants, and architectural concepts appear to be of particular interest for the understanding of crown construction in trees. Nevertheless, numerous problems are still unsolved. At the present time, architectural studies of crown construction in trees continue, but research has also spread to the understanding of the flowering process in tropical plants (Barthélémy, 1986b) and to the architectural analysis of root systems (Kahn, 1983; Atger, 1986). On the other hand, the quantitative knowledge of plant architecture linked with mathematical approaches has also led to the development of a mathematical model for computer simulation of every kind of tree architecture known at the present time and their evolution during tree growth (Reffye *et al.*, 1986).

Literature cited

Atger, C. (1986). "Quelques liens architecturaux entre réitérations aériennes et racinaires (le cas de quatre espèces guyanaises." *Mém. D.E.A., Botanique Tropicale Appliquée. Univ. Montpellier II.*

Barthélémy, D. (1986a). "Establishment of modular growth in a tropical tree: Isertia coccinea Vahl. (Rubiaceae)." pp. 89-94 *In* Harper J. L., Rosen B.R. and White J. (eds.), "The growth and form of modular organisms." *Philos. Trans. R. Soc. London, sér. B,* **313.**

Barthélémy, D. (1986b). "Relation entre la position des complexes réitérés sur un arbre et l'expression de leur floraison : l'exemple de trois espèces tropicales." pp. 71-100 *In C. R. Colloq. Int. L'arbre, Montpellier, 9-14 sept. 1985, Naturalia Monspeliensia, n° hors-sér.*

Corner, E. J. H. (1949). "The Durian theory, or the origin of the modern tree." *Ann. Bot., N. S.,*13, **52,** 367-414

Corner, E. J. H. (1953, 1954). "The Durian theory extended I, II, III." *Phytomorphology.* 3, 465-475, 4, 152-165, 263-274.

Edelin, C. (1977). "Images de l'architecture des Conifères." *Th. Doct. 3e Cycle, Biologie végétale, Univ. Montpellier II.*

Edelin, C. (1984). "L'architecture monopodiale : l'exemple de quelques arbres d'Asie tropicale." *Th. Doct. Etat, Univ. Montpellier II.* 258.

Edelin, C. (1986). "Stratégie de réitération et édification de la cime chez les Conifères." pp. 139-158 *In C. R. Colloq. Int. l'Arbre, Montpellier, 9-14 Sept. 1985, Naturalia Monspeliensa, n° hors-sér.*

Hallé, F. (1978). "Architectural variation at specific level of tropical trees." pp. 209-221 *In* Tomlinson, P. B. and Zimmermann, M. H. (eds.), "Tropical trees as living systems." *Cambridge Univ. Press, Cambridge.*

Hallé, F. (1986). "Modular growth in seed plants." pp. 77-87 *In* Harper J. L., Rosen B. R. and White J. (eds.), "The growth and form of modular organisms." *Philos. Trans. R. Soc. London, sér. B,* **313** .

Hallé, F. and Ng, F. S. P. (1981). "Crown construction in mature Dipterocarp trees." *Malays. For.* 44, 222-223.

Hallé, F. and Oldeman, R. A. A. (1970). "Essai sur l'architecture et la dynamique de croissance des arbres tropicaux." *Masson, Paris.*

Hallé, F., Oldeman R. A. A. and Tomlinson, P. B. (1978). "Tropical trees and forests." *Springer, Berlin.*

Kahn, F. (1983). "Architecture comparée de forêts tropicales humides et dynamique de la rhizosphère." *Th. Doct. Etat, Univ. Montpellier II.*

Oldeman, R. A. A. (1974). "L'architecture de la forêt Guyanaise." *Mém. n°73, ORSTOM, Paris.*

Prevost, M. F. (1967). "Architecture de quelques Apocynacées ligneuses." *Bull. Soc. bot. Fr.,* lett. bot. **114,** 24-36.

Prevost, M. F. (1978). "Modular construction and its distribution in tropical woody plants." pp. 223-231 *In* Tomlinson, P. B. and Zimmermann, M. H., (eds.), "Tropical trees as living systems." *Cambridge Univ. Press, Cambridge.*

Reffye, Ph. de, Edelin, C., Jaeger, M. and Cabart, C. (1986). "Simulation de l'architecture des arbres." pp. 223-240 *In C. R. Colloq. Int. L'Arbre, Montpellier, 9-14 sept. 1985, Naturalia Monspeliensia, n° hors sér.*

Temple, A. (1975). "Ericaceae. Etude architecturale de quelques espèces." *D.E.A. Botanique Tropicale, Univ. Montpellier II.*

Torquebiau, E. (1979). "The reiteration of the architectural model. A demographic approach to the tree." *Mém. D.E.A., Ecologie générale et appliquée, Univ. Montpellier II.*

Population dynamics of tree species in tropical forests

M. D. SWAINE

University of Aberdeen, Scotland, UK

In this paper, I summarize the current understanding of forest dynamics in terms of tree populations. Differences between species and changes in species populations are discussed. The urgent needs of forest management suggest that research priorities should concentrate on understanding ecological differences between species.

Forest dynamics

Natural and semi-natural tropical rain forests are structurally stable, maintaining an approximately logarithmic decline in numbers of trees with increasing size. This kind of size-class distribution is the consequence of forest dynamics in which the available space constrains the number of trees that can be accommodated in any size class. Continual tree mortality (at about 1-2% annually) permits further growth of the surviving trees and recruitment of new trees (Swaine and Lieberman, 1987).

Mortality rates among trees greater than about 5 cm dbh are independent of tree size - large trees are no more at risk of death than small trees. Because small trees are more numerous, however, most deaths create small gaps without opening the upper canopy. Studies of forest dynamics based on gaps with openings in the upper canopy, therefore, ignore the great majority of disturbances in forest. Plants perceive gaps very differently from people, and we need to reassess our idea of the "gap".

Mortality rates of seedlings and saplings are typically much higher than for larger trees (Fig. 1). The mortality of seedlings may approach 100 % in the first year (Lieberman and Lieberman, 1987; Swaine and Hall, 1986); those which survive longer are subject to a declining risk. Seedling populations are thus ephemeral and highly variable over short periods.

Trees >10 cm dbh, and probably smaller ones, are much less likely to die if they are growing rapidly. The slowest-growing (suppressed) trees in a plot at Bukit Lagong in Peninsular Malaysia were six times more

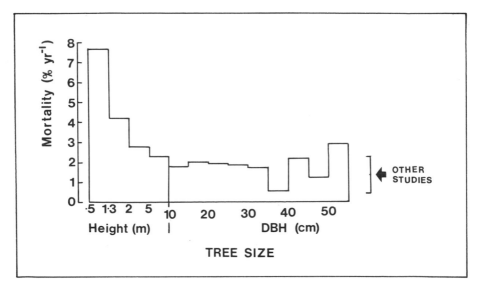

Figure 1. Annual mortality of trees in different size classes at Kade, Ghana. Trees ≥ 10 cm dbh were monitored on 2 hectare between 1970 and 1982; those <10 cm dbh on 0.25 hectare between 1979 and 1987. The range for other studies is taken from Swaine and Lieberman (1987).

likely to die than the fastest-growing trees (Fig. 2; Manokaran, 1988).

Growth rates of trees are highly variable. There are large differences between species, tree sizes, sites, and even between the same sized individuals of the same species growing at the same site. In contrast, growth of an individual tree during successive periods is much less variable. Trees which are growing fast continue to do so, slow-growing individuals remain slow, sometimes for more than 30 years (Manokaran, 1988) but are more likely, ultimately, to die (see above). Such autocorrelation of growth was detectable for over 30 years in data from Sungei Menyala in Peninsular Malaysia (Fig. 3; Manokaran, 1988). The diameter increase of a tree tends to be linear for long periods so that the future growth of individual trees is somewhat predictable and offers potential for forest management on a tree-by-tree basis.

Autocorrelation of growth and the link between growth rate and mortality means that fast-growing trees are more likely to reach maturity, and that the largest trees may not necessarily be very old. A fast-growing *Shorea macrophylla* in Peninsular Malaysia, for example, reached 42 m height in 32 years (Ng and Tang, 1974).

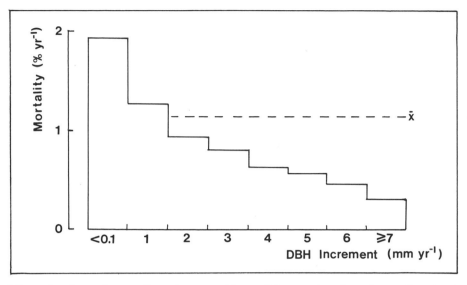

Figure 2. Annual mortality of trees ≥ 10 cm dbh as a function of growth rate at Sungei Menyala Forest Reserve, Peninsular Malaysia between 1947 and 1985. Only trees with at least two measured increments are included. Data from N. Manokaran, Forest Research Institute, Malaysia

Species populations

The features of forest dynamics outlined above also appear to apply to individual species populations and to groups of ecologically similar species. Growth autocorrelation and growth related mortality occur in small stature understory species as well as in those capable of forming the upper canopy (Fig. 3). However, it is the differences between species in these dynamic processes which determines the floristic composition of the forest.

Forest composition and species population sizes are determined in part by species tolerance of current environmental conditions (rainfall, soil conditions, *etc.*) and in part by site history, especially on a local scale (Swaine and Hall, 1988).

I offer here a comparison between two upper canopy tree species from West Africa as a means to discuss some of the issues in the determination of forest composition. *Celtis mildbraedii* and *Strombosia glaucescens* (Table 1) have overlapping distributions, *Strombosia* extending into wetter forest, *Celtis* into drier. Both species coexist over large areas; both have shade tolerant seedlings and saplings, often at high density.

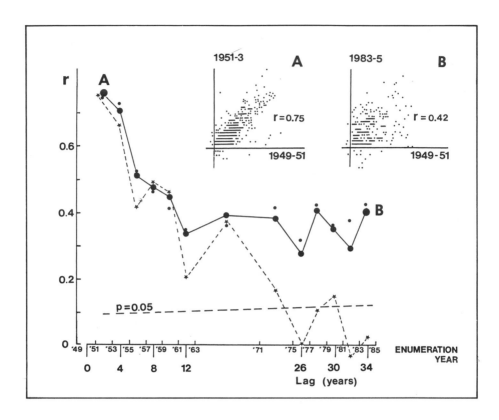

Figure 3. Autocorrelation of diameter growth in individual trees at Bukit Lagong Forest Reserve, Peninsular Malaysia. The correlation coefficients (r) are calculated between growth in the first increment period (1949-51) and in each successive period up to 1983-5; two examples are inset. The approximate value of r at 0.05 probability (P) is shown. O, canopy species; *, understory species; o, all trees. Data from N. Manokaran, Forest Research Institute, Malaysia.

At one site, Kade in Ghana, *Celtis* is very abundant in the upper canopy, *Strombosia* less so; however among the juveniles *Strombosia* is much more common (Table 1). Can *Celtis* maintain its high adult abundance with so few juveniles? Data on mortality differences (Fig. 4) between the species suggest one possibility. Mortality is much higher in *Strombosia* in all size classes - perhaps *Celtis* can maintain its adult population with relatively few juveniles because it has such low mortality. In contrast, persistent high mortality in *Strombosia* demands a high stocking of juveniles.

This kind of mechanism may be proposed as a means by which local forest composition remains in equilibrium. Unfortunately, we do not know if the life-table differences between these species persists at other sites where they coexist. Certainly, *Strombosia* juveniles are not always more abundant than those of *Celtis* (Swaine and Hall, 1988).

Table 1. Comparison between two shade tolerant canopy tree species from West Africa. Both species are evergreen, bird-dispersed, and have drupaceous fruits and epigeal germination. Population dynamics data are from Kade, Ghana (other sources: *Taylor, 1960; + Hall and Swaine, 1981).

	Celtis mildbraedii	*Strombosia glaucescens*
Family	Ulmaceae	Olacaceae
Architectural model	Troll	Roux
Buttresses	Yes	No
Max tree height (m)	60	>30
Seed weight (g)*	0.5	0.7
Max seedling growth (cm yr^{-1})*	31	24
Distribution	Ivory Coast to Angola	Sierra Leone to Ghana
Forest types+	Moist Evergreen to	Wet Evergreen to
	Dry Semi-deciduous	Moist Semi-deciduous
Flowering months	Erratic	Aug-Sep
Fruiting months	Erratic	Dec-Jan
Mean tree density (hectare^{-1})		
50 to 130 cm tall	126	834
130 cm tall to 29 cm gbh	48	392
≥30 cm gbh	84	14
Mean relative growth rate (% yr^{-1})		
Saplings (0-29 cm gbh)	5.03	4.33
Trees (≥30 cm gbh)	0.89	0.77
Recruitment (% yr^{-1})		
to 130 cm height	8.33	6.55
to 30 cm gbh	0.62	3.67
Mortality (% yr^{-1})		
50 to 130 cm tall	2.60	3.59
130 cm tall to 29 cm gbh	1.24	3.07
30 to 99 cm gbh	0.25	1.91
≥100 cm gbh	0.31	4.07

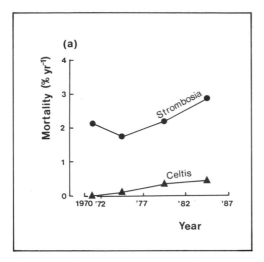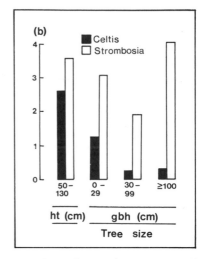

Figure 4. Comparison of annual mortality rates in *Strombosia glaucescens* and *Celtis mildbraedii* at Kade, Ghana, (a) between 1970 and 1987 (trees ≥30 cm gbh), and (b) between size classes.

Such subtleties of species population dynamics may be irrelevant to the control of composition if the forest is subject to heavy disturbance. There is increasing evidence that the details of species population size and structure at any site could be decisively influenced by infrequent natural catastrophes like drought, fire, and windstorms (Beaman *et al.*, 1985; Cousens, 1965; Johns, 1986; Sanford *et al.*, 1985). Such catastrophes need occur only once a century or less for them to exert a permanent influence on forest composition. Most of us may never experience these important events.

Whether or not forest composition is in equilibrium on a local scale is a question which I suggest cannot be properly answered. We may see powerful evidence of change in population size at any site, the more so the longer we observe change. At Sungei Menyala in Malaysia, for example, eight species in a two-hectare plot showed clear changes over 34 years (Manokaran and Kochummen, 1987). The conclusions we can draw about compositional stability from such results, however, are limited: 34 years is only a small part of the life-span of most forest trees, and the future of these species cannot be predicted without knowledge of the cause of the changes in their population. Furthermore, a two-hectare (or even 50 hectare) sample probably only includes a small fraction of the breeding population of a species - what is occurring in other parts is unknown. In any study of natural changes in forest populations, the changes we are bound to see will be open to contrary interpretations.

Of more interest and greater utility, I suggest, is to concentrate our efforts as botanists on inherent differences between species and the effects these have on dynamic processes in forest.

Understanding how a species survives in a forest, expands its population, or displaces another will not come from watching the changes occur but from a detailed knowledge of its ecology. Important differences include fecundity, seed germination, growth, and mortality, all of which are under some degree of genetic control. Environmental factors which influence the expression of these inherent traits include light, moisture, nutrients, and biota. Of these, light intensity is possibly the most important single factor for forest trees, for a variety of reasons.

Differences between species in mortality rate, insofar as they are determined genetically, are closely tied with species tolerance to shade. Trees, and especially their seedlings, are much more likely to die from disease, drought, or herbivory if they are living at or near their light compensation point.

Light intensity, through the High Irradiance Response (HIR), is probably the commonest means by which dormancy of soil seed bank species is broken. The relative frequency of HIR or LED (Low Energy Reaction) as breakers of dormancy among species, however, has not been specifically examined.

Changing the light environment of a *Pentaclethra macroloba* seedling can cause a nine-fold increase in the relative growth rate (Oberbauer and Strain, 1985). Tree species differ by a factor of at least five in their maximum photosynthetic rate (Oberbauer and Strain, 1984).

Ignoring other environmental influences has its dangers. Adding nutrients can sometimes give dramatic increases in growth (Okali and Owusu, 1975, but see Tanner *et al.*, in press) though this may not be fully realized without adequate light. Rare droughts are undoubtedly important to species which are usually well-provided (Beaman *et al.*, 1985; Woods, 1985). From the point of view of forest management, however, the light environment can be manipulated, nutrients and water cannot.

Forest management

Most of us are concerned for the future of tropical rain forests. I believe that their survival depends particularly on developing sustainable yield management methods which can predict the consequences for floristic composition of various exploitation practices. Foresters need the help of ecologists and botanists to link more closely the processes of logging and regeneration. If all Meliaceae above a certain size were removed from a forest in West Africa, which of the remaining species would benefit ?

What will be the composition of the future crop ? To answer these questions we must know the characteristics of each potential timber tree species in some detail. At present, our knowledge of species differences is uneven and very incomplete. Very often the ecologist cannot give a satisfactory answer to pertinent questions posed by foresters, government economists, and aid agencies. If we fail in what is expected of us, we will be seen as irrelevant and self-indulgent, and the forests we like to study will be converted to low-diversity systems that non-biologists can manage.

Research needs on species differences

We need greater understanding of the frequency, annual pattern, and population synchrony of seed production and its dispersal. Does the time of logging operations in relation to seed availability affect forest regeneration ?

How do species differ in their germination requirements ? A recent study in Malaysia (Raich and Gong, in press) recognizes three classes of species: those which will germinate equally in forest shade or in full sunshine (14%), those requiring canopy openings (33%), and those whose germination is depressed in large clearings (53%). These and other differences in germination have important implications for forest composition after logging.

The growth rate of trees is one of the most important aspects of species differences in the determination of forest composition and for management. Many important timber species (*e.g.* in the Meliaceae and Dipterocarpaceae) appear to need relatively open conditions for maximum growth, but this has rarely been quantified. Over-enthusiastic thinning or extraction, however, may succeed only in encouraging climbers or less valuable pioneer trees. In contrast, some species appear to grow as fast under 25% as 100% full sunlight (Oberbauer and Strain, 1985). We should also remember that the growth of many cocoa varieties (and presumably of other understory species) is depressed by exposure to full sunshine (see, for example, Okali and Owusu, 1975).

Although we have examples of all these effects from different parts of the tropics, in no single area can we describe the characteristics of all timber-sized species. This should be our aim. Much can be achieved by simple experimental work in the nursery using, for example, neutral shade treatments (Okali, 1972; Okali and Owusu, 1975). For forest management purposes we do not need to measure net carbon-dioxide uptake, nor are large scale silvicultural experiments (Maitre, 1987) an efficient approach for any but the commonest species (Philip, in press).

Acknowledgment

I am grateful to N. Manokaran and the Forest Research Institute, Malaysia, for permission to use unpublished data.

Literature cited

Beaman, R. S., Beaman, J. H., Marsh, C. W. and Woods, P. V. (1985). "Drought and forest fires in Sabah in 1983." *Sabah S. J.* **8**, 10-30.

Cousens, J. E. (1965). "Some reflections on the nature of Malayan lowland rainforest." *Malayan Forester* **28**, 122-128.

Hall, J. B. and Swaine, M. D. (1981). "Distribution and ecology of vascular Plants in a Tropical Rain Forest." *W. Junk, The Hague.*

Johns, R. J. (1986). "The instability of the tropical ecosystem in New Guinea." *Blumea* **31**, 341-371.

Lieberman, D. and Lieberman, M. (1987). "Forest tree growth and dynamics at La Selva, Costa Rica (1969-1982)." *J. Trop. Ecol.* **3**, 347-358.

Maitre, H. F. (1987). "Natural forest management in Cote d'Ivoire." *Unasylva* **30**, 53-60.

Manokaran, N. (1988). "Population dynamics of tropical forest trees." *Ph. D. thesis, University of Aberdeen.*

Manokaran, N. and Kochummen, K. M. (1987). "Recruitment, growth and mortality of tree species in a lowland dipterocarp forest in Peninsular Malaysia." *J. Trop. Ecol.* **3**, 315-330.

Ng, F. S. P. and Tang, H. T. (1974). "Comparative growth rates of Malaysian trees." *Malaysian Forester* **37**, 2-23.

Oberbauer, S. F. and Strain, B. R. (1984). "Photosynthesis and successional status of Costa Rican rain forest trees." *Photosynthesis Research* **5**, 227-232.

Oberbauer, S. F. and Strain, B. R. (1985). "Effects of light regime on the growth and physiology of Pentaclethra macroloba (Mimosaceae) in Costa Rica." *J. Trop. Ecol.* **1**, 303-320.

Okali, D. U. U. (1972). "Growth rates of some West African forest tree seedlings in shade." *Ann. Bot. (London)* **36**, 953-959.

Okali, D. U. U. and Owusu, J. K. (1975). "Growth analysis and photosynthetic rates of cocoa (Theobroma cacao L.) seedlings in relation to varying shade and nutrient regimes." *Ghana J. Agric. Sci.* **8**, 51-67.

Philip, M. S. (in press). "Managment systems in the tropical moist forest of anglophone Africa." *FAO, Rome.*

Raich, J. W. and Gong, W. K. (in press). "Effects of canopy openings on tree seed germination in a Malaysian dipterocarp forest." *J. Trop. Ecol.*

Sanford, R. L., Saldarriaga, J., Clark, K. E. , Uhl, C. and Herrera, R. (1985). "Amazon rain-forest fires." *Science* **227,** 53-55.

Swaine, M. D. and Hall, J. B. (1986). "Forest structure and dynamics," pp. 47-93 *In* Lawson, G. W. (ed.), "Plant Ecology in West Africa." *John Wiley, Chicester.*

Swaine M. D. and Hall, J. B. (1988). "The mosaic theory of forest regeneration and the determination of forest composition in Ghana." *J. Trop. Ecol.* **4,** 253-269.

Swaine, M. D. and Lieberman, D. (eds.), (1987). "The Dynamics of Tree Populations in Tropical Forest." *Special Issue, J. Trop. Ecol.* **3,** 289-369.

Tanner, E. V. J., Kapos, V., Freskos, S., Healey, J. R. and Theobald, A. M. (in press). "Nitrogen and phosphorus fertilization of Jamaican montane forest trees." *J. Trop Ecol.*

Taylor, C. J. (1960). "Synecology and Silviculture in Ghana." *Nelson, Edinburgh.*

Woods, P. V. (1985). "Drought and fire in tropical forests in Sabah - an analysis of rainfall patterns and some ecological effects." *Proceedings of the 3rd Int. Round Table Conf. on Dipterocarps,*

Speciation

Speciation in tropical forests

A. H. GENTRY

Missouri Botanical Garden, USA

Biogeographers have long emphasized the generality of the latitudinal gradient in species richness. Indeed a central question for those interested in understanding the evolutionary diversification of life on earth has long been: Why are there so many species in tropical forests ? In the last few decades there has been renewed interest in the reasons for high tropical diversity from such divergent viewpoints as those of theoreticians, trying to formulate models explaining patterns of species diversity, and conservationists, increasingly concerned about the impending loss of so much of the earth's biotic diversity due to accelerating tropical deforestation.

Although some authors have focused on the possible role of uniquely tropical modes or rates of speciation in generating the high diversity of tropical forests (Ashton, 1969; Stehle *et al.*, 1969), this is not necessarily synonymous with the maintenance of high diversity in tropical forests, and much more attention has been focused on the latter (*e.g.*, MacArthur, 1964, 1965; Pianka, 1966; Terborgh, 1985; Connell, 1987). From a different perspective it has been suggested that speciation may not be unusual in the tropics, but rather that less extinction in the benign environment of tropical forests may be the key to high tropical diversity (Stebbins, 1974). Ehrendorfer (1970) has even argued that speciation may be slower in tropical than in temperate forests with high diversity due to cumulative build up through very long periods of time. Or perhaps the stochastic processes associated with the ecological dynamics of tropical forests that are now well known to play a major role in diversity maintenance (Connell, 1978, 1987; Agren and Fagerstrom, 1984; Hubbell and Foster, 1986) also help to generate some of that diversity.

The final part of this symposium will focus more broadly on the general question of high tropical diversity, while this review is largely restricted to speciation per se. Nevertheless, generation and maintenance of diversity are clearly interrelated. Therefore, a few general points on diversity will be useful to set the stage for a discussion of tropical speciation, which, after all, must ultimately be responsible for generating the fabulous diversity of tropical forests.

Some relevant general points on diversity are: (i) Tropical forests tend to be more diverse than other forest types at scales varying from the

TROPICAL FORESTS
ISBN 0–12–353550–6

level of local communities to those of continents (*e.g.*, Gentry, 1982a,c,
1988a), including the intermediate level of between-community differentiation (*e.g.*, Gentry, 1986a). The role of speciation in generating diversity at these different levels may be very different; (ii) although emphasis
has been focused on tree diversity, tropical forests are not only rich in trees
and lianas but also in herbaceous taxa (Gentry and Dodson, 1987a,b).
Indeed a wet tropical forest typically has more species of herbs (including
epiphytes) or shrubs than does any extra-tropical vegetation (Gentry and
Dodson, 1987a), and the potentially distinctive speciation processes generating high diversity in herbaceous tropical plants must also be considered
in an essay on tropical speciation; (iii) there tend to be remarkably
regular patterns of diversity between different parts of the tropics sharing
similar ecologies. Comparable individual tropical forests on all three
main continental areas have similarly high diversities (Gentry, 1988b).
Even Africa has some forests with exceptionally high α-diversity. Wetter
tropical forests tend to have more species than drier ones, forests on poor
soils fewer species than on richer soils, low altitude forests more species
than high altitude forests (Gentry, 1988b). The repetition of these patterns
by different sets of species in different parts of the world indicates that
there may also be rules underlying the speciation processes that give rise to
this diversity.

At a different level, interpretation of tropical speciation is likely to
be influenced by personal viewpoint and biases. A cladist is highly likely
to emphasize vicariance events, an island biogeographer equally likely to
focus on dispersal and adaptive radiation. Others have concentrated on
the population genetics of speciation, sometimes generating acrimonious
debates about the relative merits of allopatric, parapatric, or sympatric
speciation. Here I will try to emphasize instead that tropical speciation is
not a single phenomenon but encompasses a rich array of speciation
phenomena, each of which may have had an important role to play in the
overall process.

One final caveat needs to be made. Despite the current explosion of
interest in tropical forests, we as yet know perilously little about them.
Distribution patterns taken as evidence for a particular mode of speciation
often turn out instead to reflect collection artifact. For example, the
reputedly Inambari-refuge distribution of *Arrabidaea nicotianiflora*
Kranzl., mapped by Gentry (1979), was dramatically changed by the discovery of that species on the Río Sinu in northern Colombia the week before
this symposium. When the major timber tree of the coastal Ecuadorian
wet forest can turn out to be an undescribed and previously uncollected
species (Gentry, 1977), that species turns out to belong to a supposedly
exclusively Asian genus (Werff and Richter, 1985), and eight additional
newly collected species of that genus now convert it to a South American

genus with an Asian outlier, it is painfully evident that the tropical data base remains woefully incomplete.

Tropical vs. temperate speciation

Suggested differences between "tropical" and "temperate" speciation patterns have tended to fall into two clusters; those directly dependent on latitudinal environmental differences, and (in plants) those related to the habit differences between predominantly woody tropical floras and predominantly herbaceous temperate ones. More recently, a third cluster of suggested tropical speciation mechanisms has focused on forest dynamics in either ecological or evolutionary time. I will divide my discussion of tropical (vs. temperate) speciation into three major parts, first re-hashing some of the latitudinal environmental differences that may be involved, second focusing on peculiarities in reproductive mode related to the predominately woody habit of tropical plants, and finally examining the relative roles of allopatric, peripatric, parapatric, and sympatric speciation processes.

Latitudinal trends in speciation

Until recently tropical forests were often viewed, at least implicitly, as having high species richness not because of differences in speciation but simply because the relatively benign and constant tropical environment had provided more time for speciation to take place (Fischer, 1960; Sanders, 1968). Some authors have emphasized that niche fine-tuning, perhaps mediated through competition or predation (*e.g.*, Terborgh, 1985; Janzen, 1970; Connell, 1987), should be more feasible in an environment buffered from climatic extremes. Others have emphasized that the greater structural complexity of tropical forests facilitates niche differentiation (MacArthur and MacArthur, 1961; MacArthur *et al.*, 1966; Terborgh, 1985; Gentry, 1989a). According to such views, intensive selection for narrow and specific adaptations could thus be a general feature of tropical forest speciation (*e.g.*, Ashton, 1969; Stehle *et al.*, 1969). Bignoniaceae, for example, show very precise differentiation of pollination niches, presumably with competition for pollinators providing the evolutionary stimulus (Gentry, 1974a,b, 1976).

At the opposite extreme, some authors have suggested nearly the opposite, that tropical forests are diverse because of low extinction rates in a benign environment. Thus, Stebbins (1974) considers tropical forests to be little more than museums of botanical antiquities, of little consequence for understanding evolutionary processes. From a different, but similarly nonselective viewpoint, Federov (1966), Ashton (1977). and Kaur *et al.* (1978) have suggested that tropical forest speciation may often result from breakdown of the sexual process, perhaps accompanied by genetic drift in small or over-dispersed populations. Many tropical species thus may represent evolutionary deadends, or as Ashton (1977) phrases it, the "declining species do not fade away gradually but, by borrowing time in a Faustian pact of apomixis, regain the stage from time to time before their inevitable nemesis." In addition to the few examples of apomixis and parthenocarpy cited by Kaur *et al.* (1978) for Dipterocarpaceae, Sapindaceae (*Xerosperma*), and Guttiferae (*Garcinia*), Morawetz (1984a) has provided several recent examples of triploid species of Annonaceae, strongly implying a breakdown of sexuality. Bedell (pers. comm.) suggests that nearly all Marcgraviaceae are autogamous, despite their complex inflorescences. Such cases strongly imply that, at least in certain tropical groups, speciation may often be an almost accidental by-product of the organisms population structure and genetic processes.

Even though it has become very clear in recent years that tropical environments are strongly fluctuating and dynamic, the amount of physiological stress imposed by the rainfall fluctuations in many tropical forests is surely relatively minor when compared to that resulting from winter in temperate zone climates. Thus tropical forests, as a rule, really do have more constant and benign environments than do temperate zone regions (*e.g.*, Janzen, 1967; Gentry, 1989a). For example, the partitioning of a shared pollinator resource among six sympatric Panamanian *Arrabidaea* species which flower at staggered intervals throughout the year (Gentry, 1974b) would be patently impossible in a temperate zone growing season truncated by winter, just as would sharing of a disperser resource by sequentially fruiting Trinidadian or Colombian *Miconia* species (Snow, 1965; Hilty, 1981).

Another aspect of the tropical environment which has sometimes been implicated in tropical speciation is the greater incidence of the sun's energy near the equator along with a resultant greater stability in biological productivity (Connell and Orians, 1964; Wright, 1983). Perhaps the greater available energy could facilitate niche specialization and evolutionary fine-tuning in a manner analagous to that discussed above for a benign non-fluctuating climate. Or perhaps greater energetic input is reflected in greater biomass accumulation, giving rise to generally larger and less extinction-prone populations at ecological equilibrium (Wright, 1983; Turner *et al.*, 1987).

Even more directly, the stronger input of ultraviolet rays from a sun more directly overhead could generate more mutations and thus more raw material for speciation in the tropics. However, the generally negligible effect of mutation rate on evolutionary change makes this kind of scenario even less likely.

Speciation in woody vs. herbaceous plants

The percentage of woody plant species is generally higher in the tropics than in the temperate zone. For example terrestrial herbs constitute 65% to almost 90% of the species in a series of temperate zone florulas tabulated by Gentry (1989b), as compared to only about half of the species of tropical dry forest sites and a third or less of the species of tropical moist or wet forests (Gentry and Dodson, 1987b). For the same series of sites, trees, shrubs, and lianas together constitute between 10 and 27% of a given temperate zone florula but a third to over half of a given tropical one. The woody life form is twice as prevalent in the northern Chilean Andes as in a similar range of sites in the Great Basin (Arroyo, 1988). Tropical woodiness may be interpreted as an automatic correlate of large body size resulting from increased selection for competition for light in tropical forests (*cf.* Grime, 1979). Thus, to some extent generalizations about differences between tropical and temperate zone modes of speciation might reflect differences in the evolutionary strategies of trees and herbs. Stebbins (pers. comm.) for example, has suggested, only half in jest, that even if there are lots of species in the tropics they are mostly trees that all look about alike and are of little evolutionary interest.

In the temperate zone there are a number of broad generalities correlating reproductive mode with habit and habitat. Aneuploid reduction is characteristic of desert herbs, polyploidy of herbs of high latitudes, especially in glaciated or otherwise disturbed areas, hybridization and constant chromosome numbers of dominant clinally-differentiated trees like oaks and hickories, and autogamy of weeds (Ehrendorfer, 1980; Løve and Løve, 1949; Grant, 1975; Stebbins, 1942, 1974).

Many tropical taxa like Bignoniaceae (Gentry, 1988c; Goldblatt and Gentry, 1979) show exactly the kinds of patterns that might be expected for woody plants: constant relatively high (likely palaeopolyploid) chromosome numbers, little hybridization in nature (despite the potential for artificial hybridization), and allopatry of the most closely related taxa. Note that although many species of Bignoniaceae are interfertile in the laboratory and even several intergeneric hybrids are known, in nature the species are very consistently differentiated and hybridization virtually unknown (Gentry, 1988c). It is noteworthy that these are the kinds of patterns typical of many animal taxa and that fit relatively well into "biological species" concepts. Looking at the mostly clear-cut species of Bignoniaceae with 2n=40 in three fourths of the counted species and almost no hybridization in nature, I have long suggested to students in my phytogeography classes that tropical plants (and thus, in view of their numerical preponderance, plants in general) may evolutionarily behave more like animals than like the temperate zone (especially California) herbs whose speciation has been most intensively studied.

Nor are Bignoniaceae alone. Ehrendorfer (1970, 1986) and Ashton (1969) have summarized many of the prevailing generalities about repro-ductive mode in tropical (woody) plants, which are characterized by higher chromosome numbers, usually with high numerical constancy, little or no hybridization, more restricted recombination due to small populations of scattered individuals, narrow strongly adaptive niche differentiation, greater genetic and morphological uniformity within populations and taxa, and local mosaic-like allopatric ecogeographic differentiation. Both Ashton (1969) and Ehrendorfer (1970) also suggested that distributions tend to clumping and that genetic interchange may be very limited with autogamy prevalent.

However, Bawa (1974, 1979, 1980) has shown that dioecy is widespread and that even most hermaphrodite tropical trees are obligately outcrossing, contrary to the suggestions of Federov (1966) and Corner (1958). Not unexpectedly, Loveless and Hamrick (1984) have shown that gene flow is much greater for outcrossed (especially wind-dispersed) species than for selfing ones. More interesting, they suggest that sig-nificant gene flow over very long distances and between isolated popula-tions may occur in tropical forests. In an isolated cluster of six *Tachi-galia* trees on Barro Colorado Islands, Panama, Hamrick and Loveless (1989, pers. comm.) report that about 25% of the pollination was by pollen from at least half a kilometer away. Thus, the role of gene flow in maintaining species integrity and mediating speciation processes in tropical forests may be more like that envisioned by advocates of pan-mictic biological species (*e.g.*, Mayr, 1942, 1963) than like the very limited and local process seen in temperate zone studies (Ehrlich and Raven, 1969; Levin, 1979, 1981) or implied by advocates of autogamy as playing a major role in tropical speciation (Ashton, 1977). Perhaps specia-tion follows a different set of rules in tropical plants than in the temperate zone.

However, rapidly accumulating new data for many additional groups of tropical plants prove that many generalities about tropical plant speciation were distinctly premature. For example, a series of important studies of chromosome variation in neotropical woody plants by Morawetz (1984a,b,c, 1986a,b) has shown a rich array of chromosomal differentiation in many groups. For example, in Annonaceae 26 genera are primitively diploid with 2n=26 but 30 other genera have base numbers of x=7, 8, or 9; some genera have aneuploid series (*e.g.*, *Cymbopetalum*, 2n=14,16,18), others show polyploidy associated with ecological specialization for the relatively open *cerrado* habitat (Morawetz, 1984c), and at least four genera even have triploid species as do genera at in least nine other woody families (Morawetz, 1986a). Structural heterozygosity of the type frequent-ly found in herbaceous plants is found in *Porcelia* (Morawetz, 1984a).

Even a single species like *Duguetia furfuracea* of the Brazilian *cerrado*, can have several different caryological races (Morawetz, 1984b).

In Myristicaceae polyploidy has been so extensive as to generate counts of 2n=280 in *Osteophloeum* and 2n=102 in *Compsoneura* (Morawetz, 1986a). While it is true that in the largest genera of these woody tropical families (*e.g.*, *Virola* uniformly 2n=52: *Guatteria*, uniformly 2n=28) many of the old generalizations still seem to hold, it is also true that in many other instances they do not.

Hybridization may also be more widespread and evolutionarily important in tropical woody plants than has been assumed. In the first thorough biosystematic investigation of a large tropical woody genus, Neill (1988) found that *Erythrina* constitutes a homogamic complex with 94% of the 86 counted species diploid with 2n=42 (the five other counted species represent four independent polyploid events). All *Erythrina* species are genetically self compatible, autogamy (but not agamospermy) occurs occasionally, and all diploid species are completely interfertile, forming viable F1 hybrids. Hybrids between sympatric species occur in nature and Neill (1988) suspects that hybrid speciation via stabilization of natural hybrids has played an important evolutionary role in the genus. Thus, in many ways *Erythrina* behaves much like such temperate zone woody genera as *Ceanothus* (Nobs, 1963) and *Quercus* (Muller, 1952; Hardin, 1975). It should be noted, however, that except under human disturbance most *Erythrina* species are allopatric so natural hybridization is far less prevalent than in genera like *Quercus* with numerous sympatric species.

Furthermore, and contrary to the popular perception, very many tropical species are herbaceous. In fact, most of the tropical taxa included in this symposium are herbs or shrubs (Hansen, this volume; Chen, this volume; Andersson, this volume; Haynes and Holm-Nielsen, this volume). When allowance is made for the fact that many tropical forest herbs are epiphytic, over half the species of most neotropical local florulas are herbaceous (Gentry and Dodson, 1987b). While this may not be true in African forests (see Makokou, Gabon in Gentry and Dodson, 1987b) nor even in Central Amazonia (Gentry, 1989b), predominantly herbaceous taxa constitute at least half of the overall neotropical flora (Gentry, 1982a). Considering that tropical regions are generally much richer in plant species than temperate ones (Raven, 1976; Gentry, 1982a), on a global scale there may be almost as many herb species in the tropics as in the temperate zone, and there would surely be more if the speciose Mediterranean regions were excluded from the comparison.

The prevalence of tropical herbs is especially marked in topographically dissected, rich-soil high-rainfall areas as along the lower slopes of the northern Andes (Gentry, 1982a, 1986b). And many of the herbaceous or subwoody tropical taxa that have been studied to date show patterns that are very different from those postulated to characterize tropical trees. The palmettos that are so characteristic of tropical forest understories are a case in point. Thus *Heliconia*, highly diversified especially in northwestern South America (Andersson, this volume), is largely self compatible (Kress, 1983). Zingiberaceae, which has radiated extensively especially in tropical Asia, includes many different chromosome base numbers, much polyploidy (especially in tribe Hedychieae), even apomictic complexes (Globbeae), and perhaps speciation via somatic mutations associated with asexual reproduction (Chen, this volume). Even in ferns, apogamy is widespread in *Asplenium* (Iwatsuki, this volume). Aquatic plants, intrinsically herbaceous, are also prevalent in the tropics and have very different speciation patterns than trees: For example, Haynes and Holm-Nielsen (this volume) emphasize that neotropical Alismatidae have speciated and migrated in seasonally dry areas and are hardly present at all in Amazonian forests under today's climatic regime. Hybridization is rampant and evolutionarily important in *Fuchsia* (Berry, 1982), while pollinator specificity has characterized evolutionary diversification not only in orchids (Dodson, 1967) but also African Loranthaceae (Polhill, this volume). Even in Bignoniaceae, there are exceptions to the rule of "well-behaved" species as in Cuban *Tabebuia* where hybridization is widespread.

We may safely conclude that tropical plants cannot be characterized as evolutionarily conservative. They are not necessarily uniformly outcrossing with high constant chromosome numbers, absence of hybridization, and overwhelmingly allopatric speciation. Rather they have adopted an extraordinarily broad and diverse array of evolutionary strategies, probably including all of the different modes of speciation used by temperate zone plants.

Geographic patterns of speciation

Much debate among systematists and evolutionists has focused on the geography of speciation. More specifically, some authors (*e.g.*, Mayr, 1942, 1963) have maintained that speciation is predominately allopatric. Others (*e.g.*, Endler, 1982a,b; Benson, 1982; Beven *et al.*, 1984) favor parapatric speciation on strong environmental gradients, and still others, especially botanists, advocate various kinds of sympatric speciation. In the tropics in recent years this debate has focused on the role played by Pleistocene refugia in tropical speciation.

Allopatric Speciation. The classical idea of speciation, especially among zoologists, has been that allopatry or geographic isolation is nearly always necessary for speciation to occur. Nowadays, the foremost proponents of allopatric speciation tend to be cladists espousing what has come to be called vicariance biogeography (Platnick and Nelson, 1978; Rosen, 1978; Nelson and Platnick, 1981).

While vicariance biogeography tends to emphasize the establishment of allopatry with a vicariance event that divides a parent population (Cracraft, 1982; Wiley and Mayden, 1985; Cracraft and Prum, 1988;), dispersal across a pre-existing barrier can establish equally good allopatry, resulting in what is sometimes termed peripatric speciation (Mayr, 1982). Vicariance biographers look for congruence in geographical patterns between different clades. The model of Pleistocene refugia, sometimes referred to as "the" model for diversification in the tropics (Prance, 1982a), rapidly gained support in part because its emphasis on allopatric differentiation appealed to classic biogeographers while at the same time postulating a series of fragmentation events amenable to the analyses of vicariance biogeography.

The idea of speciation in Pleistocene forest refugia was introduced by Haffer (1969) based on the distributional patterns of Amazonian forest birds, many taxa of which show congruent centers of endemism associated with areas of relatively high rainfall and concordant contact zones between different pairs of related species. Similar distributional patterns were also found in groups as taxonomically disparate as butterflies (Brown, 1982), lizards (Vanzolini and Williams, 1970), monkeys (Kinzey and Gentry, 1979; Kinzey, 1982), and woody plants (Prance, 1981, 1982a). These patterns were interpreted as representing allopatric speciation in isolated forest islands, or refugia, during dry periods, associated with glacial advance. As these species expanded their ranges with forest expansion during mesic interglacials they formed the zones of secondary contact observed today. Independent evidence for dry periods, when today's continuous Amazonian forest might have been fragmented into small isolated patches, came from palynological studies which indicated widespread cycles of wet and dry climate during the Pleistocene, though mostly from outside the Amazon basin (*e.g.*, Hammen, 1974, 1982; Absy and Hammen, 1976), and from geomorphology, with the presence through much of Amazonia of stonelines similar to those that are formed by wind erosion in dry areas. The Pleistocene refugia hypothesis had the advantage of providing a neat explanation for a whole series of biogeographic phenomena. Moreover, it provided a model capable of predicting such observed phenomena as high Amazonian species diversity and the relatively depauperate biota of Africa as compared to South America.

Not surprisingly, an attractive model in a field as ripe with special cases as biogeography was almost instantaneously accepted, often quite uncritically. Thus, many distribution patterns that hindsight show to have other explanations were interpreted as deriving from Pleistocene forest fragmentation. For example, concentrations of as few as two endemic plant species in Venezuela were interpreted as indicating Pleistocene refugial centers (Steyermark, 1982). My paper in the Refuge Book (Gentry, 1982b) is an even worse example. The three apparent centers of plant endemism in the northern, central and southern Chocó region that I interpreted as evidence for three distinct Pleistocene refuges turned out to be largely due to collection artifact. Geographically intermediate areas of the Chocó that have subsequently been collected have just as much local endemism as do the original three "centers," which, not at all coincidentally, had also been the main centers of plant collecting activity.

Attempts to establish a more rigorous mathematical framework in which to test whether the apparently shared distributional patterns of unrelated taxa are actually congruent have met with mixed success. For example, Beven *et al.*'s (1984) computer analysis of Haffer's Amazonian bird distribution patterns showed that the apparent overlap between sets of refugia proposed from different taxa does not differ from random placement.

In the last few years, such problems have led to a growing wave of anti-refuge papers. Many of the supposedly refuge associated distributional patterns turn out to have simpler alternate explanations based entirely on present day ecology (Gentry, 1985, 1986b). The marked areas of high boundary segment density between distributional ranges found by Beven *et al.* (1984) mostly lie along obvious present day range discontinuities such as major rivers, the base of the Andes, or the coast. Recently, Gentry (1986a) has emphasized that most of the plant distributional patterns of endemic Bignoniaceae in Amazonia that he himself had earlier (Gentry, 1979) interpreted in the light of Pleistocene refugia occur in patches of distinctive substrate and are more likely to reflect habitat specificity on marked present day ecological gradients with resultant interrupted gene flow.

New palynological evidence from the Andean foothills of Ecuador suggests that during the last glacial advance montane *Podocarpus - Alnus - Hedyosmum* forests extended much lower altitudinally than previously suspected (Liu and Colinvaux, 1985) and that glacial age temperature depression in the Amazonian lowlands may have been closer to 6°C than to the 2°C previously suggested by the CLIMAP results (Colinvaux, 1987). Thus, at least some of the area postulated as a Pleistocene refuge for lowland forest trees was apparently covered by montane forest instead. All other palynological data showing drier periods during glacial advances are extra-Amazonian except an undated core from rather peripheral Rondonia (Absy and van der Hammen, 1976) and oxbow lake cores from the Manaus area (Absy, 1986) which are likely to reflect successional rather than climatic change (Colinvaux, 1987).

Problems have also turned up in the stone line data that had been considered (Ab'Saber, 1982) to provide independent evidence for extensive Amazonian savannah formations during dry periods. According to Irion (this volume) these were more likely formed under forest cover like that of today. Salo *et al.* (1986) and Colinvaux (1987) emphasize that erosional disturbance by rivers has reworked much of the Amazonian land surface over a time scale similar to the lives of individual trees; the extremely dynamic and heterogeneous vegetational kaleidoscope that results may in itself be capable of fomenting active speciation (Salo, 1987) and at the very least provides optimal levels of disturbance for maintenance of high species diversity under today's climatic regime via Connell's (1987) "intermediate disturbance" hypothesis. Salo (1987) goes so far as to conclude that there is no geological or biostratigraphical support for the refuge hypothesis from anywhere in the Amazon basin.

Another kind of attack on the importance of Pleistocene refuges for tropical forest speciation comes from reconsideration of the time scale involved. For trees that live a hundred or more years, there have been uncomfortably few generations since the last glacial advance. If a generation time of 100 years is assumed, then only 100 or so generations may have elapsed since the last glacial advance, perhaps even fewer during the putative refugial confinement at maximum advance. Either rather strong selection or tiny refugial populations must be involved to account for speciation in so few generations. Heyer and Maxson (1982) suggest that immunological distance between South American leptodactylid frogs indicate divergence times much older than the Quaternary. Moreover, a recent reanalysis of some of the bird distributional data using the techniques of vicariance biogeography and reinforced by cladistic analysis of relationships (Cracraft and Prum, 1988) come to very different conclusions than did Haffer (1969). Cracraft and Prum (1988) find a primary trans-Andean disjunction in unrelated bird lineages suggesting that Andean uplift rather than Pleistocene refugia constituted the most important vicariance event in the evolutionary diversification of these taxa. Instead of a single recent vicariance event associated with Pleistocene refugia, the patterns of congruence and non-congruence shown by different clades suggest a sequential series of vicariance events occurring at different times during the Cenozoic as well as in the Quaternary.

Some taxa probably did speciate in Pleistocene forest refugia but many others surely did not. Far more basic biogeographic data are needed before we can asses the overall importance of this mechanism in producing the high species richness of Amazonia, much less of other tropical areas. In conclusion, while no one now doubts that there have been major climatic upheavals during the Pleistocene in the tropics as well as in the temperate zone, the role of these disturbances on speciation and the distributional patterns seen today remains hotly debated.

Parapatric Speciation. Endler (1982a,b), Benson (1982), and others have argued that parapatric speciation may be the prevalent speciation mode in tropical forest animals. According to this interpretation, interruption of gene flow on habitat gradients by disruptive selection has been strong enough to lead to speciation. In Amazonia, many locally endemic plant species are edaphic specialists. For example, Prance (1982a) pointed out that in the Guiana-Venezuela region more than 2/3 of the endemic Chrysobalanaceae species are habitat specialists. In Bignoniaceae, most species with restricted distributions are habitat specialists; only a quarter of the locally endemic Amazonian species occur in *terra firme* forest (Gentry, 1986a).

While it is possible that some of the edaphic specialist bignon species could have arisen in forest refugia and subsequently invaded their specialized habitats (cf. Gentry, 1979), it is much simpler to suppose that they originated in situ. *Phryganocydia* is illustrative of this process (Gentry, 1983). The genus is well-defined and consists of three accepted species. A wide-ranging wind-dispersed species, *P. corymbosa*, gave rise to two locally endemic water-dispersed derivatives with wingless seeds in the swamps of the Magdalena Valley (*P. uliginosa*) and the Pacific coast mangroves of northern Chocó and southern Central America (*P. phellosperma*). A third derivative, with relatively thick sub-winged seeds and white instead of magenta flowers, occurs in the seasonally inundated *várzea* and tahuampa forests along the Amazon. The taxonomic status of this unnamed form remains unresolved, but whether or not it is specifically distinct it clearly represents incipient speciation similar to that which gave rise to *P. phellosperma* and *P. uliginosa*. Typical *P. corymbosa* also occurs in Amazonia but not in seasonally inundated forest. Clearly, selection for water-dispersed seeds in the *várzea* habitat is taking place under today's climatic regime and without intervention of Pleistocene refugia nor any other kind of allopatry.

The kind of edaphic specialization shown by *Phryganocydia* seems to have been a prevalent mode of speciation in Amazonia and to be responsible for much of today's Amazonian floristic richness. The role of ß-diversity and habitat specialization is especially prominent in northwestern Amazonia where the edaphic mosaic is apparently more complex than elsewhere (Gentry, 1986a, 1986b). In many cases, high species richness in this area reflects a series of related species that replace each other in different vegetation types. For example, four species of the *Passiflora vitifolia* complex (Gentry, 1981) occur together around Iquitos, one each restricted to alluvial soil, seasonally inundated forest, lateritic *terra firme* forest, and white-sand forest. The Iquitos-area forests on all these substrates are species rich but with very little species overlap between different forest types. Only 3-20 species are shared out of the ca. 200 species

> 2.5 cm dbh. that occur in 0.1 ha. samples (Gentry, 1986a). Since all these forests have remarkably similar compositions at the familial and generic levels (Gentry, 1986b), it seems highly probable that many of the species restricted to different habitats represent speciation associated with edaphic specialization. If so, the role of parapatric speciation associated with habitat specialization may have been of paramount importance in Amazonian plants. Geesink (this volume) suggests that parapatric speciation may also be prevalent in tropical Asia, *e.g.*, in *Millettia* which seems to present taxonomically difficult cases of parallel parapatric speciation events.

In Bignoniaceae, presumed parapatric speciation usually involves vegetative changes associated with adaptation to specialized substrates or, in the case of species of seasonally inundated habitats, a switch from wind-dispersed to water-dispersed seeds. Perhaps 180 or almost a third of all neotropical Bignoniaceae seem to represent this kind of speciation (Gentry, 1988c).

We may conclude not only that active parapatric speciation has occurred, but that it is taking place today. To judge from Bignoniaceae, perhaps 3/4 of the locally endemic species of woody taxa in Amazonia represent this kind of speciation.

Explosive speciation

A very different kind of speciation seems to be taking place in the cloud forests along the base of the northern Andes and in southern Central America. In this region there is a pronounced concentration of species of taxa that are predominantly epiphytic, shrubby, or herbaceous (Gentry, 1982a,b). Local endemism is very high in these "Andean-centered" groups and the individual genera tend to be unusually large and speciose (Gentry, 1982b, 1986a). I have interpreted this concentration of species as due to very active, often essentially sympatric, "explosive" speciation in the broken terrain and complex juxtaposition of different vegetation types associated with the Andean orogeny. This pattern is poorly documented because, almost by definition, the genera involved are large and taxonomically difficult, with many undescribed species. However, the available evidence strongly suggests that there is endemism on an extremely local scale in ecologically isolated cloud forests as small as 5-10 sq. kilometers (Gentry, 1986a). In such ridge-top forests as Centinela, Ecuador, which lies only 300 m above the adjacent valley that separates it from the Andean cordillera and Cerro Tacarcuna on the border of Panama and Colombian, the level of local endemism runs between 10% and 24% (Gentry, 1986a).

A good example of this kind of explosive speciation is herbaceous

Gasteranthus with a quarter of its world total of 24 species endemic to the environs of Centinela (Gentry and Dodson, 1987b). It is likely that a number of the patterns of speciation in specific taxa documented in this symposium represent similar phenomena. Dransfield (this volume) emphasizes the difference between the relictual large-tree Madagascar palms with few species per genus and the undergrowth Dypsid palms which have radiated extensively into over 100 taxa characterized by vegetative differentiation and a plethora of leaf forms. Dransfield notes that selective advantages of the different leaf forms are not obvious, and that the taxonomic problems engendered by lack of clear cut boundaries between Dypsid genera and species are scarcely paralleled elsewhere in the family, exactly what we might expect if these palms, which are concentrated in that part of Madagascar which ecologically most closely parallels the Andean foothill cloud forests (Gentry, 1988b), are undergoing explosive speciation.

We have suggested several possible reasons why local speciation rates may be so high in northern Andean cloud forests (Gentry and Dodson, 1987b). Ultrafine niche partitioning may be possible under relatively constant cloud forest conditions. At a slightly larger scale, microgeographic differentiation is no doubt accentuated by the broken terrain of mountainous areas where adjacent slopes with different orientations often have completely different microclimates. Finally, essentially accidental, perhaps often suboptimal, genetic transilience (Templeton, 1980) associated with founder effect phenomena in a kaleidoscopically changing landscape may have been a dominant evolutionary mode in the epiphytes, understory shrubs, and herbs that constitute the bulk of the Andean foothill cloud forest species (Gentry, 1982a, 1986a). The small and localized populations associated with constant recolonization in a dynamic habitat partitioned by mountains, local rainshadows, and frequent landslides should lead to optimal conditions for genetic drift with random fixation of different allele combinations in different founder populations (Wright, 1977; Templeton, 1980). While a recent review (Carson and Templeton, 1984) focused on this kind of founder-effect mediated speciation on islands and isolated mountaintops and considers it rather rare, I suspect that it may have been extremely important in the evolutionary diversification of continental neotropical cloud forest plants. We suspect that in orchids, where this kind of process is complemented by intricate coevolutionary interactions, speciation may take place in as little as 15 years (Gentry and Dodson, 1987b).

This kind of explosive and essentially sympatric speciation has apparently been extremely important in generating the high overall species richness of the Neotropics. Nearly half of the neotropical flora may be accounted for by this phenomenon (Gentry, 1982a). Clearly

Stebbins (1974) conclusion that *"Comparisons between tropical and temperate regions... do not support the hypothesis that the well-known richness of tropical floras is due chiefly to extensive speciation in progress at the present time....In spite of the much greater richness of tropical flora in total numbers of species, genera and families, there is no evidence that speciation is more active in them than it is in the varied communities of the marginal temperate regions and it may even be less active."* Tropical speciation in as little as 15 years would seem very dynamic indeed.

If generally true, these conclusions on the nature of tropical speciation have profound implications, necessitating some major changes in our view of the evolutionary process. For example, as seen from this perspective, most *Heliconia* species are more likely to date back only 100-1000 years than the 30 million years suggested by Andersson (this volume). If randomly generated, often relatively nondescript and only slightly differentiated, species make up the bulk of such evolutionarily plastic genera as *Anthurium, Piper,* or *Pleurothallis,* conservationists may be faced with the philosophical question of whether preservation of each of these entities merits the same priority as does preservation of the more dramatically distinctive species of other groups. Moreover, if other unprepossessing cloud forest ridges like Centinela similarly have a hundred or so endemic species, it follows that many thousands of additional and hitherto unsuspected new species remain to be discovered; if so, the inventory task, much less the conservational task, facing tropical botanists is truly daunting. The Centinela ridge top has now been deforested and its hundred endemic species are extinct or about to become so. If Centinela was typical of the levels of endemism that can occur in tropical forests, the world has surely already undergone massive extinctions in already-deforested parts of the Andean and Central American foothills and perhaps elsewhere. Not only is mankind's biological heritage poorer as a result, but our intellectual heritage is also eroded as these uniquely active laboratories of speciation disappear from the face of the earth. Moreover, those of us interested in evolutionary processes have an added incentive for preserving our planet's dwindling remnants of tropical forest: We need them if we hope ever to truly understand the processes of speciation and evolution that have given rise to the diversity of life on earth.

Literature cited

Ab'Saber, A. N. (1982). "The paleoclimate and paleoecology of Brazilian Amazonia." pp. 41-59 *In* Prance (1982b).

Absy, M. L. (1986). "Palynology of Amazonia: the history of the forests as revealed by the palynological record." pp. 72-87 *In* Prance, G. T. and Lovejoy, T. (eds.), "Amazonia." *Pergamon Press, Oxford.*

Absy, M. L. and Hammen, T. van der (1976). "Some palaeoecological data from Rondonia, southern part of the Amazon basin." *Act. Amaz.* **6,** 293-299.

Agren, G. I. and Fagerstrom, T. (1984). "Limiting dissimilarity in plants: randomness prevents exclusion of species with similar competitive abilities." *Oikos* **43,** 369-375.

Arroyo, M. T. K. (1988). "Effects of aridity on plant diversity in the northern Chilean Andes: results of a natural experiment." *Ann. Missouri Bot. Gard.* **75,** 55-78.

Ashton, P. S. (1969). "Speciation among tropical forest trees: some deductions in the light of recent evidence." *J. Linn. Soc., Biol.* **1,** 155-196.

Ashton, P. S. (1977). "A contribution of rain forest research to evolutionary theory." *Ann. Missouri Bot. Gard.* **64,** 694-705.

Bawa, K. S. (1974). "Breeding systems of tree species of a lowland tropical community." *Evolution* **28,** 85-92.

Bawa, K. S. (1979). "Breeding systems of trees in a tropical wet forest." *N. Zealand J. Bot.* **17,** 521-524.

Bawa, K. S. (1980). "Evolution of dioecy in flowering plants." *Ann. Rev. Ecol. Syst.* **11,** 15-39.

Benson, W. W. (1982). "Alternative models for infrageneric diversification in the humid tropics: tests with passionvine butterflies." pp. 608-640 *In* Prance (1982b).

Berry, P. E. (1982). "The systematics and evolution of Fuchsia sect. Fuchsia (Onagraceae)." *Ann. Missouri Bot. Gard.* **69,** 1-198.

Beven, S., Connor, E. F. and Beven, K. (1984). "Avian biogeography in the Amazon basin and the biological model of diversification." *J. Biogeogr.* **11,** 383-399.

Brown, K. S., Jr. (1982). "Paleoecology and regional patterns of evolution in neotropical forest butterflies." pp. 255-308 *In* Prance (1982b).

Carson, H. L. and Templeton, A. R. (1984). "Genetic revolutions in relation to speciation phenomena: the founding of new populations." *Ann. Rev. Ecol. Syst.* **15,** 97-131.

Colinvaux, P. (1987). "Amazon diversity in light of the paleoecological record." *Quat. Sci. Rev.* **6,** 93-114.

Connell, J. H. (1978). "Diversity in tropical rain forests and coral reefs." *Science* **199,** 1302-1310.

Connell, J. H. (1987). "Maintenance of species diversity in biotic communities." pp. 201-218 *In* Kawano, S., Connell, J. and Hidaka, T. (eds)," Evolution and Coadaptation in Biotic Communities." *Univ. Tokyo Press, Tokyo.*

Connell, J. H. and Orians, E. (1964). "The ecological regulation of species diversity." *Amer. Naturalist* **98**, 399-414.

Corner, E. J. H. (1954). "The evolution of tropical forest." pp. 34-46 *In* Huxley, J. S., Hardy A. C. and Ford, E. B. (eds.), "Evolution as a Process." *Allen and Unwin, London.*

Cracraft, J. (1982). "Geographic differentiation, cladistics, and vicariance biogeography: reconstructing the tempo and mode of evolution." *Amer. Zool.* **22**, 411-424.

Cracraft, J. and Prum, R. O. (1988). "Patterns and processes of diversification: speciation and historical congruence in some neotropical birds." *Evolution* **42**, 603-620.

Dodson, C. D. (1967). "Relationships between pollinators and orchid flowers." *Atas do Simposia sobre a Biota Amazonica* **5**, 1-72.

Ehrendorfer, F. (1970). "Evolutionary patterns and strategies in seed plants." *Taxon* **19**, 185-195.

Ehrendorfer, F. (1980). "Polyploidy and distribution." pp. 45-60. *In* Lewis W. (ed.), "Polyploidy, Biological Relevance." *Plenum, London.*

Ehrendorfer, F. (1986). "Chromosomal differentiation and evolution in angiosperm groups." pp. 59-86 *In* Iwatsuki, K., Raven, P. H. and Bock, W. (eds.), Modern Aspects of Species. *Univ. Tokyo Press, Tokyo.*

Ehrlich, P. R. and Raven, P. H. (1969). "Differentiation of populations." *Science* **165**, 1228-1232.

Endler, J. (1982a). "Problems in distinguishing historical from ecological factors in biogeography." *Amer. Zool.* **22**, 441-452.

Endler, J. (1982b). "Pleistocene forest refuges: fact or fancy." pp. 641-657 *In* Prance (1982b).,

Federov, A. (1966). "The structure of the tropical rain forest and speciation in the humid tropics." *J. Ecol.* **54**, 1-11.

Fischer, A. G. (1960). "Latitudinal variations in organic diversity." *Evolution* **14**, 64-81.

Gentry, A. H. (1974a). "Coevolutionary patterns in Central American Bignoniaceae." *Ann. Missouri Bot. Gard.* **61**, 726-759.

Gentry, A. H. (1974b). "Flowering phenology and diversity in tropical Bignoniaceae." *Biotropica* **6**, 64-68.

Gentry, A. H. (1976). "Bignoniaceae of southern Central America: distribution and ecological specificity." *Biotropica* **8**, 117-131.

Gentry, A. H. (1977). "New species of Leguminosae, Lauraceae, and Monimiaceae, and new combinations in Bignoniaceae from western Ecuador." *Selbyana* **2**, 39-45.

Gentry, A. H. (1979). "Distribution patterns of neotropical Bignoniaceae: some phytogeographic implications." pp. 339-354 *In* Larsen, K. and Holm-Nielsen, L. (eds.), "Tropical Botany." *Academic Press,*

London.

Gentry, A. H. (1981). "Distributional patterns and an additional species of the Passiflora vitifolia complex: Amazonian species diversity due to edaphically differentiated communities." *Plant Syst. Evol.* **137,** 95-105.

Gentry, A. H. (1982a). "Neotropical floristic diversity: phytogeographical connections between Central and South America, Pleistocene climatic fluctuations, or an accident of the Andean orogeny ?" *Ann. Missouri Bot. Gard.* **69,** 557-593.

Gentry, A. H. (1982b). "Phytogeographic patterns in northwest South America and southern Central America as evidence for a choco refugium." pp. 112-136 *In* Prance (1982b).

Gentry, A. H. (1982c). "Patterns of neotropical plant species diversity." *Evol. Biol.* **15,** 1-84.

Gentry, A. H. (1983). "Dispersal and distribution in Bignoniaceae." *Sonderb. Natur. Wiss. Ver. Hamburg* **7,** 187-199.

Gentry, A. H. (1985). "Algunos resultados preliminares de estúdios botanicos en el Parqúe Nacional del Manú." *In* Rios, M. (ed.), "Reporte Manú. Centro de Datos para la Conservacion." *L a Molina, Peru, 2/1 - 2/24*

Gentry, A. H. (1986a). "Endemism in tropical vs. temperate plant communities." pp. 153-181 *In* Soule, M. (ed.), "Conservation Biology." *Sinauer Press, Sunderland.*

Gentry, A. H. (1986b). "An overview of neotropical phytogeographic patterns with an emphasis on Amazonia." *An. 10 Simposio do Tropico Umido* **2,** 19-35.

Gentry, A. H. (1988a). "Tree species richness of upper Amazonian forests." *Proc. Natl. Acad. USA* **85,** 156-159.

Gentry, A. H. (1988b). "Changes in plant community diversity and floristic composition on environmental and geographical gradients." *Ann. Missouri Bot. Gard.* **75,** 1-34.

Gentry, A. H. (1988c). "Evolutionary patterns in neotropical Bignoniaceae." *In* Prance, G. and Hammond, D. (eds.), "Reproductive biology of tropical plants." (in press).

Gentry, A. H. (1989a). "Tropical forests." *In* Keast, A. (ed.), "Biogeography and ecology of forest bird communities." *Oxford Univ. Press, Oxford.* (in press).

Gentry, A. H. (1989b). "Floristic similarities and differences between southern Central America and upper and central Amazonia." *In* Gentry, A. (ed.), "Four Neotropical Forests." *Yale Univ. Press, New Haven.*

Gentry, A. H. and Dodson, C. H. (1987a). "Contribution of non-trees to species richness of tropical rain forest." *Biotropica* **19,** 149-156.

Gentry, A. H. and Dodson, C. H. (1987b). "Diversity and phytogeography of neotropical vascular epiphytes." *Ann. Missouri Bot. Gard.* **74**, 205-233.

Goldblatt, P. and Gentry, A. H. (1979). "Cytology of Bignoniaceae." *Bot. Not.* **132**, 475-482.

Grant, V. (1975). "Genetics of flowering plants." *Columbia Univ. Press, New York.*

Grime, J. P. (1979). "Plant strategies and vegetation processes." *Wiley and Sons, New York.*

Haffer, J. (1969). "Speciation in Amazonian forest birds." *Science* **165**, 131-137.

Hammen, T. van der (1974). "The Pleistocene changes of vegetation and climate in tropical South America." *J. Biogeogr.* **1**, 3-26.

Hammen, T. van der (1982). "Paleoecology of tropical South America." pp. 60-66 *In* Prance (1982b).

Hamrick, J. L. and Loveless, M. D. (1989). "The genetic structure of tropical tree populations in association with reproductive biology." *In* Buck, J. and Linhart, Y. (eds.), "Plant evolutionary ecology." *Westview Press, Boulder.*

Hardin, J. W. (1975). "Hybridization and introgression in Quercus Heyer, W. R. and Maxson, L. R. (1982). "Distributions, relationships and zoogeography of lowland frogs: the Leptodactylus complex in South America, with special reference to Amazonia." pp. 375-388 *In* Prance (1982b).

Hilty, S. L. (1981). "Flowering and fruiting periodicity in a premontane rain forest in Pacific Colombia." *Biotropica* **12**, 292-306.

Hubbell, S. and Foster, R. (1986). "Biology, chance, and history and the structure of tropical rain forest tree communities." pp. 314-329 *In* Diamond, J. and Case, T. (eds.), "Community ecology." *Harper and Row, New York.*

Janzen, D. H. (1967). "Why mountain passes are higher in the tropics." *Amer. Naturalist* **101**, 233-249.

Janzen, D. H. (1970). "Herbivores and the number of tree species in tropical forests." *Amer. Naturalist* **104**, 501-528.

Kaur, A., Ha, C. O., Jong, K., Sands, V. E., Chan, H. T., Soepadmo, E. and Ashton, P. S. (1978). "Apomixis may be widespread among trees of the climax rain forest." *Nature* **271**, 440-441.

Kinzey, W. (1982). "Distribution of primates and forest refuges." pp. 455-482 *In* Prance (1982b).

Kinzey, W. and Gentry, A. (1979). "Habitat utilization in two species of Callicebus." pp. 89-100 *In* Sussman, R. (ed.), "Primate ecology: problem-oriented field studies." *Wiley and Sons, New York.*

Kress, W. J.(1983). "Self-incompatibility in Central American Heliconia." *Evolution* **37**, 735-744.

Levin, D. A. (1979). "The nature of plant species." *Science* **204**, 381-384.

Levin, D. A. (1981). "Dispersal versus gene flow in plants." *Ann. Missouri Bot. Gard.* **68**, 233-253.

Liu, K. and Colinvaux, P. A. (1985). "Forest changes in the Amazon Basin during the last glacial maximum." *Nature* **318**, 556-557.

Loveless, M. D. and Hamrick, J. L. (1984). "Ecological determinants of genetic structure in plant populations." *Ann. Rev. Ecol. Syst.* **15**, 65-95.

Løve, A. and Løve, D. (1949). "The geobotanical significance of polyploidy." pp. 273-352 *In* Goldschmidt, R. (ed.), "I. Polyploidy and latitude." *Portugaliae Acta Biol. Spec. Vol.*

MacArthur, R. H. (1964). "Environmental factors affecting bird species diversity." *Amer. Naturalist* **98**, 387-397.

MacArthur, R. H. (1965). "Patterns of species diversity." *Biol. Rev. (London)* **40**, 510-533.

MacArthur, R. H. and MacArthur, J. W. (1961). "On bird species diversity." *Ecology* **42**, 594-598.

MacArthur, R. H., Recher, H. and Cody, M. L. (1966). "On the relation between habitat selection and species diversity." *Amer. Naturalist* **100**, 319-332.

Mayr, E. (1942). "Systematics and the Origin of Species." *Columbia Univ. Press, New York.*

Mayr, E. (1963). "Animal species and evolution." *Harvard Univ. Press, Cambridge.*

Mayr, E. (1982). "Processes of speciation in animals." pp. 1-19 *In* Barigozzi, C. C. (ed.) Mechanisms of speciation." *Liss, New York.*

Morawetz, W. (1984a). "How stable are genomes of tropical woody plants ? Heterozygosity in C-banded karyotypes of Porcelia as compared with Annona (Annonaceae) and Drimys (Winteraceae)." *Pl. Syst. Evol.* **145**, 29-39.

Morawetz, W. (1984b). "Karyological races and ecology in Duguetia furfuracea as compared with Xylopia aromatica (Annonaceae). *Flora* **175**, 195-209.

Morawetz, W. (1984c). "Karyologie, Ökologie und evolution der Gattung Annona (Annonaceae) in Pernambuco, Brasilien." *Flora* **175**, 435-447.

Morawetz, W. (1986a). "Remarks on Karyological differentiation patterns in tropical woody plants." *Pl. Syst. Evol.* **152**, 49-100.

Morawetz, W. (1986b). "Karyosystematics and evolution of Australian Annonaceae as compared with Eupomatiaceae, Himantandraceae and Austrobaileyaceae." *Pl. Syst. Evol.* **159**, 49-79.

Muller, C. H. (1952). "Ecological control of hybridization in Quercus: a factor in the mechanism of evolution." *Evolution* **6**, 147-161.

Neill, D. A. (1988). "Experimental studies on species relationships in

Erythrina (Leguminosae: Papilionoideae)." *Ann. Missouri Bot. Gard.* **75,** 886-969.

Nelson, G. and Platnick, N. (1981). "Systematics and biogeography." *Columbia Univ. Press, New York.*

Nobs, M. A. (1963). "Experimental studies in species relationships in Ceanothus." *Publ. Carnegie Inst. Wash.* **623,** 1-94.

Pianka, E. R. (1966). "Latitudinal gradients in species diversity: a review of concepts." *Amer. Naturalist* **100,** 36-40.

Platnick, N. and Nelson, G. (1978). "A method of analysis for historical biogeography." *Syst. Zool.* **27,** 1-16.

Prance, G. T. (1981). "A review of the phytogeographic evidences for Pleistocene climate changes in the Neotropics." *Ann. Missouri Bot. Gard.* **69,** 594-624.

Prance, G. T. (1982a). "Forest refuges: evidence from Woody Angiosperms." pp. 137-157 *In* Prance (1982b).

Prance, G. T. (1982b). "Biological diversification in the tropics." *Columbia Univ. Press, New York.*

Raven, P. H. (1976). "Ethics and attitudes." pp. 155-179 *In* Simmons J. B., Beyer, R. I., Brandham, P. E., Lucas, G. Ll. and Parry, V. T. H. (eds.), "Conservation of threatened plants." *Plenum, New York, London.*

Rosen, D. E. (1978). "Vicariant patterns and historical explanation in biogeography." *Syst. Zool.* **27,** 159-188.

Salo, J. (1987). "Pleistocene forest refuges in the Amazon: evaluation of the biostratigraphical, lithostratigraphical and geomorphological data." *Ann. Zool. Fennici* **24,** 203-211.

Salo, J., Kalliola, R., Hakkinen, I., Makinen, Y. Niemela, P. Puhakka, M. and Coley, P. (1986). "River dynamics and the diversity of Amazon lowland forest." *Nature* **322,** 254-258.

Sanders, H. L. (1968). "Marine benthic diversity: a comparative study." *Amer. Naturalist* **102,** 243-282.

Snow, D. W. (1965). "A possible selective factor in the evolution of fruiting seasons in tropical forest." *Oikos* **15,** 274-281.

Stebbins, G. L. (1942). "Polyploid complexes in relation to ecology and the history of floras." *Amer. Naturalist* **76,** 36-45.

Stebbins, G. L. (1974). "Evolution above the species level." *Harvard Univ. Press, Cambridge.*

Stehle, F. G., Douglas, R. G. and Newell, N. D. (1969). "Generation and maintenance of gradients in taxonomic diversity." *Science* **164,** 949-977.

Steyermark, J. A. (1982). "Relationships of some Venezuelan forest refuges with lowland tropical floras. pp. 182-220 *In* Prance (1982b).

Templeton, A. R. (1980). "The theory of speciation via the founder principle." *Genetics* **94,** 1011-1038.

Terborgh, J. (1985). "Habitat selection in Amazonian birds." pp. 311-338 *In* Cody, M. (ed.), "Habitat selection in birds." *Academic Press, New York.*

Turner, J. R., Gatehouse, C. M. and Corey, C. A. (1987). "Does solar energy control organic diversity? Butterflies, moths and the British climate." *Oikos* **48,** 195-205.

Vanzolini, P. E. and Williams, E. E. (1970). "South American anoles: the geographic differentiation and evolution of the Anolis chrysolepis species group (Sauria, Iguanidae)." *Arq. Zool. Sao Paulo* **19,** 1-298.

Werff, H. van der and Richter, H. G. (1985). "Caryodaphnopsis Airy-Shawe, a genus new to the Neotropics." *Syst. Bot.* **10,** 166-173.

Wiley, E. and Mayden, L. (1985). "Species and speciation in phylogenetic systematics, with examples from the North American fish fauna." *Ann. Missouri Bot. Gard.* **72,** 596-635.

Wright, D. H. (1983). "Species-energy theory: an extension of species-area theory." *Oikos* **41,** 496-506.

Wright, S. (1977). "Evolution and the genetics of populations vol. 3. Experimental results and evolutionary deductions." *Univ. Chicago Press, Chicago.*

Speciation and Malesian Leguminosae

R. GEESINK AND D. J. KORNET

*Rijksherbarium and Instituut voor Theoretische Biologie,
Leiden, The Netherlands*

Immediately after the rediscovery of Mendel´s laws attempts were made to
connect the generally accepted typological species-concept with the Genetic
Theory and Darwin´s Evolutionary Theory. We are left with: a number of
species concepts, each emphasizing different aspects of species. Attempts
to combine the aspects into one "synthetic" species concept seem to obey
Gödel's theorem thus far: that the logically correct ones are incomplete.

Evolutionary Theory is the only available scientific theory
explaining the diversity of Living Nature. A species concept, consistent
with Evolutionary Theory is the most satisfactory one, but implies a choice
between relevant aspects as well. The ongoing debate, whether natural
species are to be regarded as sets (*i. e.* as structurally defined classes of
objects) or as unique, cohesive individuals has clarified the implications
of some alternative choices. The attempts towards a unified theory of
living and non-living nature, recently developed by Brooks and Wiley,
provides fundamental concepts from physico-chemical and mathematical
laws. Their major message is that the evolutionary process must be an
entropic process.

Disjunct species (*Acacia koa*/*A. heterophylla* and *Rhynchosia
rothii*), chaotic supraspecific evolution (tribe Millettieae), sympatric
uniparental (*Indigofera*), and parapatric (*Millettia* sect. *Fragiliflorae*)
speciation, are demonstrated to show some applications of these theoretical
considerations.

Species and speciation

Most readers will have a feeling what a species is. Usually, such an
implicit species concept contains some of the following aspects: a species
must be recognizable by its characteristics, it has some constancy of
progeny and does not interbreed with other species. Often additional
criteria will be mentioned, like having a particular ecology, a particular
geographical region, its own evolutionary history. Two explicitly defined

species concepts which do not mention characteristics as a criterion are Mayr´s biological species concept and Wiley´s phylogenetic concept: Mayr (1942) stated that "Species are groups of actually or potentially interbreeding populations reproductively isolated from other such groups." Later (1969), he modified the definition by deletion of the notion "potentially interbreeding." In 1982, he added "occupying a specific niche" to this definition.

Wiley (1981) presented a slightly modified version of Simpson´s (1961) well-known definition, emphasizing the historical component: *An evolutionary species is a single lineage of ancestral-descendant populations which maintains its identity from other such lineages and which has its own evolutionary tendencies and historical fate.* The species problem arises when attempting to combine the aspects mentioned in these definitions with the desire of museum/herbarium taxonomists to recognize species as distinct groupings based on monothetic sets (unique combinations) of characteristics.

Bentham (1861) explicitly stated that for practical reasons taxonomists can only deal with species with distinct characters: *Believing, however, as I do, that there exist in nature a certain number of groups of individuals, the limits to whose powers of variation are, under present circumstances, fixed and permanent, I have been in the habit of practically defining the species as the whole of the individual plants which resemble each other sufficiently to make us conclude that they are all, or MAY HAVE BEEN all, descended from a common parent.* (Emphasis in original). His idea of common parent was meant as, descended from one original plant, or pair of plants. Today, we think of ancestral populations rather than individuals. Bentham´s definition is interesting because of his emphasis on descent and real existence in nature. He even reduced distinctness to a practical criterion for common descent.

Ghiselin (1974a) tried to solve the species problem by regarding species as individuals rather than sets of similar specimens constituting classes of objects. With this he in fact selected the historical aspect as the primary one. Hull (1976) elaborated this idea and demonstrated its biological meaning. Williams (1985) claimed priority because her axiomatization of evolutionary change (Williams, 1970) implied that species are individuals with respect to Evolutionary Theory. *Biology and Philosophy* devoted its volume 2 part 2 (1987) to this controversial subject. We think the idea of species as individuals is a very fruitful one which deserves some further explanation.

Two sorts of relevant classes can be distinguished; (i) more artificial classes, like "red balls" or "all flowering plants with blue petals"; (ii) natural classes like "Helium", "Calcite", "DNA", or "marsh plants". In the same way we can distinguish two sorts of (composite) individuals; (i) artificial individuals, like "General Motors Company" and "The Netherlands;" (ii) natural individuals, like "the planet Saturnus" and "George Bentham." The term "individual" is used in a wider meaning than in its daily use as "specimen" or "particular organism." In our opinion the relevant comparison should be made between Natural individuals and Natural kinds. These are listed below:

Natural individuals	**Natural kinds**
Spatio-temporally restricted: unique; with a beginning and an end in time.	Spatio-temporally unrestricted: members can originate anywhere, at anytime.
Exist in nature as cohesive wholes, independent from man´s ability to recognize them.	Do not exist as cohesive wholes; individual members of them (*e.g.* gold atoms) are the real things.
Defined by relations between their parts. Characteristics can be described but are NOT defining properties.	Defined by underlying structural properties of their members.

A species is not so distinct an individual as a particular specimen. A new species comes into being (gets individualized) when as a population it looses the genetic contacts with other populations of the ancestral species. As an ancestral population it is not yet an individual consisting of specimens genetically interacting in a network, representing reproductive ties. After the isolation of the ancestral population, the newly originated species evolves genetically independent from other species. "Becoming an individual" emerges from the isolation event, and is expressed by independent autonomous evolution.

Artificial and natural classes are different from individuals. An artificial class is a category in the human mind. Take for example the class of red balls. Many different things are members of the class of red balls: tomatoes, some billiard-balls, the setting sun, *etc.* The class definition obtains only to red and globose objects. There is no common underlying structure which explains the shared properties. Hence, no predictions of other characters can be made. Natural kinds, on the other hand, are discovered groupings of similar things. Here, an underlying common structure explains the observed shared properties of the members. Classes cannot change their defining properties without becoming a different class. Classes have no shared historical development, hence they cannot evolve.

Natural selection works on members of a subset, the class (natural kind) of *e.g.* red-petaled flowers within a species variable in petal-colour. If such selection-pressure works on the entire species in its entire area, the results of this selection process is an evolutionary change (after many generations) of the individual species (Williams, 1970). If no splitting occurs, it remains the same individual. Selection pressure can also work on only a part of the species in only a part of its area, resulting in polytypic diversification, of still the same individual. Only cladogenesis (splitting) results in speciation; one ancestral species gives rise to two or more descendent species. Genera and other higher taxa are the passive results of several speciation events. This passive role of monophyletic higher taxa made Wiley (1981) distinguish individuals (species) from historical groups (monophyletic higher taxa). For a more elaborate treatment, see Zandee and Geesink (1987).

Biparental systems, uniparental systems and species

In figure 1 all specimens (dots and circles) have two parents. Only the reproducing offspring is depicted and each generation contains only a few specimens. Therefore it is impossible in this scheme to subdivide the lineage into different semi-isolated, interbreeding populations.

There is cohesion between the specimens which is represented by the arrows forming a network. Arrows indicate a relation (parenthood). Other phenomena can be inferred or follow from these relations, like sexual reproduction, inheritance of characteristics, and similarity. Sexual reproduction is the mechanism of cohesion. Fisher (1929, 1958-ed.: 160) wrote: "*With sexual reproduction species are not arbitrary taxonomic units such as they would with asexual reproduction only, but are bound together by sharing a very complete community of ancestry.*"

In Fig. 1 a genetically fixed character of the specimen in the corner left, below, can be inherited by any specimen of the isolated groups A, B, C, or D. This does not imply that all descendants do actually have all characters of all ancestors. Genetic segregation, dominance, and recessiveness provide specimens with different genetic information. If inheritance would be a random process without selection, all descendants would have equal chance to inherit all ancestral characters. Evolution is the channelling mechanism; first, there is one species (X); many generations later there are two species (Y and Z) and later there are four species (A-D). Some species (the lineages M and N) got extinct.

All species are variable. The character range will not be much

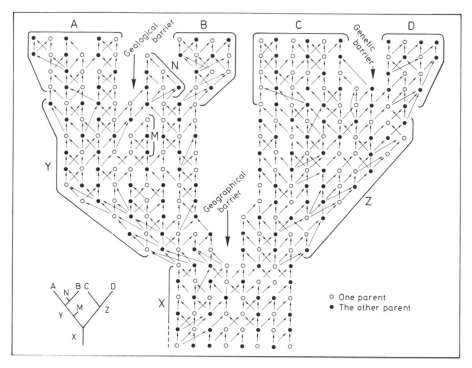

Figure 1. Idealized and simplified part of a true geneology of specimens in an outbreeding system.

different just before and just after an isolation event. The products of this splitting event develop under different circumstances and different selection pressures. So, after another few (hundreds or more) generations the ranges of variation of the lineages Y and Z may show significant differences or gaps. Because of Nature's drive to diversification (randomization of information, treated in the following section) all splittings in genealogical trees of biparental systems will most probably lead to recognizable species. Adherents to the historical species concept will draw the species demarcation line immediately above the isolation-event. Other taxonomists will be a bit hesitating, and demand something more than "just being isolated", because it cannot be predicted whether isolation will be forever. For the practising taxonomists this is more than just an academical question (section IV, example 1).

If all species consisted of systems with unproblematical cohesion, only boundary cases could spoil the picture somewhat.

Figure 2 differs from the previous one mainly in species C and its ancestors. Immediately after the first isolation event, only autogamous specimens (self-pollinating, vegetatively propagating, or variously apomictic) survived in the right hand area. Most individual lines are perfectly isolated in the uniparental system ("UPS C"). There is no genetical exchange between these lines.

If a mutation results into a changed character state, all descendants will have that changed character state. There is one significant exception: in self-pollinating systems there is still crossing-over at the meiosis, locally resulting in individual variation (see *Indigofera* p. 148).

The uniparental way of propagation is an excellent strategy of invading a new area, but such systems are susceptible to diseases and parasites like monocultures of crops.

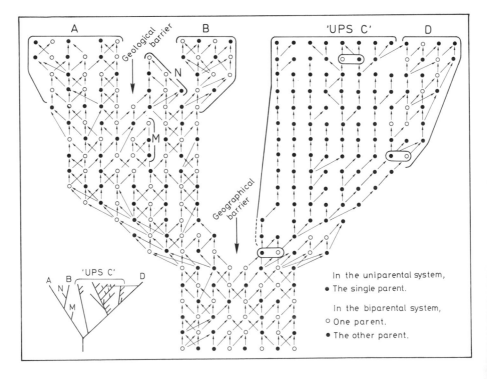

Figure 2. Idealized and simplified part of a true geneology of specimens with an autogamous branch in the right hand area.

Species as self-organising information systems

This is another controversial subject, the understanding of which we experienced as clarifying and significant. The relevant literature is Brooks and Wiley, 1986, and Wicken, 1987. Further recommended literature are Shannon and Weaver, 1949, Gatlin, 1972, and Eberling and Feisel, 1982, 1986. Collier, 1986 and Geesink, 1986 are introductions.

The basic idea of this theory is that the evolutionary proces is, like all other natural processes, an entropy increasing process, even though this is counterintuitive to those who feel that increase of complexity (evolution from simple forms to more complex ones) "must somehow" violate the second law of thermodynamics. Brooks and Wiley (1986) argue that such an intuitive idea is based on wrong analogies. They developed a theory based upon non-equilibrium thermodynamics, (see Prigogine and Stengers, 1984) and information/communication theory (Shannon and Weaver, 1949). Brooks and Wiley (1986) demonstrated by means of certain analogies that diversification, increase of complexity, and speciation are natural processes in which entropy increases, as expected. The central formula is:

$$(1)\ Hs = -\sum pi\ ld\ pi$$

in which Hs represents the measure of realized complexity of a particular system (macrostate) consisting of parts (microstates) that have chances (pi) to be realized in the system in a particular place. The logarithm in base 2 (ld) is chosen in order to obtain entropy-values that are expressed in bits (one/zero statements).
Fig. 3-a, from Schuster, (1988) shows a system with 2 possible states; either a 1 in the left hand box (state) or in the right hand box.

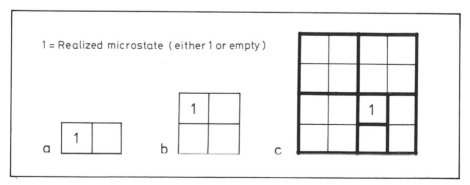

Figure 3. Entropy system with: a) two possible states, b) four possible states, c) sixteen possible states.

One question suffices to locate the one. Hence, the maximum information content is one bit. For a system with four possible states (Fig. 3-b) two questions (up/down and left/right) must be answered to locate each one-state. Its maximum information content is two bits. Similarly, 4 questions are needed for locating any one-state in the frame surrounding the 16 boxes in figure 3-c. An empty macrostate (the frame around the 16 available places) is called a "phase space". Its most applied aspect is its maximum possible information content, expressed formally as:

$$(2)\ Hmax = ld\ A$$

where Hmax is the maximum possible information content, and A the array of e.g. a matrix (the total of available spaces). In taxonomic datamatrices A is the number of taxa multiplied by the number of character states and represents the number of available places in which microstates can be realized.

Formula (1) is actually derived from formula (2) (Shannon and Weaver, 1949). Formula (2) is similar to the Boltzmann's well-known formula of statistical entropy:

$$(3)\ S = k\ ln\ \Omega$$

which is derived along different lines.

Wicken (1987) made the critical remark that the Shannon entropy (H in formulas 1 and 2) is not a true entropy, but at best an entropy-analogy. Justification of the theory is a complex matter, but philosophers of science (Collier, 1986, 1988 and contributors in Weber and Depew, 1988) took up the challenge. Several applications by Brooks and Wiley (1986) show that the "entropy-analogies" behave as true entropies would do.

The increasing system of sets of inclusive systems of size four (Fig.3) have no realization in the living nature.

The first biological replicators originated about 4 billion years ago. Small RNA-like structures containing only 2 nucleotides, in the sequence GC, were synthesised. Earlier, the 4 kinds (assuming that there were also four kinds available) of nucleotides and associated molecules were disorderly diffused (Hmax in Fig. 4a). By an endothermic reaction they became connected into the first little string of RNA of size 2. Some order is generated and the entropy level drops (Hs in Fig. 4a). There are 16 different arrangements possible with 4 kinds of nucleotides on a string of size 2, only one of which is realized. The corresponding complexity is:

$$H = -(\ 1/16\ ld\ 1/16) = 0.25\ (bits)$$

The maximum information content of a phase space size 16 possible states is:

$$Hmax = ld\ 16 = 4\ (bits)$$

The information content of this system is the difference between the maximum possible information content Hmax minus the realized information content Hs, in this case:

$$I = Hmax - Hs = 4 - 0.25 = 3.75\ (bits)$$

The first replicator "dies," the molecule desintegrates, and thermodynamic entropy is produced. Before desintegration, however, it reproduced itself, and the information of the system under consideration has been passed to a next "generation." Reproduction of DNA/RNA structures is usually accurate, and an identical second molecule is the result. Hs remains constant (the horizontal part of Hs in Fig. 4 a). The information content (Hmax minus Hs) of the system under consideration also remains the same. When the system evolves from GC to GCU the phase space also increases from 16 to 64, as a phase space of three loci with 4 states has 64 different combinations. The information content of any single system in a phase space of 64 is:

$$I = Hmax - Hs = ld\ 64 - 1/64\ ld\ 1/64 = 6 - 0.094 = 5.906\ (bits)$$

If, after many generations, one molecule reproduces by mutation into a molecule of the structure GCA instead of GCU, there are 2 different structures. Randomization of information has started. The realized information content is then:

$$I = ld\ 64 - 2(\ 1/64\ ld\ 1/64) = 6 - 0.188 = 5.812\ (bits)$$

If the process of replication was chaotic, all of the 16 possible arrangements were equally likely. After a sufficient number of chaotic "replications" the complexity of the information system would equal the maximum possible information content Hmax. But replication is not chaotic; it is a next-to-perfect copying system. The mechanism of biological heredity is the major constraint of nature's inclination of randomization. Another important mechanism of constraint is natural selection. Both mechanisms form the pillars of Darwin"s Evolutionary Theory. The randomizing mechanisms are mutations, recombinations in biparental systems and autogamous systems, differential selection, polyploidy, etc. During the course of evolution life has diversified (causing information entropy to increase) and more complex systems have evolved. The phase space of our hypothetical example was that of 2 and 3 loci long. Codons are sets of three loci, each filled in by any of the four nucleotides. Genes, units of biological information systems, consist of a few hundred up to few thousand codons. Genomes of the most complex living systems have a length of several billions loci. In figure 4 b the macroevolutionary increase of complexity (Hs) is depicted from the origin of life (t0) to the recent (t2). The corresponding Hmax-values increase more than proportionally.

Variation can be expected to increase in larger cohesive systems, but is canalized by the constraints of the mechanisms of biological heredity and natural selection. A spectacular example of hereditary conservatism is the successful induction of teeth enamel by means of epithelium of chicken by Kollar and Fisher (1980). This result indicates that the information for the development of ancestral teeth enamel is still largely intact in organisms that had this information suppressed for a geologically period of over 100 million years through many speciation events.

This evolutionary information theory has interesting implications for speciation as well. Speciation is best regarded as entropic decay of a phase space consisting of cohesive populations (Fig. 5). If the original species (A-E together) becomes less cohesive and the number of populations remains the same, the phase spaces (total number of possible connections) decrease together with their Hmax-figures (IIc in Fig. 5: entropy of cohesion). When the splitting into two species

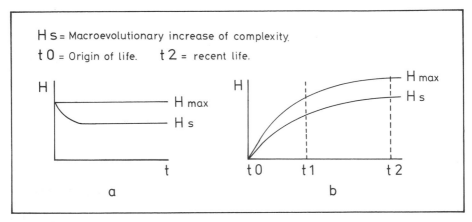

Figure 4. a) macroevolutionary increase of complexity (Hs) and maximum possible information content (Hmax) during time; b) macroevolutionary increase of complexity (Hs) depicted from the origin of life (t0) to the recent time (t2). H = Shannon entropy.

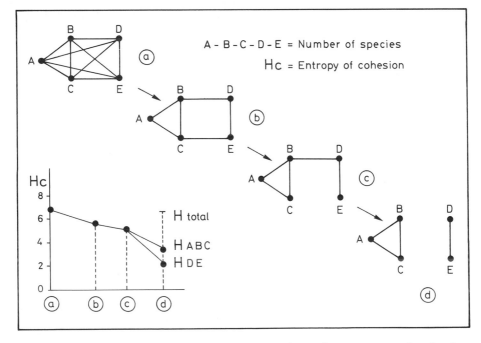

Figure 5. Speciation as entropic decay of a phase space of cohesive populations(A,B,C,D,E). The connecting lines indicate the possibility of two way gene-flow between the populations. Hc = entropy of cohesion, Htotal = total entropy.

(ABC and DE) finally takes place the sum of the two separate entropies is higher than the entropy of the least cohesive ancestral species, as can be expected from an entropic process.

The very fact that there is historical constraint enables us to reconstruct phylogenies. One practical application of the evolutionary information theory is the measure of information content of phylogenetic trees (Brooks , O'Grady and Wiley, 1986, and Geesink and Zandee, 1988, in prep.).

Some examples from Malesian Legumes

What can the philosophical considerations of species and monophyletic taxa being cohesive individuals rather than classes of similar objects, together with biological information theory contribute to the understanding of speciation patterns in Malesian Leguminosae ?

1. *Acacia koa/heterophylla* vs. *Rhynchosia rothii* : Disjunct similarity without cohesion.

Acacia heterophylla is endemic to the Mascarene Islands (Réunion). Three or four closely related (sub)species now treated as one species, *Acacia koa* (Wagner, *et al.* 1989 in press), occur in the Hawaiian islands 18 000 km away. The range of character variability of *A. heterophylla* falls entirely within the range of variability of *Acacia koa* if the latter is taken in a wide sense. If *A. koa* is taken in the narrowest sense, the Mascarene plants are not distinguishable from a recently discovered population on Kauaii Island, which combines the habit, shape, and size of the phyllodes of *A. koa sensu stricto* with the shape of the pods and orientation of the seeds of the plants sometimes distinguished as *A. koaia*. The variability within *A. koa sensu lato* as well as the distinguishing characters of it with *A. heterophylla*, as listed by Vassal (1969) and Pedley (1975), are discussed in Wagner, Herbst and Sohmer (1989, in press) in more detail.

One would expect that, in the classical, basically typological, period of taxonomy, the species would have been considered conspecific but bitopic in at least some treatments, but this is not the case; since Asa Gray distinguished the two as different species, they have been kept separate.

Areas with a rather long dry season are occupied by plants indicative for such a climate, even when these areas are as remote from each other as Burma and East Java, 3000 km. From the Legumes, *Rhynchosia rothii* is such an example, but also *Neocollettia wallichii* and *Indigofera linifolia* show a similar disjunct distribution pattern. Other species, more tolerant for a moister climate, have less disjunct patterns. Van Steenis (1961) presented some of these examples in a series of drought classes, varying from indifference to climate to the pattern of *R. rothii*.

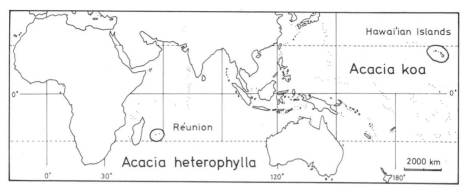

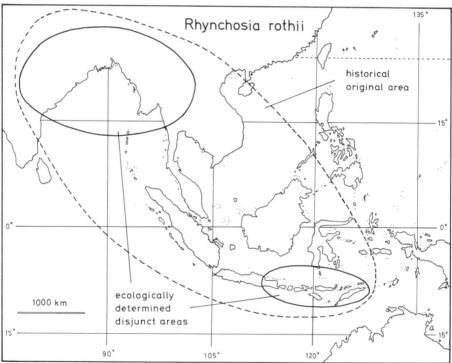

Figure 6. Distributions of *Acacia koa* and *A. heterophylla* (above), and *Rhynchosia rothii* (below).

Let us assume that, at this moment, *R. rothii* from continental S.E. Asia forms a cohesive system, apart from the other cohesive system of the plants from East Java together with the adjacent Lesser Sunda Islands. Plants from these disjunct areas have, unlike *Acacia koa/heterophylla*,

NEVER been accounted to different species. This is in contradiction to the treatment of *Acacia heterophylla/koa* , because what is the difference between an unbridgeable gap of 3000 km and one of 18 000 km ? We could argue that the populations are separate NOW, so NOW they represent different cohesive systems. Hence, the *R. rothii* specimens from continental Asia form a species apart from the specimens from East Java and the Lesser Sunda Islands. Just like the case of the two Acacias. Shouldn't we be consistent ?

The apparently inconsistent treatments of *Acacia* and *Rhynchosia* clearly show that plant taxonomists, also in the typological period, took criteria in consideration other than morphological differences, even despite the strong dictum: "Taxonomy first, then geography!"

One reason to maintain this apparent contradictory treatment is provided by our knowledge of the geological past. During the last glacial periods, West Malesia was connected to Asia mainland. The climate must have been drier than now, and this happened only 15 000 years ago. Such a period is relatively short in the life of a species (the avarage life span of most species-lineages is estimated to be a few million years). The present isolation is then only a geologically short temporary interruption. During the next glacial period the old connection may become restored, and the populations will then form one cohesive system again. If this narrative is indeed the correct one, we would argue that *R. rothii* has always remained one individual lineage, despite a major (regional) interruption of its cohesion.

2. Chaos: Scala *Millettiearum* and *Indigofera*.

2a. Millettieae.

There are groups, like in the case of the genera of the tribe Millettieae (Geesink, 1984), where a phylogenetic analysis failed to result into one or a few alternative acceptable cladograms. The originally applied group-compatibility method (but also the later applied parsimony methods, Swofford, 1985), resulted in many equally short but poorly supported cladograms, a manyfold of even poorer supported cladograms with one more step, and a manyfold thereof with two more steps, etc. In the framework of the Evolution as Entropy- theory, the quality of these sets of cladograms can be accurately quantified by calculation of the D-measure and R-value (Brooks, O'Grady and Wiley, 1986) or the Redundancy Index (Geesink and Zandee, 1988, in prep.). These calculations resulted in low values indicating a high degree of randomness of the character states assembled in the datamatrix and provide a measure for the historical constraints in the group under study. It became clear that insufficient, historically constrained characters had been taken into consideration. However, addition of more character states can also lead to more chaos. In the case in my study of the Millettieae new chemical and anatomical characters only suggested new close relationships, not yet indicated by characters used earlier. One explanation may be that much parallel (character) evolution has actually occurred in this group, and that this parallel evolution has succesfully obscured the actual historical relations.

2b. *Indigofera*.

More than 800 binomials suggest that there are many species in the genus *Indigofera* (Gillet, 1971; De Kort and Thijsse, 1984). Some of them form distinct groupings recognizable by probably unique synapomorphies. The great majority of the described species show a rather chaotic pattern of different combinations of widely distributed character states. Suspicion has arisen about the status of all those species. Ten years ago, Mary T. Kalin Arroyo drew the attention of the audience of the first International Legume Conference (Kew) to a simple indication of self-compatibility, visible even in herbarium material. With some experience one can see that in *Indigofera* nearly all full-grown flowers develop into pods. In Legume species that are known to be self-incompatible and obligatory cross-pollinating, usually relatively few flowers develop into pods. If most of the described species of *Indigofera* are indeed the doubtful products of (predominantly) uniparental systems, we deal with species as historical groups rather than cohesive systems. In *Indigofera* (De Kort and Thijsse, 1984), contrary to *Portulaca* (Geesink, 1969), it is very difficult to discover the boundaries of the cohesive ancestral species. This hampers the discovery of its natural structure.

3. *Millettia* sect. *Fragiliflorae*: A decent case ?

After the unsuccesful attempt to find the natural structure of the tribe Millettieae one of us (RG) looked for a monophyletic group of decent biparental species that would be suitable for a successful phylogenetic analysis. During the preparation of Scala Millettiearum (Geesink, 1984), it was found that the genus *Pongamia*, consisting of one widespread coastal tree species and one endemic inland species of New Guinea, actually belongs to a section of *Millettia*. The unique synapomorphy of the group is the combination of similar leaflets, characteristically curved flower buds with a closed apex, and a very peculiar pulvinate (swollen) base of the inflorescence rachis. The group has characteristically dorso-ventrally flattened axillary buds as well, but this is also characteristic for an African section, which may therefore act as a possible outgroup. Its species seem clearly different and the geographic distribution shows mainly endemic and parapatric species.

Conclusions

The term "natural" as used in "natural species", and in "natural classification" refers to different aspects of Living Nature. There is, at present, only one scientific theory available that explains the diversity of Living Nature, the Evolutionary Theory. In the context of Evolutionary Theory "natural" means organized by Nature, by means of the Evolutionary Process: Descent with modification, constrained by the replicational and developmental processes and by Natural Selection. The Evolutionary Process results in species, monophyletic groups and uniparental systems.

We agree with Ghiselin (1974a,b), Hull (1976,), Wiley (1981), and Williams (1985) who claim that the view of species and monophyletic groups as individuals is consistent with Evolutionary Theory whereas the more common (often only implicit) view based on class-definitions is not.

There are no reasons why biological theories should be exceptional with regard to general physical laws. The Evolutionary Process proceeds, as any other natural process, in accordance to the Second Law of Thermodynamics. This view may appear to be too counterintuitive for some biologists and too obvious for others. The chemical process of replication is, of course, a normally entropic process. But we want to call attention for the more interesting point of view that Information Theory offers tools for understanding the entropic nature of the evolutionary process. The theories recently developed by Brooks and Wiley (1986) and Wicken (1987) provide a model of information-flow which can directly be applied to phylogeny reconstruction. As far as we know there is no acceptable scientific alternative to this view. Rejection of this view would imply that the Evolutionary Process is not regarded as a natural process.

Acknowledgments

This paper could never have been written without the discussions with Ed Wiley and Rino Zandee. Monique Bosman and Peter van Welzen proposed several additional explanatory sentences. The correspondence with David Hull and Dan Brooks helped greatly to clarify matters.

Literature cited

Bentham, G. (1861). "On the species and genera of plants, considered with reference to their practical application to systematic botany." *Nat. Hist. Rev.*, n.s.,**1**, 133-151.

Brooks, D. R.,O"Grady, D. R. and Wiley, E. O., (1986). "A measure of the information content of phylogenetic trees, and its use as an optimality criterion." *Syst. Zool.* **35**, 571-581.

Brooks, D. R. and Wiley, E. O. (1986). "Evolution as entropy: Toward a unified theory of biology." *Univ. Chicago Press, Chicago.*

Collier, J.(1986, 1988). "Entropy in evolution." (Revised edition 1988). *Biol. Philos.* **1**, 5-24.

De Kort, I and Thijsse, G. (1984). "A revision of the genus Indigofera (Legum.-Pap.) in Southeast Asia." *Blumea* **30**,: 89-151.

Eberling , W. and Feisel, R. (eds.), (1982) (reprint 1986). "Physik der Selbstorganisation und Evolution." *Akademie-Verlag, Berlin.*

Fisher, R. A. (1929). "The genetical theory of natural selection." (Revised edition 1958). *Dover Publ., Inc., New York, and by McCLelland & Steward, Canada.*

Gatlin, L. L., (1972). "Information theory and the living system." *Columbia Univ. Press, New York.*

Geesink R. (1969). "An account of the genus Portulaca in Indo-Australia and the Pacific." *Blumea* **17**, 275-301.

Geesink, R. (1984). "Scala Millettiearum. A survey of the tribe Millettieae (Legum.-Pap.) with methodological considerations." *Leiden Bot. Ser.* **8.** *Brill/Leiden University Press.*

Geesink. R. (1986). "Bookreview" (of Brooks & Wiley, 1986). *Taxon* **35**, 905-907.

Geesink, R. and Zandee, M. (1988), (in Ms.). "Phylogeny and information theory: Redundancy Index (RI) as an optimality criterion for phylogenetic trees."

Gillet, J. B. (1971). "Indigofereae." In: *Flora of east Tropical Africa. Leguminosae (part 3) subfamily Papilionoideae (1)*, 212-330.

Ghiselin, M. T. (1974a). "A radical solution to the species problem." *Syst. Zool.* **23**, 536-544.

Ghiselin; M. T. (1974b). "The Economy of nature and the evolution of Sex." *Univ. of California Press. Berkeley.*

Hull, D. L. (1976). "Are species really individuals ?" *Syst. Zool.* **2**, 174-191.

Kollar, E. J. and Fisher, C. (1980). "Tooth induction in chick epithelium: expression of quiescent genes for enamel synthesis." *Science* **207**, 993-995.

Lotka, A. J. (1924). "Elements of physical biology." (Reprinted in 1956 as:

"Elements of mathematical biology." *Dover Publ., Inc. New York.*)

Mayr, E. (1942). "Systematics and the origin of species." *Columbia Univ. Press, New York.*

Mayr, E (1969). "Principles of systematic zoology." *McGraw-Hill, New York.*

Mayr, E. (1982). "The growth of biological thought." *Belknap, Harvard.*

Pedley, L. (1975). "Revision of the extra-Australian species of Acacia subg. Heterophyllum." *Contr. Queensl. Herb.* **18.**

Prigogine, I. and Stengers, I. (1984). "Order out of chaos." *Bantam, New York.*

Schuster, H. G. (ed.), (1988). "Deterministic chaos." *VCH Verlag, Weinheim.*

Shannon, C. E. and Weaver, W. (1949). "The mathematical theory of communication." *Univ. Illinois Press, Urbana, Chicago.*

Simpson, G. G. (1961). "Principles of animal taxonomy." *Columbia Univ. Press, New York.*

Steenis, C. G. G. J. van (1961). "Introduction in: Van Meeuwen *et al.,* Preliminary revisions of Malaysian Papilionaceae I." *Reinwardtia* **5,** 420-429.

Swofford, D. L. (1985). "PAUP. Phylogenetic analysis using parsimony." *Illinois Nat. His. Sur., Champaign, U.S.A.* Manual distributed by the author.

Vassal, J. (1969). "A propos des Acacias heterophylla et koa." *Extr. Bull. Soc. Hist. Nat. Toulouse* **105,** 443-447.

Wagner, W. L., Herbst, D. R. and Sohmer, W. L. (1989 in press). "Leguminosae" (contributed by R. Geesink, W.L. Wagner and D.R. Herbst) in: Manual of flowering plants of Hawaii, *Honolulu.*

Weber, B. H. and Depew, J. D. (eds.), (1988). "Entropy, Information, and Evolution." *MIT Press, Cambridge, Massachusetts.*

Wicken, J. S. (1987). "Evolution, Thermodynamics, and Information." *Oxford University Press, Oxford.*

Wiley, E. O. (1981). "Phylogenetics. The theory and practice of phylogenetic systematics." *John Wiley & Sons, New York.*

Williams, M. B. (1970) "Deducing the consequences of evolution: A Mathematical Model." *J. Theor. Biol.* **29,** 343-385.

Williams, M. B. (1985). "Species are individuals: Theoretical foundations for the claim." *Phil. Sci.* **52,** 578-590.

Zandee, M. and Geesink, R. (1987). "Phylogenetics and Legumes: a desire for the impossible ?" In C.H. Stirton (ed.), Advances in Legume Systematics, Part 3. *Royal Botanic Gardens, Kew,* 131-167.

Speciation patterns in the palms of Madagascar

J. DRANSFIELD

Royal Botanic Gardens, Kew, England, UK

It is well known that the palm flora of Madagascar is particularly rich and interesting when compared with that of adjacent Africa. Although an enumeration of the palm flora has not been completed, recent fieldwork has indicated the presence of new genera and many new species, further emphasizing the diversity of the palm flora. Most of the palms are confined to the humid rain forests of the east coast. Four of the six palm subfamilies have representatives on the island, *i.e.* Coryphoideae (4 genera, 6 species), Ceroxyloideae (2 genera, ca. 14 species), Arecoideae (16 genera, ca. 113 species) and Calamoideae, only represented by *Raphia*, a possible introduction. Within and between the tribes numbers of species are very uneven, some tribes or subtribes being represented by few morphologically very isolated taxa in contrast to the arecoid subtribe Dypsidinae which accounts for about three quarters of the total number of species. Within Dypsidinae there is a range of forms from tall tree palms to minute undergrowth palms, among the smallest of all palms. Genera have traditionally been separated on number of stamens, shape of anthers and nature of the endosperm, but the use of such characters seems to separate manifestly closely allied taxa. The diversity of the Dypsidinae suggest differentation of species within Madagascar with little extinction providing the disjunctions that allow the separation of genera.

Research into the palm flora of Madagascar has a long history and the names of of many botanists are associated with Madagascar palms, *e.g.* Martius, Baillon, Beccari, Perrier de la Bâthie, Jumelle, Humbert, Guillaumet, and H. E. Moore. Two floristic accounts have been published (Beccari, 1914a; Jumelle and Perrier de la Bâthie, 1945), yet the palm flora remains poorly known. Even the delimitation of genera is sometimes confused, mostly owing to the small number of herbarium collections - many taxa are known only from their types which tend to be poorly prepared and incomplete. With representatives of four of the six palm subfamilies (Uhl and Dransfield, 1987), Madagascar has a rich and diverse palm flora (Table 1), particularly so when contrasted with Africa.

TROPICAL FORESTS
ISBN 0–12–353550–6

Table 1. Subfamilies, tribes, subtribes, and genera of palms recorded in Madagascar. For each genus its number of species is given. Endemic genera are marked with an *. Numbers in parenthesis are endemics.

	Genera	Species
Coryphoideae		
Phoeniceae *(Phoenix-2)*	1(0)	2(0)
Borasseae		
Lataniinae *(Borassus-2)*	1(0)	2(2)
Hyphaeninae *(Hyphaene-1, Bismarckia*-1)*	2(1)	2(1)
Calamoideae		
Calameae		
Raphiinae *(Raphia-1)*	1(0)	1(0)
Ceroxyloideae		
Ceroxyleae *(Ravenea-10, Louvelia*-4)*	2(1)	14(14)
Arecoideae		
Areceae		
Oraniinae *(Orania-1, Halmoorea*-2)*	2(1)	3(3)
Dypsidinae *(Vonitra*-4, Chrysalidocarpus-20,*		
Neodypsis*-14, Neophloga*-35, Phloga*-3,		
Dypsis*-24, Antongilia*-1)	7(6)	101(101)
Unnamed subtribe 1 *(Masoala*-1, Marojejya*-3)*	2(2)	4(4)
Unnamed subtribe 2 *("Lemurophoenix"*-1)*	1(1)	1(1)
Cocoeae		
Beccariophoenicinae *(Beccariophoenix*-1)*	1(1)	1(1)
Butiinae *("Voanioala"*-1, Cocos-1)*	2(1)	2(1)
Elaeidinae *(Elaeis-1)*	1(0)	1(0)
Totals 23(14)	134(128)	

We know sufficient about this diverse flora to be aware of how poorly understood and incomplete it is, at the same time the palms in Madagascar are as severely threatened with extinction as any other group and in some degree more so. There is a very real danger that some palms will disappear before they are even represented in herbaria.

In the present paper, I shall describe some aspects of the diversity of the palms in Madagascar in the context of the suprageneric classification of the family (Uhl and Dransfield, 1987) and attempt to account for this diversity.

Subfamily Coryphoideae

Of tribe Phoeniceae two species of *Phoenix* are present. The widely cultivated *P. dactylifera* has been introduced and the widespread African *P. reclinata* appears to be native in the north-west of the island. The Coryphoid tribe Borasseae is represented by three genera: *Hyphaene* with one species, *H. coriacea*, which is also present in coastal East Africa; *Borassus* with two endemic species, *B. madagascariensis* and *B. sambiranensis;* and *Bismarckia*, the only species of which, *B. nobilis*, is endemic to Madagascar. The four Borassoid genera are all plants of the drier western part of the island where they may be conspicuous components of the landscape. The species of *Borassus* and *Bismarckia* are tall single-stemmed tree palms while *Hyphaene coriacea* is a shrubby palm with dichotomously branched trunks.

Subfamily Calamoideae

Subfamily Calamoideae is represented only by *Raphia farinifera*. It is also found in East Africa where it is widespread and so variable in size, clustering behavior, and fruit shape that several species have been described in the past. In marked contrast, *R. farinifera* in Madagascar appears to be very uniform; it is apparently always solitary and there is very little variation in fruit size and shape. The palm is not encountered in primary vegetation but always seems to occur near areas of cultivation or human settlements. These facts seem to suggest that the Madagascar populations of *R. farinifera* are descended from palms introduced from the African mainland during the human settlement of the island. The absence from the rain forests of Madagascar of any other members of this subfamily, so characteristic and well developed in the humid rain forests of West Africa, south-east Asia, the Malesian region, and of parts of northern South America, is noteworthy (see Uhl and Dransfield, 1987).

Subfamily Ceroxyloideae

Of subfamily Ceroxyloideae, tribe Ceroxyleae is represented in Madagascar by two genera, *Ravenea* with eight described species and two varieties and *Louvelia* with three described species. There are three other genera, viz: *Ceroxylon* in the Andes of South America, *Juania* on the island of Juan Fernandez, and *Oraniopsis* in Queensland, Australia. *Louvelia* is endemic to Madagascar while *Ravenea* is also represented in the Comores by two endemic species (Dransfield and Uhl, 1986).

Ravenea hildebrandtii in the Comores and one or two undescribed taxa on Marojejy and Tsaratanana in northern Madagascar are small slender palms with stems rarely exceeding about two meters tall. All other species of *Ravenea* are tree palms, some such as *R. rivularis* and *R. robustior* being robust canopy palms. All species are single-stemmed and bear pinnate leaves with numerous single-fold leaflets. The staminate and pistillate inflorescences of *Ravenea* are superficially similar (Fig. 1A). There is a basal prophyll that incompletely sheathing a single or several more or less equal peduncles that bear up to seven sheathing peduncular bracts. The rachis is well exserted from the leaf sheaths and the rachillae are usually numerous and widely spreading at anthesis. The presentation of the flowers and their sweet scent is suggestive of pollination by bees but there are no critical observations to confirm this. Although certain taxa can be recognized from afar, the whole genus is rather uniform; species of *Ravenea* can usually be instantly identified to genus. The problems arise while naming them to species. In contrast, *Louvelia* is more heterogeneous (Table 2). The type of the genus, *Louvelia madagascariensis* (Fig. 1B) is a rather robust squat palm of the forest undergrowth; it is acaulescent or erect with a short trunk and a crown of finely pinnate leaves. *Louvelia albicans,* known only from its type, is apparently a plant of similar habit while *L. lakatra* has a habit similar to that of most species of *Ravenea*. The inflorescences of the first two species are highly condensed and borne hidden among the leaf-sheaths, thus very different from the aspect of inflorescence presented by *Ravenea,* the inflorescence of *L. lakatra,* however, is open and lax just as in *Ravenea,* and yet the morphology of the leaf bears a striking resemblance to that of *L. madagascariensis.*

Table 2. *Ravenea* and *Louvelia* compared

	Ravenea	*Louvelia*
Inflorescence	sometimes multiple lax, exserted from leaf sheaths	solitary condensed, scarcely exserted from leaf sheaths
Peduncular bracts	coriaceous to sub-woody	rather fleshy to membranous
Sepals and petals	almost always rather rounded, truncate or bluntly triangular	where known, narrow triangular, acute, spreading
Fruit	usually 1-seeded	usually more than 1-seeded
Eophyll	bifid (? always)	pinnate (? always)

A

B

Figure 1. A - The multiple staminate inflorescence of *Ravenea amara*, with several, more or less equal axes borne within a common incomplete prophyll (hidden) and maturing centrifugally. B - The condensed staminate inflorescence of *Louvelia madagascariensis* borne among leaf litter and petiole bases. (Photographs: J. Dransfield).

Unfortunately, the only known extant population of *L. lakatra* at Perinet has been pruned into a perpetually juvenile state by the constant harvesting of young leaves for fiber so it is unlikely that fertile material will be collected. The type consists of material in fruit, and as far as I am aware the palm has never been collected since except for sterile voucher specimens I made in 1986. An undescribed palm from the Masoala Peninsula seems to combine features of habit and flowers of both *Louvelia* and *Ravenea*, yet fruiting material which seems to belong to this same taxon collected on Marojejy is single-seeded and scarcely different from most species of *Ravenea*. The condensed inflorescences of *L. madagascariensis* have a musty scent and are visited by coleopterans (Dransfield and Henderson, pers. obs.). *Louvelia albicans*, despite being searched for recently, has not yet been refound. At present, it does seems unlikely to maintain two genera in the Ceroxyleae in Madagascar, but any formal taxonomic decisions will have to wait until more collections are available.

If the Madagascar representatives of the Ceroxyleae are taken as a whole, diversity in the group in terms of habit is rather low. There are no caespitose species, almost all are tree palms of moderate to robust stature with a few squat palms of the undergrowth and one or two slender undergrowth palms, and the species can usually be identified from a distance by their distinctive crown and leaf form. There is diversity in inflorescence structure and presentation. Inflorescences in *Louvelia* and some species of *Ravenea* are solitary, a leaf subtending a single inflorescence bearing a prophyll at its base. In contrast, many species of *Ravenea* have so-called multiple inflorescences (Fisher and Moore, 1977; Fig. 1A) where a leaf apparently subtends several inflorescences. These inflorescences are borne within a single common prophyll but otherwise each are equivalent in structure to a solitary inflorescence. The most central inflorescence is the first to mature followed successively outwards. Multiple inflorescences are confined to staminate individuals except in *R. madagascariensis* var. *monticola* in which the pistillate as well as the staminate individuals have multiple inflorescences. The presence or absence of multiple inflorescences has been used to distinguish taxa. As already indicated above, there are well-exserted, open and spreading inflorescences, apparently associated with bee-pollination and in other taxa, congested inflorescences scarcely exserted from the leaf-sheaths, that are probably associated with beetle-pollination. However, the group has diversified into a wide range of habitats, perhaps the widest range for any group of Madagascar palms. Species such as *R. robustior* and *R. latisecta* are palms of the mid-elevation rain forests in the eastern part of Madagascar. There are three as yet unidentified taxa in rain forest at lower elevations in the Masoala Peninsula. There is a coastal taxon, apparently exclusive to the forest developed on the white sands of the coastal raised beaches; it keys to *R. madagascariensis* but is definitely not that species. *Ravenea madagascariensis* itself is a forest palm ranging from 900-1400 m above sea-level. Two varieties have been recognized but whether they can be maintained will require further study. *Ravenea rivularis* and *R. amara* are riverine palms of the plateau, the former at relatively low elevations, the latter at about 1000-1500 m above sea-level. *Ravenea glauca* occurs in the limestone gorges of Isalo; it has been confused with *R. rivularis*, and may not be distinct from *R. xerophila* which occurs in dry Didieriaceous forest in the extreme south. This last species, or at least the population which was described as *R. xerophila*, is probably the most drought-resistant of all palms in Madagascar. It is on the verge of extinction, having suffered the effects of habitat destruction and also the attentions of palm enthusiasts.

Subfamily Arecoideae

The largest of the palm subfamilies, Arecoideae, is represented in Madagascar by two of the six tribes. From tribe Areceae there are representatives in four subtribes (two of which as yet undescribed) and of tribe Cocoeae there are, apart from the coconut and the African oil palm which are probably introduced, representatives from two subtribes.

Tribe Areceae. Subtribe Oraniinae. Subtribe Oraniinae comprises two genera, *Orania* distributed from Thailand and the Philippines south and east to New Guinea, with one species, *O. longisquama* in Madagascar. The second genus, *Halmoorea,* was recently described based on material collected by Moore in Madagascar and confused with *Orania longisquama* (Dransfield and Uhl, 1984b). *Halmoorea* was separated from *Orania* on the basis of a number of characters discordant in the latter genus. However, recent field work has turned up a third palm belonging to this group, effectively intermediate between the two described genera, and so in the future it may be necessary to sink *Halmoorea* in synonymy. The three taxa are tree palms, *Orania longisquama* and *Halmoorea trispatha* contributing to the forest canopy while the undescribed taxon rarely is taller than about five meters. The last two are remarkable in bearing crowns of distichously arranged leaves reminiscent of *Ravenala.* All three palms are said to be poisonous and are thus left while other palms are destroyed for their edible apices. Subtribe Oraniinae is of particular interest because of the apparently rather unspecialized triovulate gynoecium (most other members of the subtribe are pseudomonomerous), a character which Oraniinae shares with the New World genera, *Manicaria, Leopoldinia,* and *Reinhardtia.* At present, *Orania longisquama, Halmoorea trispatha,* and the undescribed taxon seem to be rather rare palms occuring in the most humid areas of lowland rain forest in Madagascar. They seem to represent the remnants of an early radiation of tribe Areceae.

Tribe Areceae. Subtribe Dypsidinae. This subtribe accounts for 75% of all Madagascar palm species. Additionally, it has two representatives in the Comores, one on the island of Pemba off the coast of Tanzania and two species known only from cultivation but presumably originating in Madagascar. The subtribe is thus quintessentially a Madagascar palm group. I believe it to be monophyletic as presently circumscribed, but it must be admitted that one of its defining features is that it occurs in Madagascar. Its relationships with other subtribes of Areceae are still not understood. It has much in common with Euterpeinae and there are even some shared, unusual features such as small size and similar form of flower, basal stigmatic remains, and presence of "piassava" in common between one of its genera, *Vonitra*, and the monogeneric Leopoldiniinae of Amazonia. The relationships will need to be clarified by further morphological and developmental studies, and DNA sequencing studies just begun, which may add to our understanding. Within Madagascar, however, it does seem likely that all the dypsids belong to the same group and that if a sister group is to be searched for, we should look outside Madagascar rather than among the remaining subtribes (but see discussion under "*Lemurophoenix*").

Having established that the bulk of the Madagascar palm flora seems to belong to a single line of evolution, we may then examine the relationships among these dypsids. Here is where the real problems lie in the delimitation of genera. For although seven genera have been recognized, some of the characters traditionally used to define these genera are known to be unreliable elsewhere in the family. Furthermore they separate, within the Dypsidinae, pairs of species which are manifestly closely related. A good example is provided by *Chrysalidocarpus lutescens* and *Neodypsis baronii*. The former is a common variable palm of the coastal white sand forest on the east coast, and the latter is an upland life form replacement occurring in forests at elevations of about 800-1500 m above sea-level. The two palms are superficially similar in habit, flowering, and fruiting characters. It takes a practiced eye to distinguish the two in cultivation in Tananarive. Yet they are placed in different genera because the endosperm of *C. lutescens* is homogeneous, while that of *N. baronii* is marginally ruminate. It would seem to be nonsensical to keep these two, manifestly, closely related palms in separate genera. On the other hand, palms displaying extremes of morphology in the subtribe are easily placed in familiar genera, but there are so many intermediates that the generic boundaries break down. Yet it would seem at first sight to be ridiculous to include a minute forest undergrowth palm such as *Dypsis mocquerysiana* and a tall tree palm such as *Neodypsis lastelliana* in the same genus; the former has a stem scarcely 50 cm tall, short, entire, bifid leaves, once-branched inflorescences bearing minute flowers, the staminate with three stamens, and fruit with homogeneous endosperm, while the latter is a tall, canopy palm with large pinnate leaves, massive inflorescences branched to three orders, relatively large staminate flowers with six stamens, and fruit with deeply ruminate endosperm. How can two such extremes be congeneric ? However, there are individual taxa occupying positions intermediate between all the described genera. Although I have suggested that the use of the nature of the endosperm, whether ruminate or homogeneous, may not be reliable for separating genera, it seems that there may in fact be two forms of ruminations in *Neodypsis*; in one the endosperm is shallowly and rather regularly penetrated by neat intrusions of integument (as found in *N. baronii*) while in the other the seed is deeply and irregularly penetrated in a manner similar to that found in the seeds of *Areca catechu* L. (as found in *Neodypsis lastelliana*). It may be that two different developmental processes are involved, and that it may be possible to divide species based on this character into apparently more natural groups.

Vonitra is at first sight the most easily distinguished and clearly circumscribed of the dypsid genera. The three most familiar species, *V. fibrosa*, *V. crinita*, and *V. utilis* are closely related and doubtfully distinct. They are moderate to robust palms with erect dichotomously branched trunks. There is no crown-shaft and the base of the crown is obscured by a mass of pendulous fibre formed by the disintegration of the leaf-sheath and ligules, the so-called "piassava". The inflorescences are borne on long peduncles, the flowers are minute, and the staminate with six stamens borne in two series with didymous anthers. The fruit is relatively large and the endosperm deeply ruminate. *Vonitra fibrosa* is an abundant palm, certainly one of the commonest in lowland forest. On the Masoala Peninsula, besides growing on hill-slopes and ridge-tops, it will also grow as a rheophyte on the margins of fast-flowing, frequently flooding, rocky rivers. So far, this seems to be the only rheophytic palm in Madagascar. A fourth, undescribed species has recently been discovered on the Masoala Peninsula where it is sympatric with *V. fibrosa*. It is a dwarf with a short erect stem to one meter only, covered with short broad fibres and has an inflorescence branched to only one order. A fifth species, *V. nossibensis*, transferred to *Vonitra* from *Chrysalidocarpus* by Perrier de la Bâthie, lacks the "piassava" and the branching trunk and has large flowers. It has ruminate endosperm so it must be misplaced in *Chrysalidocarpus*, but it seems equally misplaced in *Vonitra*.

The refinding of the genus *Antongilia* and the finding of a strange palm which seems to combine features of *Vonitra*, *Antongilia*, and *Neodypsis*, together with the difficulty of assigning some taxa to any described genus, further emphasises the difficulty of delimiting genera. *Antongilia perrieri* is a massively constructed squat palm of the rain forest surrounding the Bay of Antongil. It has an appearance superficially similar to that of an undescribed pinnate-leaved species of *Marojejya* (see below), with a shuttlecock-like crown of marcescent leaves, but the petioles are long and the inflorescence has a long peduncle. In inflorescence structure and fruit, the genus is similar to *Vonitra* but the plant lacks "piassava". The flowers also are much larger than the minute flowers of *Vonitra* and the stamens are scarcely biseriate. However, without using the distinctive habit this genus cannot be keyed out from *Neodypsis*.

Most species of *Chrysalidocarpus* and *Neodypsis* are moderate to large palms. Some, as in *Vonitra*, display aerial branching. *Neodypsis decaryi* is famed for the tristichous arrangement of the leaves, though this unusual leaf arrangement is also present in *Chrysalidocarpus madagascariensis* and another as yet unnamed species of *Neodypsis*. Trunks may be single or densely tufted even in the same species. In *Chrysalidocarpus decipiens,* a palm of the high plateau, the stem appears consistently to divide dichotomously below ground level while still in the rosette stage, then the resulting stems grow upwards and do not branch further. Often only one of the resulting trunks grows to its splendid bottle-shaped maturity, but double trunks do occur.

The small undergrowth palms are included in the genera *Phloga, Neophloga* and *Dypsis*. Given flowers and fruit it is usually easy to assign such small palms to one or other of these genera. Palms with six stamens with didymous anthers and ruminate endosperm belong to *Phloga*, those with six stamens with sagittate anthers and homogeneous endosperm belong to *Neophloga* while those with three stamens and homogeneous endosperm are included in *Dypsis*. There are particularly robust species of *Neophloga (e.g. N. scottiana)* which really cannot be separated generically from small species of *Chrysalidocarpus. Phloga nodifera,* without flowers and fruit, cannot reliably be distinguished from *Dypsis pinnatifrons,* the two palms being astonishingly similar in habit. Yet in this instance the two taxa are strikingly different in flower and fruit structure.

Dypsis has been divided into three separate genera. *Dypsis sensu strictu* has three antesepalous fertile stamens with short distinct filaments, anthers adnate to the filament, and no staminodes. *Trichodypsis* has three antepetalous fertile stamens alternating with antesepalous staminodes, the filaments connate in a ring, and the anthers adnate to the filament. *Adelodypsis* has three antesepalous fertile stamens with linear distinct filaments, and pendulous anthers, staminodes being absent. *Dypsis* and *Trichodypsis* comprise the smallest palms. These two segregate genera parallel each other remarkably. *Adelodypsis* on the other hand includes considerably larger palms, some such as *Dypsis pinnatifrons* with its trunk to five meters tall approaches species of *Neodypsis* and *Chrysalidocarpus* in habit. To add further confusion, a palm from Masoala recently collected has the habit of *Adelodypsis* species while displaying between three and six stamens. The three different forms and arrangements of stamens suggest that the reduction of six stamens to three may have arisen independently in each of the segregate genera, further suggesting that *Dypsis sensu lato* may be an artificially defined genus representing three different evolutionary lines. As yet, nothing is known of the floral biology of these palms. The flowers are among the smallest in the family and in the instance of species such as *D. hirtula*, the flowers scarcely open, and pollen production from the three pairs of tiny anthers must be very low. Yet mature fruits are often produced in abundance.

The genus *Neophloga* parallels *Dypsis sensu lato* in habit, leaf shape, and degree of branching of the inflorescence, but staminate flowers and fruit are clearly different. In some forest types in Madagascar it is possible to find up to five or six different species of *Dypsis* and *Neophloga* growing sympatrically. The two genera are similarly represented by species with entire bifid leaves, species with finely pinnate leaves or almost any intermediate leaf condition imaginable, species with spicate inflorescences or with branched inflorescences. Guillaumet (1973) and Koechlin *et al.* (1974) have commented on the plethora of leaf forms to be found in the Dypsidinae and have suggested that the entire-leaved forms have arisen through a process of neoteny. The selective advantages of the different leaf forms are not understood, though in the New World genus *Geonoma*, progress has been made towards solving this problem (Chazdon 1985, 1986). Chazdon suggests that the differences in leaf shape and size in *Geonoma* are related to the biomass costs of light interception in the forest undergrowth and, hence, shade tolerance, and changes in plant size may have played an important role in the adaptive radiation of *Geonoma*. The same principles probably apply in the Dypsidinae and I suggest that there is an even greater range of leaf size and shape in this group making it a very promising field for further research. Other genera of undergrowth palms displaying great diversity in leaf size and shape are *Licuala* in Malesia, *Salacca* in West Malesia, *Chamaedorea* in New World Tropics, *Ptychosperma* in New Guinea, *Gronophyllum*, *Pinanga*, and *Areca* all in Malesia (especially Bornean species), *Iguanura* in West Malesia, and *Bactris* and *Geonoma* in the New World Tropics.

Such is the diversity of dypsids in Madagascar that the subtribe gives the impression of having radiated on the island into a wide range of habitats with very little extinction, so that generic and sometimes specific boundaries appear artificial.

Tribe Areceae. Unnamed subtribe 1. Marojejya and *Masoala* are endemic genera of peculiar habit and morphology and which until recently were rather poorly known. *Marojejya insignis* was described from the Marojejy Massif by Humbert (1955) and refound by Moore on the eastern side of the Masoala Peninsula. The genus further includes two species, *M. darianii* (Dransfield and Uhl, 1984a), and a third species which was discovered near the type locality of *M. darianii* in 1986 (Dransfield, in prep.). *Masoala madagascariensis*, the only species in its genus and until recently known only from its incomplete type collected by Perrier de la Bâthie, was rediscovered in 1986 on the Masoala Peninsula (Fig. 2 A,B). The systematic position of *Masoala* was in doubt when *Genera Palmarum* went to press (Uhl and Dransfield, 1987) and the genus was placed as "Areceae incertae sedis". Because of the habit, condensed form of the inflorescence, and numerous incomplete peduncular bracts, *Marojejya* was included tentatively with the West African endemic genus *Sclerosperma* in the Sclerospermatinae. I now believe this to be mistaken, and that the relationships of *Marojejya* are not with *Sclerosperma* but with *Masoala;* a new subtribe will need to be delimited. *Masoala* is still not completely known. We have only one half-eaten mature fruit at our disposal and pistillate flowers at anthesis are not yet known. Despite this, the affinity of the two genera seems certain. *Marojejya* is a genus of rather massively constructed, squat palms of the forest undergrowth or middle canopy, occasionally becoming tall enough to reach the upper canopy. The leaves are marcescent, large, pinnate or partially or wholly entire and are more or less held erect giving the crown a characteristic shuttle-cock appearance with an untidy skirt of dead leaves. The petioles are short or absent and the base of the leaf is expanded into a pair of auricles. The shortness or absence of petioles together with the auricles results in the crown accumulating leaf-litter and debris and adventitious roots grow into this litter. The inflorescences are highly condensed and are unisexual but with both staminate and pistillate inflorescences being borne on the same individual - an unusual situation found in only a few palm genera (*e.g. Elaeis, Wallichia*). There is a remarkable superficial similarity in the morphology of the rachillae of *Marojejya* and that of *Elaeis, Salacca,* and *Eleiodoxa,* and, although pollination studies have not been made, it is very tempting to suppose that as in the three other genera (at least in the wild) pollination is effected by beetles.

   Masoala has a habit similar to that of *Marojejya*. It too is a litter accumulator but its leaf bases are even more extraordinary than those of *Marojejya*; each auricle continues distally into the respective rein (the true leaf margin of the unopened, unsplit leaf) forming a broad band of tissue which hangs untidily from the crown. The inflorescences are hardly condensed and bear flowers of both sexes, arranged in triads of a central pistillate and two lateral staminate in shallow pits. As in *Marojejya* there are several incompletely sheathing small peduncular bracts distal to the first large peduncular bract.

Figure 2. A - *Masoala madagascariensis* growing in primary forest on the Masoala Peninsula . B - The lax inflorescence of *Masoala madagascariensis* (Photographs: J. Dransfield).

The fruit has terminal stigmatic remains and apparently homogeneous endosperm. In the lax branching and the presence of flowers of both sexes on the same rachillae, the inflorescence of *Masoala* is less specialized than that of *Marojejya*, but the two genera are otherwise strikingly similar. It is this similarity which suggests that the sister group of *Marojejya* is *Masoala* rather than *Sclerosperma* and that previously assumed affinity between *Sclerosperma* and *Marojejya* was based on superficial similarities which I now assume to be the result of parallel evolution. However, this still does not account for the affinities of *Masoala* and *Marojejya* with the rest of the tribe Areceae where we assume they belong. There appear not to be obvious relationships with the Dypsidinae or the Oraniinae, the only other subtribes in Areceae present in Madagascar, and the implications are that the two genera represent an independent line of evolution in Madagascar.

Tribe Areceae. Unnamed subtribe 2. The fourth subtribe of Areceae in Madagascar is as yet undescribed. It consists of a single monotypic genus newly discovered in the hills northwest of Maroantsetra at the head of the Bay of Antongil. It is one of the largest palms in Madagascar and although clearly belonging to tribe Areceae because of the presence of triads and two bracts in the inflorescence, the gynoecium being pseudomonomerous and the fruit lacking a bony endocarp with three pores, its affinities are not obvious. It possesses so many unusual features that its inclusion in an already described subtribe seems precluded. "*Lemurophoenix*" (Dransfield, in prep.) is a palm of the forest canopy with a trunk to 35 m tall and to about one meter in diameter at the base (Fig. 3A). The staminate flowers are in bud superficially like those of the larger members of the Dypsidinae but unlike the dypsids, there are over 50 stamens. The fruit is to five cm in diameter and is covered by corky warts (Fig. 3B), a type of pericarp otherwise unknown in Madagascar palms and known only in a few palm genera scattered through the family (*e.g. Sommieria, Pelagodoxa, Manicaria, Ammandra, Palandra, Phytelephas, Johannesteijsmannia,* some species of *Licuala, Pholidocarpus* and *Chelyocarpus*). The endocarp bears a curious heart-shaped button at its base, a structure also unusual in the family. The endosperm is superficially ruminate and in marked contrast to the Dypsidinae, the embryo is apical. Although all these peculiarities seem to preclude the inclusion of this spectacular palm in any described subtribe in Areceae and necessitate the description of a new taxon, there remains the problem of where the palm's affinities lie. Of the groups occurring in Madagascar, it seems to share some characters with Dypsidinae, though as stated above, there are too many aberrant features to allow its inclusion there. The problem may not be resolved until the completion of a cladistic analysis of the family which is at present in progress (Dransfield and Uhl, in prep.).

Tribe Cocoeae. Tribe Cocoeae is presently predominantly a New World tribe with a few Old World representatives, although there is evidence of a wider distribution in the past (Uhl and Dransfield, 1987). Elaeidinae is represented in Madagascar by the African oil palm, *Elaeis guineensis.* According to Jumelle and Perrier de la Bâthie (1945), two varieties occur on the island; the typical variety which is said to be an escape from cultivation and *E. guineensis* var. *madagascariensis,* indigenous and at one time thought to represent a separate species (Beccari, 1914b). The indigenous variety is supposed to have red fruit and harder and sharper petiole spines. In fact it does not differ significantly from the wild type of African *E. guineensis* and I do not believe it worthwhile to recognize a separate variety. The oil-palm may have arrived relatively recently through natural means or have been introduced by man.

A   B

Figure 3. A - *"Lemurophoenix"* with several infructescenses. B - Immature fruit, about 3 cm in diameter, of *"Lemurophoenix"* showing the pericarp cracking to produce corky warts. (Photographs: J. Dransfield).

There are, however, two cocoids which are peculiar to Madagascar. *Beccariophoenix madagascariensis* is the sole representative of an endemic subtribe (Dransfield, 1988). It is a robust, single-stemmed tree palm contributing to or extending above the canopy of montane forest developed on ridge-tops at about 1000 m above sea-level. As a juvenile rosette palm it bears an unexpected resemblance to juveniles of *Louvelia*, the lack of petioles allowing it to accumulate litter. The inflorescence of *Beccariophoenix* is unlike that of any other cocoid; the massive peduncular bract is borne at the tip of the peduncle at the base of the rachis, and is circumscissile, falling to leave a broad collar. Nothing is known of the floral biology or other aspects of the natural history of this splendid palm.

The other endemic cocoid is a new genus belonging to subtribe Butiinae. *"Voanioala"* (Dransfield, in press; Fig. 4 A,B) is at present known with certainty from two mature trees and a juvenile. There is

circumstantial evidence of other populations but this is undoubtedly an
extremely rare palm. It is a massively constructed, single-stemmed tree
contributing to the forest canopy and known only from the interior of the
Masoala Peninsula at about 450 m above sea-level. The staminate flowers
have about 12 stamens, and the fruit is unusual in the heavily and
unevenly thickened endocarp. The ground beneath the two mature trees
seen by myself was in 1986 carpeted with a layer of fallen endocarps, few of
which had germinated. An impression is gained of a limited dispersal.
Indeed, it is difficult to imagine which extant animal agent might be
responsible for dispersal. As with *Jubaeopsis* in southern Africa, also a
member of Butiinae, "*Voanioala*" seems to be an isolated relic, surviving
on the verge of extinction.

Figure 4. A - Tree of "*Voanioala gerardii.*"   B - Part of the inflorescence of
"*Voanioala gerardii*" showing the split, grooved peduncular bract and the triads
of flowers within (Photographs: J. Dransfield).

Discussion

Some generalizations can be made about the palms of Madagascar. Spines are of very rare occurrence. Where they do occur (in *Hyphaene coriacea, Raphia farinifera, Elaeis guineensis* and *Phoenix reclinata*) the palms are elsewhere known in Africa and are assumed to be relatively recent arrivals in Madagascar. This is in contrast to the palms of the Seychelles and Mascareignes where most palms are spiny at least in the juvenile state. There seem to have been no pressures selecting for the presence of spines in the endemic Madagascar palm groups. There are no true climbing palms, although a slender species of *Neophloga* from Tsaratanana has been reported to be not self-supporting (Razafindratsira, pers. comm.), and there is a rather remarkable abundance of big litter-accumulating species. There are also several palms which have aerially-branching stems, two distichously-leaved species, and at least three strikingly tristichous palms, these being unusual features. It seems that all the endemic palm genera which I have interpreted as being relictual are massively constructed tree species or squat massive palms of the understory.

If we assume that the accepted ultimate suprageneric categories (subtribes, tribes, or subfamilies where there is no division into subtribes) each represent separate evolutionary lines, then there is evidence of ten different palm groups having reached the island Madagascar, excluding the subtribe Raphiinae which, as argued above, seems to be represented in Madagascar by a palm which has been introduced and is widely cultivated. In comparison, out of a total of 46 ultimate suprageneric categories (Uhl and Dransfield, 1987), continental Africa has representatives of 11 lines, the Malay Peninsula 16 and Colombia 17. Of the ten lines indigenous in Madagascar, nine are represented by one or two genera only, the tenth group, the Dypsidinae, has seven genera recognized at present. The Dypsidinae also account for an overwhelming ca. 75% of the native palms. Whereas groups such as Butiinae seem to be represented by relictual taxa or groups such as Phoeniceae by widespread African taxa which may have reached Madagascar recently, the Dypsidinae have diversified into over 100 taxa. Furthermore, the subtribe has taxa adapted to most vegetation types from sea-level to over 2000 m above sea-level and comprises some of the largest canopy palms as well as minute forest undergrowth palmlets among the smallest in the whole family. This peculiarly successful group seems to have radiated in a way scarcely paralleled elsewhere in the family. Yet nowhere in the family are generic boundaries more in dispute. It is as if there have been few extinctions resulting in major disjunctions in morphology allowing for the separation of genera. Tribe Ceroxyleae with its ca. 14 taxa in two poorly defined genera is the only other extant palm group for which there is evidence for significant radiation in Madagascar.

Acknowledgments

I thank Dr. Voara Randrianasolo, M. Armand Rakotozafy and M. Gerard Jean for logistic help and guidance. P. P. Lowry II and George Schatz (Missouri Botanical Garden) have helped and encouraged me in the field.

Literature cited

Beccari, O. (1914a). "Palme del Madagascar." *Istituto Micrografico Italiano, Firenze,* 59 pp.

Beccari, O. (1914b). "Contributo alla conoscenza della palma a olio (Elaeis guineensis)." *Agric. Colon.* **8,** 5-37, 108-118, 201-212, 255-270.

Chazdon, R. (1985). "Leaf display, canopy structure and light interception of two understory palm species." *Amer. J. Bot.* **72,** 1493-1502.

Chazdon, R. (1986). "Physiological and morphological basis of shade tolerance in rain forest understory palms." *Principes* **30,** 92-99.

Dransfield, J. (1988). *"Beccariphoenix madagascariensis." Principes* **32,** 59-68.

Dransfield, J. (in press). "Voanioala (Arecoideae: Cocoeae: Butiinae), a new palm genus from Madagascar." *Kew Bull.*

Dransfield, J. and Uhl, N. W. (1984a). "A magnificent new palm from Madagascar." *Principes* **28,** 151-154

Dransfield, J. and Uhl, N. W. (1984b). "Halmoorea, a new genus from Madagascar, with notes on Sindroa and Orania." *Principes* **28,** 163-167.

Dransfield, J. and Uhl, N. W. (1986). "Ravenea in the Comores." *Principes* **30,** 156-160.

Fisher, J. B. and Moore Jr., H. E. (1977). "Multiple inflorescences in the palms (Arecaceae): their development and significance." *Bot. Jahrb. Syst.* **98,** 573 - 611.

Guillaumet, J.- L. (1973). "Une éspèce de palmier nain de Madagascar." *Adansonia ser.* 2, **13,** 341 - 349.

Humbert, H. (1955). "Palms, in Une merveille de la nature à Madagascar: première exploration botanique du Massif du Marojejy et des satellites." *Mém. Inst. Sci. Madagascar. Sér. B.*

Biol. Veg. **6,** 1-210.

Jumelle, H. and Perrier de la Bâthie, H. (1945). "Palmiers." *In* Humbert, H. (ed.), *Flore de Madagascar et des Comores,* **30.** Tananarive

Koechlin, J., Guillaumet J.-L. and Morat, P. (1974). "Flore et vegetation de Madagascar." *J. Cramer, Vaduz.*

Uhl, N. W. and Dransfield, J. (1987). "Genera Palmarum. A Classification of Palms based on the work of H. E. Moore Jr." L. H. Bailey Hortorium and the International Palm Society. *Allen Press, Kansas.*

An evolutionary scenario for the genus Heliconia

L. ANDERSSON

University of Göteborg, Sweden

This paper is based on recent taxonomic revisions of *Heliconia* (Andersson, 1981a, 1985a, 1985b, unpubl.; Kress 1984). In these, species are circumscribed by traditional phenetic criteria. However, the paper is based also on the assumption that the biological nature of species is best understood in the light of Van Valen's (1976) ecological species concept. This concept implies that for each phenetically recognizable, discrete species, there is a corresponding discrete niche. The driving force of speciation is thus supposed to be a drive towards niche separation. Inversely, the possibility to diversify is dependent on the availability of empty or divisible niches and the ability to utilize them. Although most aspects of this hypothesis still need further theoretical development, it is accepted for the present purpose because it is in good agreement with my field experience. In primary vegetation, closely related species mostly occupy definable habitats (Andersson, 1985a; Berry, 1982; Burger, 1974). On this rather general basis, the secret of having evolutionary success may be stated using an old slogan from military pedagogics: the secret of success is a matter of being; (i) in the right place; (ii) at the right time; and (iii) provided with the correct equipment. This paper is essentially an attempt to identify times, places and "equipment" that have been crucial for the evolutionary success of *Heliconia*.

The genus

The number of species in the genus is a matter of some controversy between splitters (Santos, 1978; Abalo and Morales, 1982, 1983a, 1983b) and lumpers (Andersson, 1981a, 1985a, 1985b). In my view, a reasonable figure is about 120. These have been distributed among four subgenera (Andersson 1985b) and 19 sections and species groups (Andersson, 1985a, 1985b, unpubl.; Kress, 1984). A vast majority of the species is neotropical, but eight (Kress, 1985) occur in the South Pacific islands from Samoa to Indonesia. The Pacific exclave is supposedly the result of a long distance dispersal (Raven and Axelrod, 1974) and will not be further considered in this context.

Although widely distributed in the Neotropics, the genus is heavily centered along the Andean-Central American orogenic axis. A minority of the species is strictly montane, occurring mainly between 1000 and 2000 m altitude, while a majority is found in the Andean foothills and adjacent lowlands. The area along the Andes and the Central American mountains constitutes a pronounced center of diversity both at the species (Fig. 1) and section (Fig. 2) levels. High endemism occurs only in southern Central America and along the northern Andes (Fig. 3). The genus, thus, constitutes a good example of the Andean-centered floristic element shown by Gentry (1982) to account for nearly half the neotropical flora. Facts and hypotheses that shed light upon the evolution of sizable groups belonging to this element are, thus, of considerable interest in neotropical biogeography and biohistory.

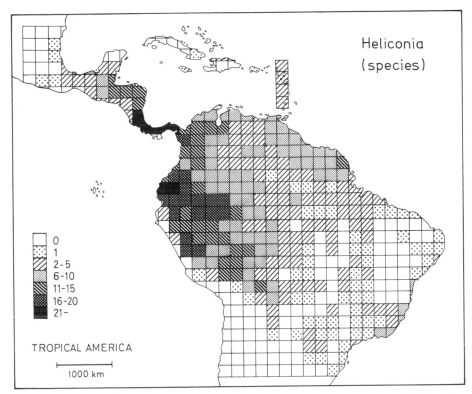

Figure 1. Geographical variation in species level diversity in *Heliconia*. Diversity is probably underestimated in NW Colombia due partly to poor sampling, partly to poor taxonomic understanding of the subgen. *Griggsia* in this area.

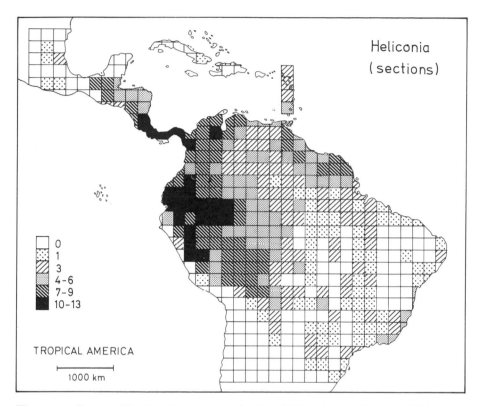

Figure 2. Geographical variation in section level diversity in the genus *Heliconia*.

Relationships

Heliconia clearly belongs to the Musaceae in a wide traditional sense (Winkler, 1930). This family consists of three quite discrete groups, which are nowadays usually (*e.g.* Cronquist, 1981; Dahlgren *et al.*, 1985) treated as families: Heliconiaceae (*Heliconia*), Musaceae (*Ensete*, *Musa*), and Strelitziaceae (*Phenakospermum*, *Ravenala*, *Strelitzia*). Of these, the Strelitziaceae appears to be the most primitive group, judging from, its distichous phyllotaxy, bracteolate, bisexual flowers with, in *Ravenala*, six fertile stamens, and many-seeded, dehiscent fruits. One may thus assume that *Heliconia* was derived from a strelitzioid ancestor.

Heliconia differs from the Strelitziaceae in two respects that I think

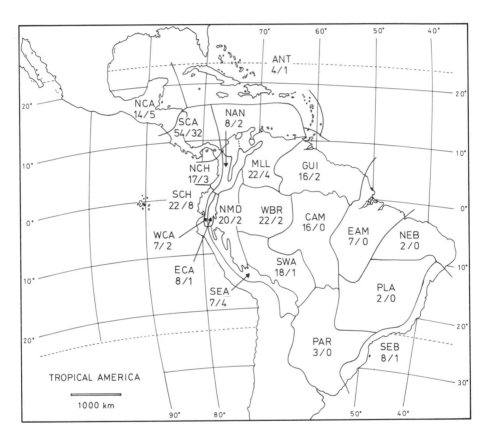

Figure 3. Number of species (left of slash) and endemics (right of slash) in phytogeographic regions of the neotropics. Regions were defined from an analysis of distributions of the Zingiberales (Andersson, unpubl.) and the delimitation differs more or less from such subdivisions used by other authors. ANT-Antilles, CAM-central Amazon, EAM-eastern Amazon, ECA-east central Andes, GUI-Guianas, MLL-the Magdalena-*Llanos* area, NAN-northern Andes, NCA-northern Central America, NCH-northern Chocó area, NEB-northeastern Brazil, NMD-Napo-Marañón drainage, PAR-Río Paraguay drainage, PLA-Brazilian Planalto, SCA-southern Central America, SCH-southern Chocó area, SEA-southeastern Andes, SEB-southeastern Brazil, SWA-southwestern Amazon, WBR-western Brazilian Amazon with adjacent areas in Peru and Colombia, WCA-western central Andes.

are very significant when trying to trace its evolutionary history: firstly in floral structure and secondly in robustness and anatomical organization.

In both *Heliconia* and the Strelitziaceae, the ground plan of floral construction is the same: epigynous, pentacyclic, trimerous, heterochlamydeous. In both, perianth segments and filaments are fused at the base to form a short tube. Above the tube, perianth segments are free in Strelitziaceae (but highly modified in *Strelitzia*), but fused in *Heliconia* to form an open tube with a posterior slit (Fig. 4). This slit is tightened at the base by a more or less laminar staminode. This construction allows a considerable amount of nectar to accumulate in the tube, but the tube may still be dilated without rupturing. A bird's beak may thus be repeatedly inserted in the tube without causing destruction. Also, the petal margins lining the slit are thickened and reinforced by fibrous tissue.

As a broad generalization, it may be said that species of *Heliconia* are more slender than those of the Strelitziaceae. Furthermore, the tallest and most robust species of *Heliconia* belong to subgenus *Griggsia*, which by virtue of its pendent inflorescence is probably a derived group. The smallest species of the Strelitziaceae belong to the genus *Strelitzia*, which is characterized by highly specialized flowers. It may thus be assumed that extreme robustness is a derived character state in *Heliconia*. Anatomical differences between *Heliconia* and the Strelitziaceae may be generalized in the statement that anatomical organization is less complex in *Heliconia*: polyarch root stele, absence of vein buttresses, one layered hypodermis, simple system of air canals in leaf axis (for details, see Tomlinson, 1959, 1960).

Prehistory

The Musoid stock, like the Zingiberales as a whole, belongs to the West Gondwana element (Andersson, 1981b; Raven and Axelrod, 1974). The rise of this stock may thus be dated to before early Paleocene, when Africa and South America had become effectively separated. At this time, the Strelitziaceae had probably acquired approximately its present family characteristics, since it is represented by endemic genera in both Madagascar (with adjacent islands), South Africa, and tropical South America. All fossils seem to be disputable as to generic identification, and the fossil record does not give any further details or corroboration.

The floral characteristics of *Heliconia* strongly suggest coevolution with, or at least adaptation to flower-visiting birds: the origin of the sophisticated construction of the floral tube could hardly be otherwise envisaged. All modern neotropical species of the genus seem to be pollinated by hummingbirds (*e.g.* Stiles, 1979). The few Pacific species

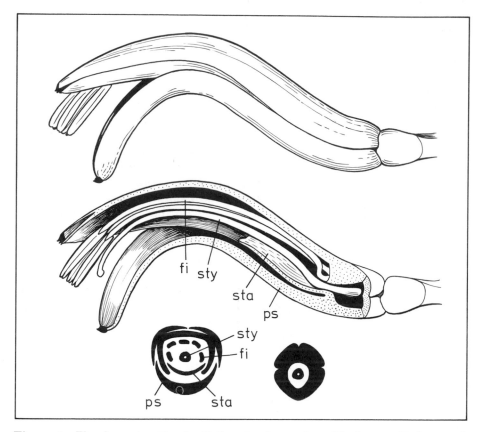

Figure 4. Floral construction in *Heliconia*, drawn from *H. obscura* Dodson and Gentry. fi-filament, ps-posterior sepal, sta-staminode, sty-style.

are, as far as known, bat-pollinated (Kress, 1985), but this is supposedly a secondary phenomenon. The Musaceae (s.str.), *Ravenala*, and *Phenakospermum* are largely pollinated by bats (Kress, pers. comm.), but effective pollination of *Musa* by tree shrews have also been reported (Nur, 1976). The genus *Strelitzia* seems to be pollinated mainly by sunbirds (Frost and Frost, 1981). Specialized bat and bird pollination syndromes both seem to have arisen from syndromes involving more primitive Paleogene vertebrates (Sussman and Raven, 1978). While the origin of modern, specialized, flower-visiting bats is dated by the fossil record to the upper Oligocene and early Miocene (Sussman and Raven, 1978), the origin of hummingbirds is rather obscure. Sussman and Raven state that provided hummingbirds were derived from swift-like ancestors, their origin could be no earlier than Eocene. However, still according to Sussman and Raven, there is "no strong evidence for widespread flower visiting among birds until Miocene."

Floral characteristics, thus, suggest that *Heliconia* was differentiated from strelitzioid ancestors and that the driving force was adaptation to a new kind of pollinator, viz. flower-visiting birds. This identifies both the "right time" and the "correct equipment". The time is somewhere between upper Eocene and lower Miocene. The "correct equipment" was the earlier primitive vertebrate pollination syndrome, including such features as a basal floral tube and rich nectar production. This piece of equipment turned out to be "correct", because it allowed the heliconiaceous ancestor to utilize a novel kind of pollinator that offered a least resistance method to achieve reproductive isolation.

The Strelitziaceae, with their caulescent growth habit, robust plant body, and long life cycles, depend on more or less permanently light-open habitats. *Strelitzia* grows in the macchia-like vegetation along the coast of the eastern Cape Province, *Phenakospermum* grows mainly in savanna scrub, *campina,* and Amazonian *caatinga.* In primary vegetation *Ravenala* grows mostly along ridgelines (Gentry, pers. comm.). *Heliconia,* on the other hand, occurs predominantly in forest understory, riverside thickets, and rainforest light-gaps of comparatively short duration. Only few and clearly derived species occur mainly in savanna scrub (*H. psittacorum*), gallery forest (*H. hirsuta*), or swamps (*H. marginata*). The rise of the genus *Heliconia* thus seems to have been coupled with an invasion of rainforest habitats, possibly from savannas or cerrados. In the new set of habitats, the genus has become an "intermediate disturbance" element in the sense of Connell (1978). This required a shorter life cycle (due to unstable habitat conditions), which in turn led to reduction in plant volume and simplification of anatomical structure.

Present day diversity patterns suggest an origin in northwestern South America. Today, this area is characterized by high precipitation and comparatively nutrient-rich soils derived from Cretaceous and, mainly, Tertiary rocks. Although minor land masses have existed in this area since late Cretaceous, topography remained rather level until the first major phase of Andean uplift in Eocene and Oligocene (Simpson, 1975). Still at the Miocene-Pliocene boundary, a major part of the northern Andes were under 1500 m altitude (Hammen, 1974). The final uplift of the Isthmus of Panama has been dated to middle (Raven and Axelrod, 1974) or late (*e.g.* Marshall *et al.,* 1982) Pliocene, islands were certainly present in the Panamanian portal long before that (*e.g.* Savage, 1982). Present day characteristics of climate and soils in northwestern South America, thus, came gradually into being during Oligocene and lower Miocene.

While it has been widely supposed that northern South America has been more or less extensively covered with tropical rainforest throughout the Tertiary (*e.g.* Solbrig, 1976), this is contradicted by the (admittedly scrappy) palynological record. Of the families of tall dicotyledonous trees recorded by Flenley (1979), and which are more or less restricted to lowland rainforests, the first to appear is the Caryocaraceae (middle Eocene), the second one the Dipterocarpaceae (middle Oligocene). This contrasts sharply to the records of temperate tree genera. For example, *Alnus, Ilex, Juglans,* and *Nothofagus* all occur in upper Cretaceous (Senonian) and *Betula, Carya,* and *Liquidambar* in the Paleocene. Although many families which are richly diversified in the rainforest belong to the Gondwana element and were no doubt distinct at the end of the Cretaceous, these families are often ecologically diverse and there is no strong evidence that early representatives were really rainforest organisms. Genera represented both in present day rainforests and in the Paleocene pollen record are all ecologically diverse, occuring also in dry forests, marsh-lands, savannas, *etc.* Such records can clearly not be taken as evidence for the existence of rainforest vegetation in the Paleocene. In other words, while temperate and subtropical forests are likely to have existed in an approximately modern aspect already by early Paleocene, the tropical rainforest possibly did not arise in its present appearance until well into the Oligocene.

The reasons for the considerable evolutionary success of the genus *Heliconia* could be summarized in this way: the strelitzioid ancestor happened to be in the right place, viz. northwestern South America, at the right time, about Oligocene. At this time, a new habitat was rapidly developing through the combined effects of the first major uplift of the Andes and the development of the tropical rainforest in its modern aspect. The ancestor also happened to possess the "correct equipment." One important piece of equipment was the palmetto habit which, with a limited amount of alteration, made it possible to take advantage of a set of intermediate disturbance niches. Another piece of equipment was adaptation to vertebrate pollinators, which made it possible to utilize a class of pollinator in rapid development, viz. the hummingbirds. The hummingbirds offered a "least resistance" way to achieve reproductive isolation. The rapid rise of both a new habitat and a new class of pollinators caused a rare situation of low competition that was favorable for rapid diversification.

Later diversification

Although all the phytogeographic subdivisions of the northwestern Neotropics have a fair to high degree of endemism at the species level (Fig. 3), this is not true at the sectional level. This may be taken to mean that early diversification took place in a restricted area, *e.g.* in the northern Andes, from where various early representatives of *Heliconia* spread gradually. In all probability, most major groups of the genus had reached Central America already before the land connection was completed. The possibility that certain sections arose in Central America and spread into northwestern South America cannot be ruled out. In fact, modern diversity patterns suggest that such was the case for sect. *Tortex* and the *Heliconia pogonantha* group. However, accepting the ecological species concept, one is rather inclined to think that present peaks in diversity correspond to peaks in number of available niches. That makes all historical interpretations based on modern diversity patterns rather dubious. It would explain, however, why the genus did not diversify much in other parts of South America. It is still true to its original set of niches and therefore, dependent on very high precipitation and young soils.

The role of Pleistocene climatic fluctuations for the formation of modern species has been rather differently viewed (see Whitmore and Prance, 1987, for discussions). The distribution areas of most *Heliconia* species with a restricted range are usually centered around one or the other of the "consensus" refugia recognized by Brown (1987). This would seem to suggest that modern species are usually of Pleistocene age, having arisen in rainforest refugia as a consequence of isolation and reduced population size. However, present distributions coincide even better with modern patterns in environmental factors. Thus, the "Guiana" element is restricted to areas with soils derived from the Guiana shield and with a high precipitation. The "Olivença" element is restricted to areas where extremely high precipitation (over 3000 mm per year) coincides with sedimentary soils of the Amazon Valley. The "Napo" element is largely restricted to areas where extremely high precipitation coincides with soils derived from the Andes. Only a better understanding of niche separation in rainforest plants can resolve this conflict. However, the view that modern species distributions are determined essentially by modern patterns in the distribution of environmental factors is a more meaningful one in the framework of the ecological species concept. On the other hand, the vegetation history of the Pleistocene seems rather clearly reflected in the distribution patterns of some widely distributed species, some instances of vicariance, the variation pattern of some widespread polymorphic species, and reticulate relationship patterns characteristic of the floras in some areas.

Widespread species (present in more than five of the areas of Fig. 3) are all species occurring mainly in habitats that may have persisted over vast areas even at the height of Pleistocene aridity: swamps (*H. marginata*), swamp and gallery forest (*H. hirsuta*) or savanna scrub (*H. psittacorum*). They have a patchy distribution today that could hardly be explained except as relictual and resulting from a once wider and more continuous distribution of drier and more seasonal climate.

Unambiguous cases of vicariance in ecologically equivalent taxa are rare in *Heliconia* and are mostly in the category of cis- and trans-Andean disjunctions, which is not directly relevant in this context. The best examples are *H. dasyantha* (eastern Guiana Shield) and *H. velutina* (western Amazon Basin), and *H. pendula* (eastern Guiana Shield and southeastern Brazil) and *H. revoluta* (western Guiana Shield, coastal range of Venezuela, and Sierra Nevada de Santa Marta).

Disjunctions clearly relating to Pleistocene climatic change are also scarce. One is *H. aemygdiana ssp. aemygdiana*, disjunct between the southwestern Amazon and southeastern Brazil.

Some polymorphic species such as *H. bihai* (Andersson, 1981a), *H. hirsuta* and *H. acuminata* (Andersson, 1985a) show variation patterns which are probably best interpreted as secondary contact phenomena. At least in *H. hirsuta* and *H. acuminata*, the patterns are further complicated by ecotypical differentiation and extreme complexity.

The southern part of the Chocó area and southern Central America are areas where species complexes with reticulate relationship patterns are particularly conspicuous. In the former area, such complexes are the *H. obscura* group and the *H. longa* group (Andersson, 1985b). In the latter area, intricate reticulate patterns are present in the *H. pogonantha* group (Kress, 1984) and sect. *Tortex* (Andersson, unpubl.). This may be interpreted as a result of severe reduction of the habitat and the extinction of a large number of species. The reestablishment of a wider array of suitable niches may have caused a situation where the least resistance way to form new genotypes was by means of introgressive hybridization.

To sum up, the Pleistocene climatic fluctuations may have had discernible consequences but mainly below the species level. They have not much affected the diversity of the genus as a whole. This in turn leads to the conclusion that modern species may largely be of Pliocene age, some perhaps upper Miocene. There seems to be few reliable criteria to decide which are younger and which are older, except that it may be suspected that many species in areas and complexes showing reticulate relationship patterns are of Pleistocene age.

Literature cited

Abalo, J. and Morales, G. (1982). "Veinticinco (25) Heliconias nuevas de Colombia." *Phytologia* **51**, 1-61.

Abalo, J. and Morales, G. (1983a). "Doce (12) Heliconias nuevas del Ecuador." *Phytologia* **52**, 387-413.

Abalo, J. and Morales, G. (1983b). "Diez (10) Heliconias Nuevas de

Colombia." *Phytologia* **54,** 411-433.

Andersson, L. (1981a). "Revision of Heliconia sect. Heliconia (Musaceae)." *Nord. J. Bot.* **1,** 759-784.

Andersson, L. (1981b). "The Neotropical genera of Marantaceae. Circumscription and relationship." *Nord. J. Bot.* **1,** 218-245.

Andersson, L. (1985a). "Revision of Heliconia subgen. Stenochlamys (Musaceae-Heliconioideae)." *Opera Bot.* **82,** 1-123.

Andersson, L. (1985b). "Musaceae." *In* Harling, G. and Sparre, B. (eds.), *Flora of Ecuador* **22,** 1-86.

Andersson, L. (unpubl.). "Revision of Heliconia subgen. Taeniostrobus and subgen. Heliconia."

Berry, P. E. (1982). "The Systematics and evolution of Fuchsia sect. Fuchsia (Onagraceae)." *Ann. Missouri Bot. Gard.* **69,** 1-198.

Brown, K. S. (1987). "Conclusions, synthesis, and alternative hypotheses." pp. 173-196 *In* Whitmore, T. C. and Prance, G. T. (eds.), "Biogeography and quaternary history in tropical America." *Clarendon Press, Oxford.*

Burger, W. C. (1974). "Ecological differentiation in some congeneric species of Costa Rican flowering plants." *Ann. Missouri Bot. Gard.* **61,** 297-306.

Connell, J. H. (1978). "Diversity in tropical rain forests and coral reefs." *Science* **199,** 1302-1310.

Cronquist, A. (1981). "An integrated system of classification of flowering plants." *Columbia University Press, New York.*

Dahlgren, R., Clifford, H. T., and Yeo, P. F. (1985). "The families of the Monocotyledons." *Springer, Berlin.*

Flenley, J. (1979). "The equatorial rain forest: a geological history." *Butterworths, London and Boston.*

Frost, S. K. and Frost, P. H. G. (1981). "Sunbird pollination of Strelitzia nicolai." *Oecologia* **49,** 379-384.

Gentry, A. H. (1982). "Neotropical floristic diversity: phytogeographical connections between Central and South America, Pleistocene climatic fluctuations, or an accident of the Andean orogeny." *Ann. Missouri Bot. Gard.* **69,** 557-593.

Hammen, T. van der (1974). "The Pleistocene changes of vegetation and climate in tropical South America." *J. Biogeogr.* **1,** 3-26.

Kress, W. J. (1984). "Systematics of Central American Heliconia (Heliconiaceae) with pendent inflorescences." *J. Arnold Arbor.* **65,** 429-532.

Kress, W. J. (1985). "Bat pollination of an Old World Heliconia." *Biotropica* **17,** 302-308.

Marshall, L. G., Webb, S. D., Sepkoski, J. J. and Raup, D. M. (1982). "Mammalian evolution and the great American interchange." *Science* **215,** 1351-1357.

Nur, N. (1976). "Studies on pollination in Musaceae." *Ann. Bot. (London)* **40,** 167-177.

Raven, P. H. and Axelrod, D. I. (1974). "Angiosperm biogeography and past continental movements." *Ann. Missouri Bot. Gard.* **61,** 539-673.

Santos, E. (1978). "Revisão das espécies do género Heliconia L. (Musaceae s. l.) espontâneas na região fluminense." *Rodriguesia* **45,** 99-221.

Savage, J. M. (1982). "The Enigma of the American Herpetofauna: Dispersals or Vicariance ?" *Ann. Missouri Bot. Gard.* **69,** 464-547.

Simpson, B. B. (1975). "Pleistocene changes in the flora of the high tropical Andes." *Paleobiology* **1,** 273-294.

Solbrig, O. T. (1976). "The origin and floristic affinities of the South American temperate desert and semidesert regions." pp. 7-49 *In* Goodall, D. W. (ed.), "Evolution of desert biota." *University of Texas Press, Austin and London.*

Stiles, F. G. (1979). "Notes on the natural history of Heliconia (Musaceae) in Costa Rica." *Brenesia* **15** *(supl.),* 151-180.

Sussman, R. W. and Raven, P. H. (1978). "Pollination by Lemurs and Marsupials: An archaic coevolutionary system." *Science* **200,** 731-736.

Tomlinson, P. B. (1959). "An anatomical approach to the classification of the Musaceae." *J. Linn. Soc. (Bot.)* **55,** 779-809.

Tomlinson, P. B. (1960). "The Anatomy of Phenakospermum (Musaceae)." *J. Arnold Arbor.* **41,** 287-297.

Van Valen, L. (1976). "Ecological species, multispecies, and oaks." *Taxon* **25,** 233-239.

Whitmore, T. C. and Prance G. T. (eds.), (1987). "Biogeography and quatenary history in tropical America." *Oxford Monographs in Biogeography Vol. 3. Clarendon Press, Oxford.*

Winkler, H. (1930). "Musaceae." pp. 505-541 *In* Engler, A. and Prantl, K. (eds.), "Die Naturlichen Pflanzenfamilien." Ed. **2,** Vol. **15a.** *Engelmann, Leipzig.*

Evolutionary patterns in cytology and pollen structure of Asian Zingiberaceae

CHEN Z-Y.

Academia Sinica, Guangzhou, China

The pantropical family Zingiberaceae, includes 52 genera and 1500 species, of which 46 genera and 1300 species are distributed mainly in tropical Asia. There Zingiberaceae is one of the constituents of the undergrowth of the tropical rain forest and monsoon forest. The plants are perennial and grow mostly in damp and humid shady places. They are also found infrequently in secondary forest. Some species can stand full exposure to the sun, harsh conditions, and high elevations. Only a few species are epiphytic. In southern and southwestern China there are about 21 genera and 180 species. Having finished the revision of Zingiberaceae for *Flora Republicae Popularis Sinicae*, Wu *et al.* (1981) initiated systematic research, work on resource utilization, and the conservation of germ plasm of Zingiberaceae in China. This project was supported by the Chinese Academy of Science. Up to now we have planted about 18 genera and 120 species of Chinese Zingiberaceae at the South China Botanical Garden. They have been studied by a multidisciplinary approach including chemotaxonomy, cytology, anatomy, and palynology. The preliminary results can now be presented. The evolutionary patterns of Zingiberaceae in tropical Asian forest are now better understood.

Cytology

The cytological data show that Zingiberaceae is a natural group. Chen and Chen (in press) discussed the taxonomic significance of chromosome numbers and showed that the genera of Zingiberaceae have different basic numbers reflecting different evolutionary levels within each genus (Fig. 1). The family has been divided into two subfamilies and four tribes. Zingiberoideae and Costoideae may both have been derived from the same ancestor, but each developed along distinct evolutionary lines. The tribe Globbeae (mainly *Globba*) with unilocular ovary and parietal placenta, is an apomictic complex. The tribes Zingibereae and Alpinieae have trilocular ovaries with axile placentation and seem to have proceeded largely at diploid and tetraploid evolutionary levels, respectively. The

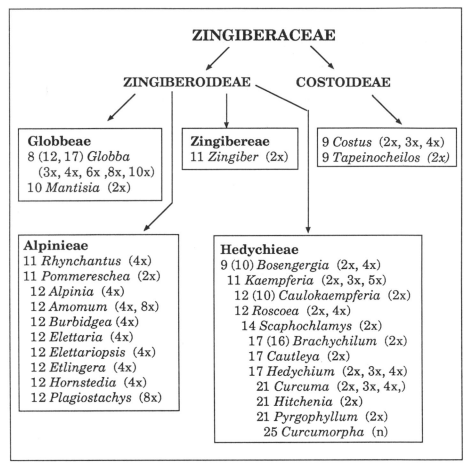

Figure 1. The basic chromosome numbers in different genera of Asian Zingiberaceae

tribe Hedychieae has a wide range of basic chromosome numbers and is at an active stage of evolution. Twelve is the most frequent basic chromosome number in Zingiberaceae. Mahanty (1970) suggested that 12 is the original basic number for the family and presumably derived from 11 which is the basic number for the Zingiberales.

Venkatasubban (1946), Chakravorti (1952), and Sharma and Bhattacharya (1959) showed that polyploid, aneuploid, and structural changes of chromosomes are important in the evolution of Zingiberaceae. Several polyploids exist in tropical Asian Zingiberaceae, especially in Globbeae and Alpinieae. Larsen (1972) showed that *Globba* is a polyploid complex in the sense of Stebbins (1950). Lim (1972) provided strong evidence for a polyploid complex in Malayan *Globba* which are all allotetraploids and heteroploids. The connection between apomixis, hybridization, and polyploidy is so closely related that the situation in the growth of Globbeae is complicated and many patterns are evident. Almost all species in Alpinieae are polyploids. Only *Amomum austrosinense* and *Plagiostachys austrosinensis*, which occur at the distributional limits of their respective genera, are secondary octoploids.

Polyploids are scattered in different genera of Hedychieae. The genera with higher basic chromosome numbers such as *Hedychium* (x=17), *Cautleya* (x=17), *Curcuma* (x=21), *Hitchenia* (x=21), *Pyrgophyllum* (x=21) and *Curcumorpha* (x=25) are all in this tribe and evolved euploidy and aneuploidy.

Palynology

Palynological data also demonstrate differences between the subfamilies and tribes. Liang (1988) studied the pollen morphology of 89 species and three varieties in 19 genera of Chinese Zingiberaceae using LM and SEM. The ultramicrostructure of pollen walls for 18 species in 12 genera were observed under TEM. Presence or absence of aperture and different ornamentations resulted in two main types and six subtypes. Costoideae has aperturate pollen. The wall has a thin intine and a thick exine which is resistant to acetolysis. The exine surface ornamentation is colpate, porate, or forate. The Zingiberoideae has non-aperturate pollen. Its wall has 2-4 layers of intine and a thin exine which is not very resistant to acetolysis. The ornamentations of the exine surface are psilate, spinate, cerebroid, or striate. The tribal distribution of the pollen types is shown in Figure 2.

Vegetative reproduction

Zingiberaceae usually have prostrate or tubular rhizomes which are sympodial and promote the formation and development of axillary buds. Most Zingiberaceae can be propagated vegetatively. Asexual reproduction is predominant in some species although they produce seeds. This is true in *Curcuma, Zingiber, Kaempferia,* and *Stahlianthus.* Moreover, in some species of *Globba* the flowers are often replaced by bulbils. Sexual reproduction is largely weakened or even lost with increasing ploidy levels. In *Zingiber officinale*, which has been in cultivation for several thousand years, Wu (1985) suggested that the area of origin could be somewhere between the Yangtze and Yellow Rivers. Because of great changes in the natural environments, wild ginger disappeared from its original range. It is suggested that vegetative reproduction by rhizomes

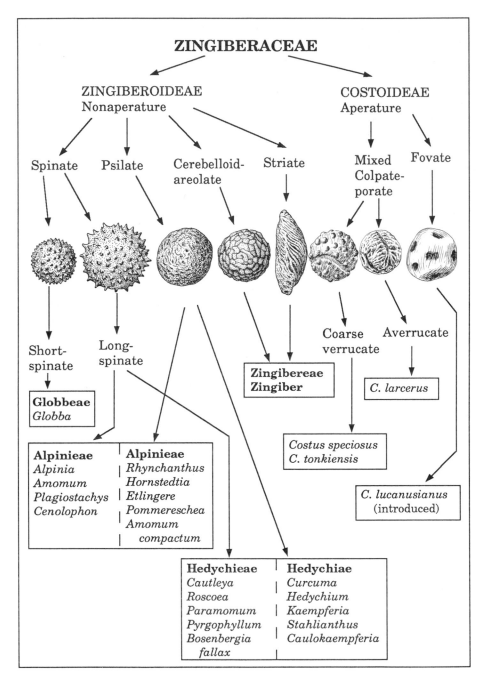

Figure 2. Pollen types of Zingiberaceae

could overcome the sterility caused by incompatibility and chromosome structural hybridity for interchanges and inversions, and therefore they could overcome the sterility caused by incompatibility and chromosome genotype could be preserved.

Asexual reproduction can produce large numbers of individuals with the same genotype. Sharma and Bhattacharya (1959) suggested that the origin of different basic chromosome numbers in various genera is connected with asexual reproduction. Evidently, somatic mutation plays a distinct role in the origin of new individuals in the species through its participation in the formation of daughter shoots. Ramachandran (1961) considered that the vegetative mode of propagation and the small size of the chromosomes in *Curcuma* have favored the perpetuation of polyploid types.

Habitat specialization

Natural selection for habitat specialization is another cause for evolution in Zingiberaceae in tropical Asia. The genus *Roscoea* is found in the Himalayan mountain range from Kashmir in the west to Assam in the east, northwards to Tibet and Sichuan, and southwards to Yunnan of China. It differs from other genera in its habitat. *Roscoea* can withstand exposure to the sun, harsh conditions, and high elevations. Adapted to these habitats, the species of *Roscoea* have closed leaf sheaths, reduced rhizomes, and they develop roots which reach deeply into the soil. These features all have adaptive significance in the plateau environment because there is no room to accomodate the fleshy rhizomes in the surface of the rocky soil. The tufted fleshy tuber-roots presumably store nutrition and are able to withstand the bad weather and low temperatures during the winter. It seems, therefore, that this adaptation to changeable weather must have been induced by the Himalayan orogeny.

Conclusion

Tropical Asia is the distribution center of Zingiberaceae. In this area, due to stable, damp and hot climate during its long history, changable habitats would favor the development and differentiation of Zingiberaceae. There we find number of monotypic and small genera such as *Brachychilum* (Java), *Siliquamomum* (Indo-China), *Nanochilus* (Sumatra), *Cyphostigma* (Sri Lanka), *Cenolophon* (Thailand), *Paracautleya* (India), *Pommereschea* (Burma and China), *Pyrgophyllum* (China) *etc.* During the last few years, dozens of new species and several new genera have been described from tropical Asia. Chinese taxonomists have also described 38 new species and two new genera, *Paramomum* (Tong, 1985) and *Pyrgophyllum* (Wu and Chen, in press), in China since 1982. *Paramomum* is monotypic and intermediate between *Costus* and *Amomum*. It has a petaloid stamen but also superior glands, spiral leaf arrangement, and open leaf-sheaths. These studies show that the Zingiberaceae in tropical Asia is obviously still actively evolving.

Acknowledgment

This study was supported by the Science Fundation of the Chinese Academy of Science. I am grateful to Dr. Wu Te-lin for revising the manuscript.

Literature cited

Chakravorti, A. K. (1952). "Cytological studies in Zingiberaceae." *Ind. Sci. Congr. Assoc. 39th Proc.* **3**, 30-31.

Chen, Z-y., Chen, S-j., Huang, X-x. and Huang, S-f. (1982-88). "The reports of chromosome numbers in Zingiberaceae (1-5)." *Guihaia* 2 **(3)**, 153-157, 4 (1), 13-18, 7(1), 39-44, 8(2), 143-147. *Acta Bot. Austro Sin.* **3,** 57-61.

Chen, Z-y. and Chen, S-j. (in press). "The taxonomic significance of chromosome numbers in the Zingiberaceae." *Proc. Sino-Jap. Symp. Pl. Chrom.*

Larsen, K. (1972). "Studies in the genus Globba in Thailand." *Notes Roy. Bot. Gard. Edinburgh* **31,** 229-241.

Liang, Y-h. (1988). "Pollen morphology of the family Zingiberaceae in China I. Pollen types with reference to the taxonomy of the family and their significance in the taxonomy." *Acta Phytotax. Sin.* **26(4),** 265-281

Lim, S-n. (1972). "Cytogenetics and taxonomy of the genus Globba L. (Zingiberaceae) in Malaya II: Cytogenetics." *Notes Roy. Bot. Gard. Edinburgh* **31,** 271-284.

Mahanty, H. K. (1970). "A cytological study of the Zingiberales with special reference to their taxonomy." *Cytologia* **35,** 13-49.

Ramachandran, K. (1961). "Chromosome numbers in the genus Curcuma Linn." *Curr. Sci.* **30,** 194-197.

Sharma, A. K. and Bhattacharya, N. K. (1959). "Cytology of several members of Zingiberaceae." *La Cellule* **59,** 299-346.

Stebbins, G. L. (1950). "Variation and evolution in plants." *Columbia Univ. Press, New York.*

Tong, S-q. (1985). "Paramomum, a new genus of Zingiberaceae from Yunnan." *Acta Bot. Yunn.* **7(3),** 309-312.

Venkatasubban, K. R. (1946). "A preliminary survey of chromosome numbers in Scitamineae of Bentham and Hooker." *Proc. Indian Acad. Sci.* **23B,** 281-300.

Wu, T-l., Chen Sen-jen, Tsai, H-t., Tong, S-q., Chen, P-s., Zhao, S-w. and Li, H-w. (1981). "Zingiberaceae" *Fl. Rep. Popularis Sin.* **16(2),** 26-152.

Wu, T-l. (1985). "The origin of Zingiber officinale." *Agricultural Archeology* **2,** 247-250.

Wu, T-l. and Chen, Z-y. (in press). "Pyrgophyllum, a new genus of Chinese Zingiberaceae." *Acta Phytotax. Sin.*

Speciation in the Asplenium unilaterale complex

K. IWATSUKI

Botanical Gardens, University of Tokyo, Japan

When we investigate the diversity of organisms, the difficulty is that the process of diversification differs according to the taxa concerned. The present paper aims to introduce the reader to the evolution of various phenetic features in a particular group of plants and to the principles inferred by the observed facts.

In this paper the fern group, *Asplenium* sect. *Hymenasplenium*, will be used as an example of species diversification in tropical forests. *Asplenium* is a genus of some 700 species, found mostly in the tropics. A small number of species belong to temperate areas, and these have been carefully studied applying a variety of techniques. The reticulate evolution of Appalachian *Asplenium* species, the cytogenetics of the *A. trichomanes* complex, and speciation in the *A. varians* complex are well-known examples of such studies. In contrast, most of the tropical species are known just in descriptive terms. As the systematics of the genus are less known in the tropics, the interrelationships of the species are thus, hardly elucidated.

Asplenium sect. *Hymenasplenium* was the subject of two earlier papers (Iwatsuki and Kato, 1975; Iwatsuki, 1975). In the former, it was shown that the Old World species with dorsiventral rhizomes and pinnate fronds could be referred to one taxon corresponding to the genus *Hymenasplenium* (Hayata, 1927). This taxon was reestablished as a section of the vast and heterogeneous genus *Asplenium* in connection with a revision of all the Asiatic species (Iwatsuki, 1975). Since then the anatomy, cytology, chemotaxonomy, speciation and ecological adaptation of *Asplenium* sect. *Hymenasplenium* has been carefully studied by various techniques. The result of these recent studies of sect. *Hymenasplenum* are briefly summarized below and will be discussed on the basis of the information available.

TROPICAL FORESTS
ISBN 0 12 353550 6

Adaptation to wet habitats

Most Hymenaspleniums grow in moist or wet habitats. Even the epipetric *Asplenium unilaterale* var. *saxicola*, which is endemic to limestone cliffs in Seram, belongs in humid tropical forests where the limestone cliffs are constantly wet and often covered with water drops. The lamina of this variety is simple and bistratose or tristratose, with small intercellular spaces (Kato and Iwatsuki, 1986).

The lamina of many members of the section are 4 - 8 cell layers thick and slightly differentiated, although with a distinct epidermis with many abaxial stomata. A semiaquatic form, *A. obliquissimum*, grows under constant spray. The lamina of this species are bistratose without intercellular spaces and stomata, or with some incomplete scars of stomata. The epidermial cells have a tetragonal, not sinuate, outline in surface view (Iwatsuki, 1975).

Another particular ecotype from Seram is apogamous and aquatic. It grows submerged more than one meter under water in a spring. Constant bubbling seems to give the plants airing, and they reproduce only vegetatively by means of frequent gemmae. Roots become long and slender, forming an entangled rhizome-root system, often trapping detached leaves and/or pinnae-bearing gemmae (Kato and Iwatsuki, 1985). The same plant under cultivation in nearly moisture saturated air, produced no gemmae. On the other hand, Japanese epipetric plants could be grown submerged in water with constant bubbling, some of them for several months. One of these plants bore gemmae on the pinnae. These facts indicate that the *A. unilaterale* complex adapts easily to moist conditions, and has the potential to develop gemmae in moisture saturated air, or when submerged.

Chromosomes, apogamy, and speciation.

Chromosome numbers of *Asplenium* are known from more than 100 species. The prevailing number is n=36 or multiples of it. We have found the numbers n=38 or 39, or multiples in sect. *Hymenasplenium*, although n=40 was formerly reported. These numbers are unique in *Asplenium* and suggest the distinctness of sect. *Hymenasplenium*. However, the section needs to be compared to the American *A. obtusifolium* complex and one species referred to the genus *Boniniella* in order to elucidate its distinctness. Apogamy was observed in three species and some forms of *A. unilaterale sensu lato* These can be distinguished easily as different species from the sexually reproducing ones by a combination of phenetic features (Murakami and Iwatsuki, 1983; Murakami and Hatanaka, 1988a).

The genetic diversity of *A. hondoense,* an apogamous Japanese form, was examined by means of electrophoretic analysis, and four biotic types recognized. One of these was assumed to have arisen through hybridization (Watano and Iwatsuki, 1988). Chromosome counts support this hypothesis as this biotype is tetraploid (2n=ca. 156), and the others are triploid apogamous (n=117, 2n=117). In contrast, *A. cataractarum,* a sexual Japanese form, is diploid (n=39, 2n=78) (Mitsui *et al.,* in prep.). Former studies on the gametophytes of the apogamous forms (Iwatsuki, 1975) failed to demonstrate the presence of gametangia, but recently, Yoroi (pers. comm.) found antheridia on the gametophytes of *A. hondoense* after two years in cultivation.

The three other biotic types are highly differentiated from *A. cataractarum* (Nei's genetic distance (D)=0.50) and share no alleles with the latter among four of eleven loci that were observed. These observations suggest that the three biotypes of *A. hondoense* were not direct allies of *A. cataractarum.* Interrelationships among the species of sect. *Hymenasplenium* should be traced after examination of all the species included in this section.

Chemical characters

Murakami and Hatanaka (1983, 1985, 1988b; Murakami *et al.,* 1985) observed nonprotein amino acids in the species of sect. *Hymenasplenium.* They found D-2-aminopimelic acid (D-APA) and *trans*-3,4-dehydro-D-2-aminopimelic acid (D-delta-APA), both of which belong to the D-series of amino acids that are rarely found in plants as free forms. They also detected 4-hydroxy-2-aminopimelic acid (OH-APA) which is a usual L-amino acid. Among 29 Japanese *Asplenium* species screened, 23 had OII-APA and 17 had APA. Delta-APA was found only in three species of sect. *Hymenasplenium,* and in *A. wilfordii* which has no close similarity to sect. *Hymenasplenium.* Observations of the metabolic processes of these amino acids suggest that glutamate and asperate are probable precursors. First 4-hydroxy-L-2-aminopimelic acid is synthesized, then two D-amino acids are derived from it. Although knowledge of the metabolic pathway of the D-amino acids is still insufficient to establish conclusions about the species relationships in sect. *Hymenasplenium,* these compounds are interesting for future studies of the section. Also studies, in progress, of chloroplast DNA are expected to yield an effective contribution to the elucidation of species relationships in the section.

Circumscription of sect. Hymenasplenium

The species of sect. *Hymenasplenium* are characterized by the chromosome number n=38 or 39, or their multiples. This is unique in *Asplenium* which is otherwise characterized by n=36 or multiples. However, the American *A. obtusifolium* complex and the Asian genus *Boniniella* (= *Asplenium cardiophyllum*) warrant careful study. *Asplenium repandulum*, a close ally or synonym of *A. obtusifolium*, has n=39 II or 40 II (Smith and Mickel, 1977). Both taxa have long, creeping, dorsiventral rhizomes, comparable to sect. *Hymenasplenium* (Iwatsuki, 1975; Mitsuta *et al.*, 1980; Smith, 1976; Tryon and Tryon, 1982). These similarities suggest a close relationship, however more information is needed about the American species in order to make a final conclusion.

Boniniella Hayata is a monotypic genus, known from the Bonins, southern Ryukyus and Hainan, and represented by the species *B. ikenoi* (= *A. cardiophyllum*). Kurita (1972) reported its chromosome number as n=76. Its long creeping dorsiventral rhizome, including the phyllopodia, is similar to that of sect. *Hymenasplenium* (Hayata, 1927), but it differs by its simple frond. Also, Hayata recorded the presence of abscission at the base of stipes in *Boniniella*; abscission is unknown in Sect. *Hymenasplenium*. This feature and the difference in frond construction, are the only features distinguishing *Boniniella* from sect. *Hymenasplenium*. Although our knowledge is still incomplete, several sources of information suggest that the Old World species allied to *A. unilaterale*, *Boniniella*, and the American *A. obtusifolium* complex are better treated as a distinct genus *Hymenasplenium*.

Discussion

The type of reproduction, apogamous or sexual, is a helpful character in distinguishing the East Asian species of sect. *Hymenasplenium*. *Asplenium hondoense* and *A. cataractarum* are distinguished on the basis of these characters, as well as phenetic features. The origin of apogamous reproduction has hardly been elucidated in any particular fern group. Walker (1962) suggested that apogamous reproduction is inherited by hybridization between apogamous and sexual species. Evidence for such a case was given by electrophoretic analysis by Gastony and Gottlieb (1985). We have no evidence that it arose in this way in sect. *Hymenasplenium*, and we still do not know how it was introduced.

Vegetative reproduction seems to have played an important role in the development of apogamous ferns. In sect. *Hymenasplenium*, vegetative reproduction is distinct especially in aquatic forms. The species of sect. *Hymenasplenium* commonly form large colonies by means of the long creeping rhizomes, and this structure, thus, is important for the expansion of their habitats.

Gemmae are known to occur on the leaves of various species of *Asplenium*. In the case of our aquatic form of *A. unilaterale*, the gemmae are luxuriantly formed in submerged conditions, but never in cultures above water. *Asplenium hondoense*, an epipetric Japanese form, also produces gemmae when grown in water, but no gemmae have been observed in its natural habitat. Only in the semiaquatic species *A. obliquissimum* do the gemmae occur subspontaneously. The potential of forming gemmae is not unique for sect. *Hymenasplenium*, but occurs extensively in other fern groups (Iwatsuki and Kato, 1985). The occurrence of gemmae seems to be important for the ability of ferns to introduce apogamy, and thus for the diversification of ferns.

Hymenasplenium share several special features with the American *A. obtusifolium* complex. Even the aquatic adaptation is represented by the Venezuelan *A. obtusifolium* var. *aquaticum*, a semiaquatic form, with bi- or tristratose laminae lacking intercellular spaces and with few incomplete stomata (Giesenhagen, 1892).

In conclusion, it may be repeated that speciation in the tropics is not a matter of brief discussion, but from the observations made in such a diverse group as *Asplenium* sect. *Hymenasplenium*, we get information of importance to the general patterns of speciation.

Acknowledgments

Biosystematic analysis in *Asplenium* Sect. *Hymenasplenium* has been carried out in collaboration with M. Kato, N. Murakami, S.-I. Hatanaka, Y. Watano, the late K. Mitsui, S. Nakato, R. Yoroi and others, and this review is the result of our joint efforts. We are indebted to the local botanists who have assisted us in obtaining the material for study.

Literature cited

Gastony, G. J. and Gottlieb, L. D. (1985). "Genetic variation in the homosporous fern Pellaea andromedifolia." *Amer. J. Bot.* **72**, 257-267.

Giesenhagen, K. (1892). "Über hygrophile Farne." *Flora* **76**, 157-181.

Hayata, B. (1927). "On the systematic importance of the steler system in the Filicales. 1." *Bot. Mag. Tokyo* **41**, 697-718.

Iwatsuki, K. (1975). "Taxonomic studies of Pteridophyta X." *Acta Phytotax. Geobot.* **27**, 39-54.

Iwatsuki, K. and Kato, M. (1975). "Steler structure of Asplenium unilaterale and its allied species." *Kalikasan* **4,** 165-174.

Iwatsuki, K. and Kato, M. (1985). "Diversity of vegetative reproduction in the ferns with reference to leaf-borne proliferation" pp. 124-131 *In* Hara, H. (ed.), "Origin and evolution of diversity in plants and plant communities." *Academia Scientific Books, Tokyo.*

Kato, M. and Iwatsuki, K. (1985). "An unusual submerged aquatic ecotype of Asplenium unilaterale." *Amer. Fern J.* **75,** 73-76.

Kato, M. and Iwatsuki, K. (1986). "Variation in ecology, morphology and reproduction of Asplenium sect. Hymenasplenium (Aspleniaceae) in Seram, Indonesia." *J. Fac. Sci. Univ. Tokyo III.* **14,** 37-48.

Kurita, S. (1972). "Chromosome numbers of some Japanese ferns (8)." *Ann. Rep. For. Stud. Coll. Chiba Univ.* **7,** 47-53.

Mitsui, K., Murakami, N. and Iwatsuki, K. (1989). "Chromosomes and systematics of Asplenium sect. Hymenasplenium (Aspleniaceae)." in prep.

Mitsuta, S., Kato, M. and Iwatsuki, K. (1980). "Stelar structure of Aspleniaceae." *Bot. Mag. Tokyo.* **93,** 275-289.

Murakami, N., Furukawa, J., Okuda, S. and Hatanaka, S.-I. (1985). "Stereochemistry of 2-aminopimelic acid and related amino acids in three species of Asplenium." *Phytochemistry* **24,** 2291-2294.

Murakami, N. and Hatanaka, S.-I. (1983). "D-2-Aminopimelic Acid and trans-3, 4-dehydro-D-2-aminopimelic acid from Asplenium unilaterale." *Phytochemistry* **22,** 2735-2773.

Murakami, N. and Hatanaka, S.-I. (1985). "Chemotaxonomic studies of the non-protein amino acids in Aspleniaceae (Pteridophyta)." pp. 142-148 *In* Hara, H. (ed.), "Origin and evolution of diversity in plants and plant communities." *Academia Scientific Books, Tokyo.*

Murakami, N. and Hatanaka, S.-I. (1988a). "A revised taxonomy of the Asplenium unilaterale complex in Japan and Taiwan." *J. Fac. Sci. Univ. Tokyo III.* **14,** 183-199.

Murakami, N. and Hatanaka, S. -I. (1988b). "Chemotaxonomic studies of Asplenium sect. Hymenasplenium (Aspleniaceae)." *Bot. Mag. Tokyo,* **101,** 353-372.

Murakami, N. and Iwatsuki, K. (1983). "Observation on the variation of Asplenium unilaterale in Japan with special reference to apogamy." *J. Jap. Bot.* **58,** 257-262.

Smith, A. R. (1976). "Diplazium delitescens and the Neotropical species of Asplenium sect. Hymenasplenium." *Amer. Fern J.* **66,** 116-120.

Smith, A. R. and Mickel, J. T. (1977). "Chromosome counts for Mexican ferns." *Brittonia* **29,** 391-398.

Tryon, R. M. and Tryon A. F. (1982). "Ferns and allied plants with special reference to tropical America." *Springer, New York.*

Walker, T. G. (1962). "Cytology and evolution in the fern genus Pteris L.." *Evolution* **16,** 27-43.

Watano, Y. and Iwatsuki, K. (1988). "Genetic variation in the Japanese apogamous form of the fern Asplenium unilaterale Lam." *Bot. Mag. Tokyo* **101,** 213-222.

Justicia and Rungia (Acanthaceae) in the Indo-Chinese Peninsula.

B. HANSEN

University of Copenhagen, Denmark

A necessary basis for understanding speciation in the tropics is a firm taxonomic foundation in which the species are properly described and defined. Many tropical plant species remain undescribed, but still more problematic is the fact that others have been overdescribed. Knowledge of tropical plant species has often developed as an accumulation of regional studies. Many widespread species have been described as new in different floras or in revisions covering parts of their distributional area. In other cases, the limited amounts of available material has hampered the understanding of variation within a species - again resulting in the description of "new species" which are merely morphological variants. This paper treats some examples of overdescription in local treatments and also overdescription based on the lack of understanding of variation in the Acanthaceae from the Indo-Chinese peninsula. Clarke (1908) enumerated 19 Justicias and two Rungias for the Malay Peninsula. Ridley (1923) treated 41 Justicias, 20 of which were new. Benoist (1935) treated 39 Justicias, 28 new, and four Rungias for Cambodia, Laos, and Vietnam. Imlay (1938) enumerated 48 Justicias, 14 new, and seven Rungias from Thailand. Bremekamp (1961, 1965, 1966, 1969) added 19 new Justicias *sensu lato* and one *Rungia* to the flora of Thailand. By 1970 there were a total of 67 Justicias and eight Rungias recorded for Thailand.

After having revised these treatments and the corresponding herbarium material, I now accept a total of 50 Justicias and 18 Rungias in the Indo-Chinese Peninsula. What has happened to the rest of them ? The number of Rungias has been raised because Benoist and Imlay each treated four Rungias under *Justicia*. Moreover, Imlay's *Justicia muticitheca* is really the widespread *Leptostachya wallichii*, and two of Benoist's Justicias now belong in *Ptyssiglottis* and four in *Isoglossa*. Only about half of Craib's, Imlay's, and Benoist's 47 new Justicias stand, even including those which have now been transferred to other genera. Only three of Ridley's 20 new Justicias stand. Only two of Bremekamp's 19 new Justicias, most of them published in segregate genera, stand. Most new species in *Justicia* have become synonyms of older names.

TROPICAL FORESTS
ISBN 0–12–353550–6

Justicia sect. *Rostellaria* (Nees) T. Anders.

This section has spicate inflorescences and 5-partite calyces with one lobe strongly reduced or missing. Variation in shape and size of leaves and bracts is considerable. Calyx and bract lengths are of little taxonomic value because these structures continue growing until the fruit is mature. The calyx grows much more than the bracts and to a size significantly larger than in the flowering stage (Fig. 1). Nevertheless, the bract/calyx ratio is important in distinguishing *Justicia diffusa* from *J. procumbens* but only in early anthesis or late fruiting stages. Species of this section were treated in the genus *Rostellularia* by Bremekamp (1948), who stated that the differences between most of them were small but constant, and he predicted the discovery of many new species, none of which would prove to be widespread. Enamored by his own prediction, he described 18 new species from the Indo-Chinese Peninsula and west Malesia over the next 20 years.

Justicia procumbens L., 15. 1753. [Syn.: *Rostellularia procumbens* (L.) Nees, 101. 1832; *R. linearifolia* Brem., 5. 1957; *R. chiengmaiensis* Brem., 84. 1961; *R. cradengensis* Brem., 85. 1961; *R. elegans* Brem., 85. 1961; *R. neglecta* Brem., 86. 1961.] Bremekamp (1957) described *Rostellularia linearifolia* from Laos as having linear, glabrous leaves; bract, bracteoles and calyx lobes ciliate on midnerve and margins, and calyx 5-lobed. The general statement (Bremekamp, 1948) on the variability and distribution of *Rostellularia* species is repeated verbatim. Later Bremekamp (1961,1965,1969) published four species from Thailand, *R. chiengmaiensis, R. cradengensis, R. elegans,* and *R. neglecta,* none of which were compared to *R. procumbens,* which Bremekamp confined to India as evidenced by his protologue to *R. neglecta.* The differences between them are certainly small, mostly involving number of calyx lobes, *i.e.* 4-5; pubescence of bracts, especially presence or absence of capitate hairs; and size and shape of the leaves. However, 5-lobed calyces occur in all entities as do capitate hairs on bracts and bracteoles. Even the type specimen of *R. chiengmaiensis* has 5-lobed calyces and bracts with capitate hairs. This leaves only leaf shape to distinguish the various microspecies, but in a large sample of material leaf variation becomes continuous. My conclusion is that all these entities are local forms of the widespread *Justicia procumbens.* The number of more or less well distinguishable forms is endless, and I can see no reason to name them. The usual form with elliptic to elliptic-lanceolate leaves occurs in forests. Narrow-leaved forms occur in streambeds as rheophytes and in grasslands.

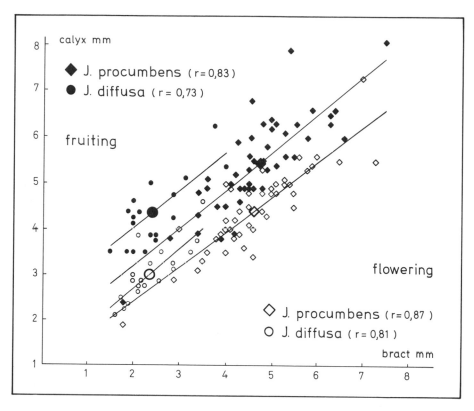

Figure 1. Linear regression graphs and scatter diagram of the bract-calyx length relationship in flowering and fruiting stages of *Justicia diffusa* and *J. procumbens*. The four large symbols denote calculated mean values. Correlation coefficient, r. Differences in calyx length between flowering and fruiting stages are significant at p=0.001, differences in bract length are not significant even at p=0.1. Differences between corresponding stages of *J. diffusa* and *J. procumbens* are significant at p=0.001. The bract/calyx indices in corresponding stages of *J. diffusa* and *J. procumbens* are significantly different at p=0.001.

Justicia diffusa Willd., 87. 1797. [Syn.: *Rostellularia diffusa* (Willd.) Nees, 100. 1832; *R. ramosissima* Brem., 87. 1965; *R. bankaoensis* Brem., 222. 1965; *R. palustris* Brem., 223. 1961; *R. rachaburensis* Brem., 84. 1969.] Bremekamp (1961) described *Rostellularia ramosissima* and found that it differed from *R. diffusa* by absence of cystoliths on the leaf margins, narrower and longer bracts, and longer calyx lobes. In a note appended to the protologue he also stated that *R. diffusa* seemed to be confined to India. Bremekamp (1965) published *R. bankaoensis* and *R.*

palustris, both with bracts considerably shorter than the calyx and with capitate hairs present although not mentioned in protologues. Finally, Bremekamp (1969) published *R. rachaburensis* which he distinguished from *R. diffusa* by its erect habit and larger leaves. In the same paper, he identified other specimens from the type locality of *R. rachaburensis* as *R. diffusa*, thereby admitting that *R. diffusa* does occur outside India. The last four new species are distinguished from each other and from *R. diffusa* by a few, very variable characters. Three of them are known from the type only, belying the claim that the differences are constant. In my opinion, they clearly belong to the widespread, variable species *Justicia diffusa*, which also has many local forms.

Justicia quinqueangularis Koenig ex Roxb., [Hort. Beng. 86. 1814] 134. 1820. [Syn.: *Rostellularia quinqueangularis* (Koenig ex Roxb.) Nees 101. 1832.] While *Justicia procumbens* and *J. diffusa* are widespread and closely related, a third species in the section *Rostellaria* also resides in our region, namely *J. quinqueangularis* Roxb. A rather common species in India, it has been collected only a few times in northern Thailand, always in moist places near water. It is clearly distinguished from the other two species by being entirely glabrous and by having lanceolate, obtuse bracts.

Justicia sect. *Rhaphidospora* (Nees) T. Anders.

Nees (1832) proposed the genus *Rhaphidospora* [Gr.: "spiny seeds"] to contain *Justicia glabra* Koenig ex Roxb., the only distinguishing character of the genus being its spiny seeds. In all other respects it is a perfectly typical *Justicia* of which Anderson (1867) treated it as a section. Unfortunately, he also merged into this section a number of species which do not belong such as *J. decussata* Roxb., *J. wynaadensis* (Nees) T. Anders., *J. vasculosa* (Nees) T. Anders., *J. collina* T. Anders., *J. dichotoma* Bl., and *J. edgeworthii* T. Anders.

Justicia scandens Vahl, 15. 1791. [Syn.: *J. glabra* Roxb., 132. 1820; *Rhaphidospora glabra* (Roxb.) Nees, 115. 1832.] *Justicia scandens* Vahl and *J. glabra* Roxb. are based on the same collection of Koenig from Malabar. Although the latter name is used in a large number of floras it must go into synonymy.

Justicia kampotiana R. Ben., 118. 1936. [Syn.: *Rhaphidospora lanceolata* Brem., 218. 1965; *Mananthes prachinburensis* Brem., 82. 1969.] Although the protologue does not mention it, the type specimen of *J. kampotiana* has seeds with spiny testa for which reason it clearly belongs in this section. Bremekamp (1965, 1969) published *Rhaphidospora lanceolata* and *Mananthes prachinburensis* from Thailand. A thorough analysis of the type material of these two entities clearly shows that they belong to *Justicia kampotiana*. Bremekamp (1948) created *Mananthes* as a segregate from *Justicia*, characterized by spiciform, sparsely branched but rather elongated inflorescences with small bracts and with rugulose seed coats. To *Mananthes* he also referred *Justicia sumatrana* (Miq.) Kurz, *J. flaccida* Kurz, *J. glomerulata* R. Ben., *J. vasculosa* (Nees) T. Anders., and *J. patentiflora* Hemsl., none of which are closely related to *J. kampotiana*. A total of 19 specimens available for the present study, come from an area between Khao Yai mountain in Thailand to Kampot in Cambodia, a range of about 350 km x 100 km (Fig. 2A). The plants are usually rheophytic with relatively short internodes, long, narrow leaves, and grow in streambeds in evergreen forest.

Justicia sect. *Adhatoda* (Nees) Lindau

Nees (1832) erected *Adhatoda* as one of several *Justicia* segregates. Like most of his segregates it is not well separated from *Justicia* (Clarke, 1884-1885; Burkill and Clarke, 1900). The stylar furrow in the upper corolla lip and the calcarate thecae, particularly the lower one, are present.

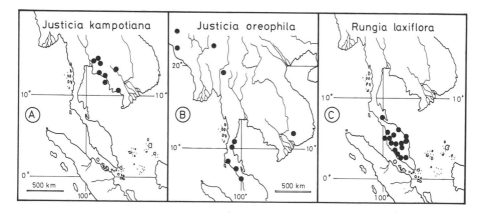

Figure 2. Distribution of A - *Justicia kampotiana*, B - *J. oreophila*, and C - *Rungia laxiflora* in the Indo-Chinese Peninsula.

Justicia adhatoda L., 15. 1753. [Syn.: *Adhatoda vasica* Nees, 103. 1832.]
In *J. adhatoda* the corolla tube usually is shorter than the lips, the peduncle
slightly longer than the inflorescences, and both bracts at a node are
fertile. The bracts are elliptic, imbricate, and palmately nerved. It is
distributed from India to Java, commonly cultivated as a hedge and
medicinal plant, and occasionally naturalized in evergreen forests.

Justicia oreophila C. B. Clarke, 526. 1884-1885. [Syn.: *Adhatoda
oreophila* (C. B. Clarke) C. E. C. Fischer, 575. 1935; *Justicia ventricosa*
var. *major* R. Ben., 120. 1936; *J. hirticarpa* Imlay, 456. 1938 without latin
description; *J. trichocarpa* Imlay, 140. 1939; *Rhyticalymma trichocarpa*
(Imlay) Brem., 83. 1961.] Most of the species treated by Nees under
Adhatoda are not closely related to *Justicia adhatoda*. I doubtfully refer
to this section *J. oreophila* C. B. Clarke (1884-1885), described from
Chittagong in India, which is strikingly similar to *J. adhatoda* in corolla
and inflorescence morphology. Fischer (1935) included it in *Adhatoda*.
The inflorescence is longer than in *J. adhatoda* and usually also longer
than the peduncle. In *J. adhatoda,* the bract pubescence is appressed while
in *J. oreophila* it is erect and has many whitish, sessile, globular glands.
The corolla is 25-36 mm long in *J. adhatoda* and 14-20 mm long in *J.
oreophila*. Imlay's (1939) *J. trichocarpa* from Peninsular Thailand is
strikingly similar but has smaller bracts than the typical northern form.
Imlay referred correctly *J. ventricosa* var. *major* R. Ben. (1936) from
Vietnam to the synonymy of *J. trichocarpa*. The floral measurements of
the Vietnamese collection are intermediate between northern and
southern populations, and I shall here consider all material to belong to
one widespread, variable species (Fig. 2B).

Rungia

Although *Rungia laxiflora* seems to be common and widely distributed in
the Malay Peninsula from where it was described by Clarke (1908), the
name is rarely used. Material of this species is usually identified with
various *Justicia* names published by Ridley and even by Clarke himself.
To clear up this situation a full synonymy is given below.

Rungia laxiflora C. B. Clarke, 698. 1908. [Syn.: *Justicia clarkeana*
Ridl., 599. 1923, Type: *Kunstler 3676* (K lectotype); *J. uber* C. B. Clarke,
688. 1908, Syntypes: *Scortechini 96, 319* (K), *Wray 386,* King's collector
10245 (K), *Ridley 8216, 9773; J. bracteata* Ridl., 56. 1909, Syntypes:
Ridley 13670 (K), *13672a,* (K); *J. secundiflora* Ridl., 86. 1911, Type:
Ridley 14530 (K); *J. pectinella* Ridl., 34. 1912, Syntypes: *Ridley 2185
(n.v.), 15738* (K); *J. breviflos* Ridl., 187. 1920, = *J. flaccida* Ridl., 85.

1911 non Kurz 1870, Type: *Ridley 1452* (K); *J. microcarpa* Ridl., 306. 1922, Type: *Ridley s.n.* (K).] Evidently Clarke (1908) did not make a transfer of *Justicia laxiflora* Bl., now *Andrographis laxiflora* (Bl.) Lindau, when he established *Rungia laxiflora*, even if he mentioned *J. laxiflora* Bl. with a question mark at the end of the description. The proof of this is in protologue: "*Justicia laxiflora* Blume, has the filament *antice barbata*, and could hardly be this. It was Zollinger, n. 1539 fide Moritzi Verz. Zoll. Pfl. 47, which I have not got. But I believe I got the name *Rungia laxiflora* from a Java plant marked *Justicia laxiflora*, which passed through my hands." To Clarke, *J. laxiflora* is not the same as *R. laxiflora* and he got the epithet from the label of another specimen than Blume's type. Of course he knew the taxon *A. laxiflora* (Bl.) Lindau very well, having treated it some thirty pages earlier in the same paper. Technically, he described a new species of *Rungia* with the valid epithet *laxiflora*. Ridley (1923), in his transfer of the species to *J. clarkeana*, gives the three most important characters for a *Rungia*: inflorescence secund, lower thecae spurred, capsule with rising placentae. Clarke (1908) also published *J. uber* characterized by 1-sided spikes and spurred lower thecae, but he overlooked the rising placentae clearly represented in the type. I have choosen his *Rungia* epithet as the species name for this entity, including seven of Ridley's Malayan *Justicia* species as shown in the synonymy. The structure of the inflorescences and the calyces and corollas are quite uniform within this entity. There is considerable variation in the shape and size of bracts and bracteoles (Fig. 3). However, when all available specimens are considered together, this variation seems to be continuous and cannot support specific distinction. So far, *R. laxiflora* is only reported from the Malay Peninsula and the adjacent peninsular part of Thailand (Fig. 2C).

Figure 3. Variation in floral details of various collections of *Rungia laxiflora* with corollas opened and cuts shown as broken lines. **a** - sterile bracts, **b** - fertile bracts, **c** - bracteoles, all x 6.5; **d** - corollas, **e** - calyces, both x 2.5; **1f** - anther, x 6.5; **5 f** - anther, x 10; **5 g** - top of inflorescence, x 2.5; **5 h** - one valve of opened capsule, x 2.5.- [1 - *Ridley s.n.* (*J. microcarpa*-type); 2 - *Ridley 14530* (*J. secundiflora*-type); 3 - *Ridley 13670* (*J. bracteata*-type); 4 - *Ridley 15738* (*J. pectinella*-type); 5 - *King's collector 10245* (*J. uber*-type).]

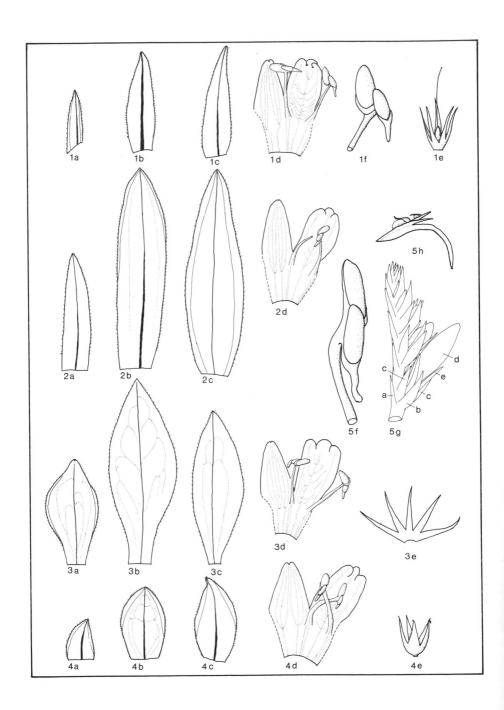

Conclusions

These examples could be supplemented by numerous other cases from the Acanthaceae, and I have no reason to doubt that the situation is much different in other tropical plant groups. Prior to the writing of flora treatments, there is a tenuous job of critical taxonomic revision. Particularly careful study of variation in the taxa involved and a comparison of their distribution areas must be emphasized. If this work is combined with a detailed study of all relevant type material, the need for taxonomic reductions will become clear to the taxonomic worker. Omission of one or more of these three steps in the past caused most of the problems demonstrated in the selected examples in the present paper.

Literature cited

Anderson, T. (1867). "An enumeration of the Indian species of Acanthaceae." *J. Linn. Soc. Bot.* **9**, 425-526.

Benoist, R. (1935). "Acanthaceae." pp. 610-772 *In* Lecomte, M. H. (ed.), Flore Générale de l'Indo-Chine **4.**

Benoist, R. (1936). "Acanthacées nouvelles d'Indo-Chine." *Notulae Systematicae* **5**, 106-131.

Bremekamp, C. E. B. (1948). "Notes on the Acanthaceae of Java." *Verh. Kon. Ned. Akad. Wetensch., Afd. Natuurk., Tweede Sect.* **45 (2),** 1-78.

Bremekamp, C. E. B. (1957). "Some new Acanthaceae and Rubiaceae from Laos (Indo- China)." *Proc. Kon. Ned. Akad. Wetensch. Ser. C,* **60 (1),** 1-8.

Bremekamp, C. E. B. (1961-1965-1969). "Scrophulariaceae Nelsonieae, Thunbergiaceae, Acanthaceae." *Dansk Bot. Ark.* **20,** 55-88, **23,** 195-224, **27,** 71-85.

Bremekamp, C. E. B. (1966). "Thunbergiaceae and Acanthaceae." *Dansk Bot. Ark.* **23,** 273-279.

Burkill, I. H. and Clarke, C. B. (1900). "Acanthaceae." Flora of Tropical Africa **5,** 1-262.

Clarke, C. B. (1884-1885). "Acanthaceae." pp. 387-558 *In* Hooker, J. D. (ed.), Flora of British India **4.**

Clarke, C. B. (1908). "Acanthaceae." Flora of the Malayan Peninsula. *J. Asiat. Soc. Bengal, Pt.* **2,** *Nat. Hist.* **74.** Extra Number, 598-628.

Fischer, C. E. C. (1935). "Contributions to the flora of Burma." *Kew Bull.* 572-576.

Imlay, J. B. (1938). "The Taxonomy of the Siamese Acanthaceae. *Unpubl. thesis, Univ. of Aberdeen.*

Imlay, J. B. (1939). "Contributions to the flora of Siam. Additamentum 51. New and re-named Siamese Acanthaceae." *Kew Bull.* 109-150.

Linnaeus, C. (1753). "Species plantarum." *Stockholm.*

Nees, C. G. (1832). "Acanthaceae Indiae Orientalis." pp. 70-117 *In* Wallich N. (ed.), Plantae Asiaticae Rariores **3,** *London.*

Ridley, H. N. (1909). "The Flora of the Telom and Padang valleys." *Fed. Malay States Mus.* **4,** 56.

Ridley, H. N. (1911). "A scientific expedition to Temengoh, upper Perak." *J. Straits Branch Roy. Asiat. Soc.* **57,** 86.

Ridley, H. N. (1912). "New and rare Malayan plants. Series VI." *J. Straits Branch Roy. Asiat. Soc.* **61,** 34.

Ridley, H. N. (1920). "New and rare species of Malayan plants. Series XI." *J. Straits Branch Roy. Asiat. Soc.* **82,** 187.

Ridley, H. N. (1922). "New and rare Malayan plants. Series XII." *J. Straits Branch Roy. Asiat. Soc.* **86,** 306.

Ridley, H. N. (1923). "Acanthaceae." Flora of the Malay Peninsula **2,** 554-610.

Roxburgh, W. (1820). "Acanthaceae." *In* Flora Indica. (*Ed. 1*) **1,** 110-136.

Vahl, M. (1791). "Symbolae botanicae." **2,** 1-108.

Willdenow, C. L. (1797) "Species plantarum." **1(1),** 1-495. *Berlin.*

Speciation of Alismatidae in the Neotropics

R. R. HAYNES AND L. B. HOLM-NIELSEN

University of Alabama, USA and Aarhus University, Denmark

The Alismatidae is a subclass of primitive monocots that occurs predominantly in aquatic or marsh habitats, if Triuridales is not included in the subclass (Haynes and Holm-Nielsen, 1985). It is composed of three orders, the Alismatidales, consisting of Alismataceae, Limnocharitaceae, and Butomaceae, the Potamogetonales, including Potamogetonaceae, Najadaceae, Juncaginaceae, Zannichelliaceae, Cymodocacaceae, and Zosteraceae, and the Hydrocharitales, with only Hydrocharitaceae. Twenty-two genera of these ten families are represented in the neotropics, four of which, *Hydrocleys, Limnocharis, Egeria,* and *Echinodorus* only occur here. Members of the subclass are characterized by having helobial endosperm development, all of which is utilized prior to seed maturation, mostly separate carpels, and trinucleate pollen (Dahlgren *et al.,* 1985). The Alismatidae has often been considered to be an ancestral group for the monocots. Clearly, members of the subclass are most similar to the Magnoliidae, but, whether the subclass is a basal group for the monocots is suspect (Cronquist, 1981; Dahlgren *et al.,* 1985). The subclass is widely dispersed on all continents. The collective distribution patterns of all neotropical Alismatidae present a rather surprising point. That is, the subclass that is almost entirely aquatic or amphibious and has several genera with distributions predominantly in South America, is almost completely absent from the Amazon basin, the region with extensive fresh water habitats and with the greatest diversity of fresh water vertebrates (Haynes and Holm-Nielsen, 1986b; Junk, this volume).

Distribution patterns

A number of aquatic plant species are widely distributed throughout the world, for example, *Ceratophyllum demersum* and *Eichhornia crassipes.* Alismatidae, similarly, have widely distributed species, as well as many that are endemic to very limited areas. Distribution patterns of neotropical species can be grouped into seven categories. These include: (i) north temperate species that extend into the tropics, as illustrated by *Sagittaria latifolia,* which is widely distributed from southern Canada to the northern coast of the Gulf of Mexico and south through Central America into Ecuador; (ii) species with distribution centers in Central America and the Caribbean Islands, as illustrated by *Sagittaria intermedia* and *Echinodorus nymphaeifolius;* (iii) bicentric species with centers in northern South America and on the Brazilean shield, for example, *Sagittaria planitiana* and *Echinodorus trialatus;* (iv) species with centers in southern South America, as illustrated by *Potamogeton gayii;* (v) south temperate species that extend into the tropics, for example, *Potamogeton ferrugineus;* (vi) species restricted to the Andes, such as *Potamogeton paramoanus* and *Zannichellia andina;* and (vii) species of wide-ranging distribution, such as *Najas guadalupensis.*

Origin and dispersal

An analysis of the distribution patterns of neotropical Alismatidae gives us an understanding of the possible patterns of speciation in some or all of the genera. Two such examples, *Hydrocleys* and *Echinodorus* sect. *Longipetali,* are illustrated below. *Hydrocleys* presently is distributed from Nicaragua south to Colombia and Ecuador then, north of the Amazon drainage, east to the Atlantic coast. The genus does not occur in the central Andes, but is disjunct to eastern Bolivia, from where it is widely distributed to the Atlantic coast in Uruguay to central Brazil (Fig. 1). When this distribution is compared to current vegetational types of South America, surprisingly, most taxa of the genus occur in the seasonal forests zone, mostly in temporary pools but a few in slow-moving rivers. *Hydrocleys* consists of five species, differing in size of flowers, whether the petals are longer than sepals, as long as sepals, or shorter than sepals, number of staminodia, number of stamens, abundance of glandular trichomes on the seed, and presence or absence of midvein on the sepals (Holm-Nielsen and Haynes, 1985). The only other genus of Limnocharitaceae in the Neotropics is *Limnocharis,* (Hooper and Symoens, 1982). *Limnocharis,* provisionaly considered an out-group to *Hydrocleys,* has large flowers with petals longer than sepals, many stamens and staminodia, no glandular trichomes on the seed, and no midvein on the sepals. We can then assume that the character states in *Limnocharis* are primitive based upon out-group comparison as illustrated by Dahlgren and Rasmussen (1983). When *Hydrocleys* is examined with cladistics, a cladogram as illustrated in Figure 2 develops.

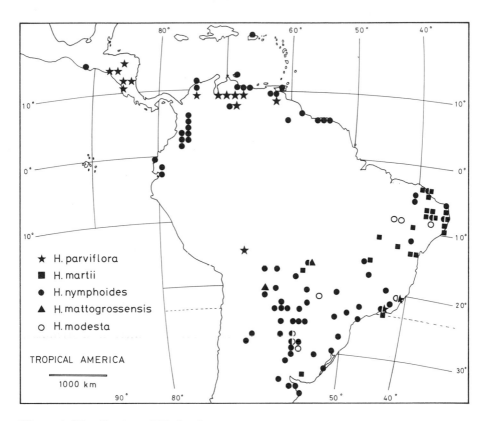

Figure 1. Distribution of *Hydrocleys*.

The first bifurcation, resulting from the development of a midvein in the sepals, produces two natural groups or clades. The more derived group, which we will call group A, consisting of *H. parviflora* and *H. martii*, with a sepal midvein, and the more primitive group, which we will call group B, containing the other three species, without a sepal midvein. The final bifurcation for group A in the cladogram is the reduction in flower size, producing what we presently know as the two species, *H. parviflora* and *H. martii*. In group B, *Hydrocleys nymphoides* separates from the remaining two, *H. modesta* and *H. mattogrossensis* at the first branch by retaining the large flower, whereas the other two species have a reduction in flower size. These two separate by reduction in stamen and staminodia number.

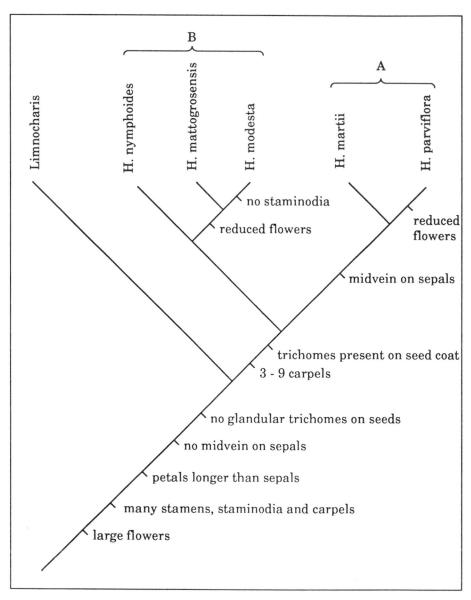

Figure 2. Cladogram of *Hydrocleys*.

The geographical region with the greatest concentration of primitive character states of *Hydrocleys* is east-central Brazil, particularly the states of Pernambuco and Paraiba.   This area is included within the distribution   of   *H . nymphoides, H . martii*, and *H . modesta*. We propose that *Hydrocleys* originated in the above area of present day Brazil and migrated southwest and then north to Central America along the belt of seasonal forests on the east side of the Andes (Fig. 3).   These areas of seasonally   dry   to   seasonal   forests   were   exceedingly   more   common (Hammen, 1979) during the Pleistocene glaciations.   Primary speciation most likely occurred early in the history of the genus, presumably in the above mentioned area, producing groups A and B.   The hypothetical ancestors (of group A) probably inhabited temporary ponds of seasonally dry forests of central Brazil, from which a speciation event resulted in two species *H . martii* and *H . parviflora*.   Group B similarly underwent speciation in the present area of central Brazil, producing three species. *Hydrocleys parviflora* and *H . nymphoides* were the only two species of the genus to become widely distributed.   They migrated north along the eastern slope of the developing Andes to their present positions of northern South   America   and   Central   America.   Intermediate   populations presumably disappeared during the continued uplift of the Andes.

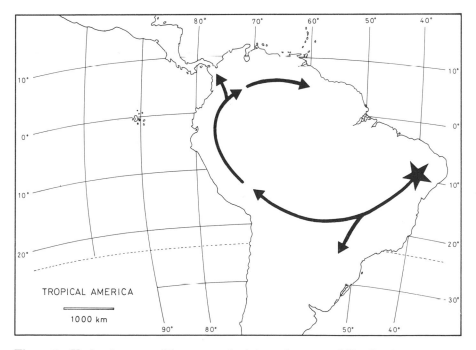

Figure 3. *Hydrocleys,* possible center of origin and routes of distribution.

Allthough the genus *Echinodorus* occur in Amazonia (Haynes and Holm-Nielsen, 1985; Junk and Furch, 1985), the section *Longipetali* is absent there and has probably had a similar history as shown for *Hydrocleys*. Cladistic data aviable to support such an hypothesis are limited. This section has a distribution pattern quite similar to that of *Hydrocleys* (Fig. 4). Individuals of section *Longipetali* are characterized by pellucid markings in a network in the leaf tissue independent of the venation and by sepals that are appressed around the flower and mature fruit. Both of these character states are considered derived, because no other *Echinodorus* has such sepals and only *E. glandulosus* has such a pellucid network. Our belief that the ancestor of this section also inhabited temporary ponds and flooded areas in the seasonally dry forests of east-central Brazil is supported by the present distribution of *Echinodorus glandulosus*, which is known only from the Brazilian states of Ceara and Pernambuco.

The ancestor to *E. glandulosus* and to section *Longipetali* most likely originated on the Brazilian shield by speciation, resulting in a species with a reticulate pellucid pattern in the leaves. A later derivation, production of appressed sepals, separated an ancestor of the section *Longipetali* from *E. glandulosus* or its ancestor.

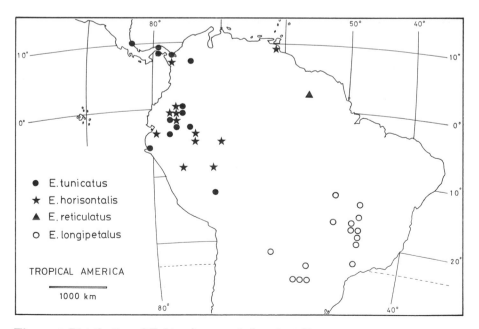

Figure 4. Distribution of *Echinodorus* sect. *Longipetali*.

Probably the ancestoral species of section *Longipetali* was most similar to *E. longipetalus*, because that is the only species of the group presently occuping the Brazilian shield. The ancestoral species became apparently widely distributed throughout the seasonally dry or seasonal forests of South America by migrating south and then west and north along the east slope of the developing Andes. These expansions of distribution occurred during times in which the dry and seasonal forests probably inhabited lower elevations than at present and probably continued until populations of plants from this section occupied areas of the present distribution. Changes in environmental conditions resulted in moister forests spreading to higher elevations, effectively separating the section into three isolated populations. One population became adapted to more humid but still seasonal conditions and eventually gave rise to two forest floor dwelling species, *E. horizontalis* and *E. tunicatus*. The northeastern population and southern population continued to exist in seasonally dry forests, the northern consisting of *E. reticulatus* and the southern of *E. longipetalus*. Such a theory is supported by comparing the present distributions with current ecological conditions. *Echinodorus longipetalus* occurs only on the forest floor of seasonally dry, central Brazil. Likewise, *E. reticulatus* is known from only one locality, this being a seasonally dry forest in Surinam (Haynes and Holm-Nielsen, 1986a). The other two species, *E. horizontalis* and *E. tunicatus*, occur on the forest floor of seasonal forests or of seasonal flooded flood plain forests at the headwaters of the Amazon River (Balslev *et al.*, 1987). Hammen (1979) indicates the major expression of the Andes uplift has occurred within the last five million years. He also notes that within the past million years, there have been periods of high rainfall intermixed with periods of lower rainfall. During these latter periods the grasslands and the forest of the high Andes occupied much lower elevations (Colinvaux, 1987), and the drier climatic conditions most likely influenced the forests too, resulting in seasonal forests in which the forest floor completely dried. These forests with seasonal ponds and flooding were habitats through which we propose that many Alismatidae migrated.

Conclusions

As one surveys the flora of South America, it becomes increasingly evident that few species of Alismatidae occur in the Amazon basin or even the headwaters of the river. A comparison of distributions for different groups in the subclass in relationship to vegetational types, as we have done, illustrates that most taxa are restricted to areas with wet and dry seasons and, in fact, to ephemeral ponds or temporarily flooded forest floors within these areas. We believe that many members of the

Alismatidae actually require two seasons, a dry period and a prolonged period of high water level, the absence of such seasonality is probably the reason for the absence of such plants in much of the Amazon basin. The speciation and distribution of the Alismatidae provide an unexpected example of how aquatic plants that were common to dry and seasonal forest areas during the Pleistocene climatic oscillations have adapted to the forest floor of the tropical rain forest. Hence, the case of Alismatidae provides further evidence for the origin of tropical forest diversity.

Literature cited

Balslev, H., Luteyn, J., Øllgaard, B., and Holm-Nielsen, L. B. (1987). "Composition and structure of adjacent unflooded and floodplain forest in Amazonian Ecuador." *Opera Bot.* **92,** 37-57.

Cronquist, A. (1981). "An integrated system of classification of flowering plants." *Columbia Univ. Press, New York.*

Colinvaux, P. (1987). "Amazon diversity in light of the paleoecological record." *Quat. Sci. Rev.* **6,** 93-114.

Dahlgren, R. and Rasmussen, F. N. (1983). "Monocotyledon evolution: Characters and phylogenetic estimation." *Evolutionary Biol.* **16,** 255-395.

Dahlgren, R. M. T., Clifford, H. T., and Yeo, P. F. (1985). "The families of monocotyledones." *Springer-Verlag, Berlin, New York.*

Hammen, T. van der. (1979). "History of flora, vegetation and climate in the Colombian Cordillera Oriental during the last five million years." pp. 25-32 *In* Larsen, K. and Holm-Nielsen, L. B. (eds.), *Tropical botany. Acad. Press, London.*

Haynes, R. R. and Holm-Nielsen, L. B. (1985). "A generic treatment of Alismatidae in the Neotropics with special reference to Brazil." *Acta Amaz.* **15** *(supl),* 153-193.

Haynes, R. R. and Holm-Nielsen, L. B. (1986a). "Notes on Echinodorus. (Alismataceae)." *Brittonia* **38,** 325-332.

Haynes, R. R. and Holm-Nielsen, L. B. (1986b). "The absence of aquatic plants along the Amazon River." (Abstract) p. 14 *In Symposium tropical botany: practice and principles. Posters and poster abstract, Rijksuniversitet, Utrecht.*

Holm-Nielsen, L. B. and Haynes, R. R. (1985). "The identity of Limnocharis mattogrossensis Kuntze (Limnocharitaceae) and its allies." *Phytologia* **57,** 421-425.

Hooper, S. S. and Symoens, J. J. (1982). "Observations on the family Limnocharitaceae Takhtajan ex Hooper and Symoens." pp. 50-60 *In* Symoens, J. J., Hooper, S. S., and Compére, P. (eds.), "Studies on aquatic vascular plants." *Royal Bot. Soc. Belgium, Brussels.*

Junk, W. J. and Furch, K. (1985). "The physical and chemical properties of Amazonian waters; aquatlic plants and animals." pp. 3-17 *In* Prance G. T. and Lovejoy T. E. (eds.), "Key Environments Amazonia." *Pergamon Press, Oxford.*

Speciation patterns in African Loranthaceae

R. M. POLHILL

Royal Botanic Gardens, Kew, England, UK

There are some 230 species of Loranthaceae in Africa, attributed to 20 genera at the present time. They grow as shrubby parasites attached to the branches of trees or shrubs and are widespread over the wooded parts of Africa. About half the species occur in forests and the number of species is concentrated somewhat in eastern Africa. Their peculiar life-history attracts considerable interest, but the biology is complex and will take many years to unravel. Nonetheless, a few studies in recent years do provide some insight into the general patterns of speciation that may be helpful to a fuller understanding of the African flora.

Dispersal

The Loranthaceae of Africa, like virtually all members of the family, depend on birds for their dispersal. In Africa the main agents are tinkerbirds, *Pogoniulus* (Capitonidae). The fruits are berry-like, normally between 5 and 15 mm long, with a thin but tough pericarp usually ripening yellow to red, or rarely blue. The skin is peeled to reveal the nutritious mesocarp that surrounds the seed and is attractively yellow, red, or white. The inner layer of the pulpy mesocarp is highly viscid and the tinkerbird is induced to regurgitate the seed shortly after ingestion, wiping it from the bill on to a branch where it readily adheres. In Asia and Australia, the principal agents are flower peckers (Dicaeidae) which often have the gut modified to allow ready passage of seeds (Docters van Leeuwen, 1954; Liddy, 1983) but in Africa, the seeds appear to be defecated only by some opportunist foragers, such as mousebirds (Godschalk, 1983).

The discoidal modified radicle is appressed to the substrate forming a holdfast from which the haustorium attempts to penetrate the host and form a connection with the host xylem. The mortality rate is high and, once established, the seedling generally takes a number of years to mature. Godschalk (1983) estimates that the range of dissemination is generally small, and that occasional long distance dispersal is probably effected by opportunist feeders that retain the seed in the gut for some time.

TROPICAL FORESTS
ISBN 0–12–353550–6

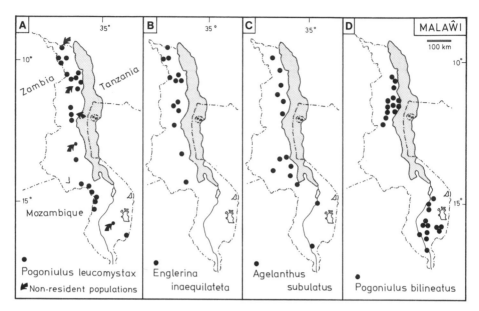

Figure 1. A - Distribution of *Pogoniulus leucomystax*, green moustached tinkerbird, in Malawi (arrows indicate non-resident populations), B - distribution of *Englerina inaequilatera* in Malawi, C - distribution of *Tapinanthus (Agelanthus) subulatus* in Malawi, D - distribution of *Pogoniulus bilineatus*, golden-rumped tinkerbird, in Malawi.

Dowsett-Lemaire (1982, 1988) found that on the Nyika Plateau of Malawi, mistletoe fruits (Loranthaceae and Viscaceae) formed a major part of the tinkerbirds' diet, supplemented by insects and small fruits of a few other plants. In areas of forest with several species of mistletoe fruiting successively through the year, the birds are resident. In patches of forest with only a single mistletoe, *Englerina inaequilatera*, pairs moved in only temporarily to breed in the fruiting season. The upper altitudinal limit of tinkerbirds and mistletoes was coincident, and no other birds seemed to eat mistletoes.

Figure 1A-B shows the range of *Englerina inaequilatera*, which is common at montane forest edges, and *Pogoniulus leucomystax*, the green moustached tinkerbird, in Malawi. Their southerly limits are almost coincident. The small population of tinkerbirds on Mt. Mulanje, east of the Rift Valley, was found only in 1983, high up where several mistletoes grow together (Dowsett-Lemaire, pers. comm.). Figure 1C shows the records in Malawi for *Tapinanthus (Agelanthus) subulatus*, a species of the Zambezian woodlands that extends into the edges of montane forests. It is utilized by the golden-rumped tinkerbird, *Pogoniulus bilineatus*, which feeds on a broad spectrum of fruit (Dowsett-Lemaire, pers. comm.) but does follow the mistletoe from the lowlands into the forests of the Viphya Plateau in the Northern Province of Malawi (Fig. 1D).

African Loranthaceae in general show little host specificity. This is particularly evident in the forests where mistletoes occur in the canopy and often more abundantly at the edges. In drier areas, the diversity of suitable hosts declines sharply and genera such as *Acacia*, *Combretum*, or *Commiphora* often become the usual host. A few species regularly parasitise other mistletoes, both Viscaceae and more usually other Loranthaceae, but in Africa these epiparasitic associations are rarely obligate (Wiens and Calvin, 1987).

Pollination

Apart from some species of *Helixanthera*, a relatively primitive Afro-Asian genus, the Loranthaceae of Africa all have flowers adapted for bird

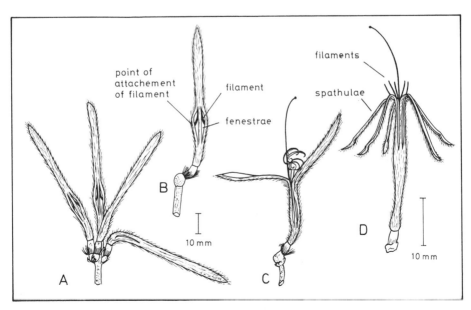

Figure 2. Pollination mechanism of *Erianthemum*. A-C *Erianthemum taborense*. A and B - Flowers just before dehiscence. C - Flower after dehiscence. D. - *Erianthemum dregei* following dehiscence. Note reflexed *spathulae* and broken ends of filaments (after Feehan, 1983).

pollination.   Only the larger sunbirds (Nectariniidae) can manipulate the
more specialized flowers.   Smaller sunbirds and other passerines, such as
white-eyes (Zosteropidae), are secondary foragers and can open simpler
flowers (Gill and Wolf, 1975; Kirkup, pers. comm.).
     Pollen is released explosively in most genera either by probing
vents (*fenestrae*) or by tapping the head of the mature bud to expose
subsidiary vents inside.   The sorts of mechanisms involved are described
by Feehan (1985).   In *Erianthemum* (Fig. 2), vents appear halfway up
mature buds marked by color signals, and abundant nectar wells up the
corolla-tube.   After imbibing nectar, the withdrawing beak tears between
the upper connivent petal-lobes (*spathulae*), which are held under tension;
the flower explodes, the spathulae reflexing and the stamens coiling
inwards with such force as to break the filaments and shower the bird's
head with pollen.   At the same time, the style inflexes to meet the
pollinator's head.

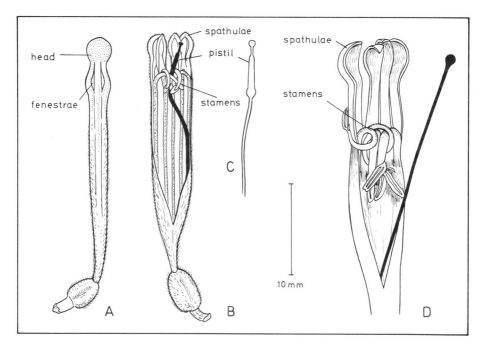

Figure 3. Pollination mechanisms of *Phragmanthera* and *Tapinanthus*
(*Agelanthus*). A-C *Phragmanthera usuiensis* (*Tapinanthus rufescens* auct.).
A - Mature bud. B - Flower after dehiscence. C - Pistil. D - *Tapinanthus
subulatus*, Flower after dehiscence (after Feehan, 1983).

In a more sophisticated genus of the same alliance, *Phragmanthera* (Fig. 3A-C), the vents are shorter and stiffer to open, the close-up signals are more refined, and the nectar reward is increased for a shorter period.

When the flower is opened by a downward V-split of the tube, the aim of the smaller pollen deposition and the stylar inflexion are more precise. Including changes of growth patterns and inflorescences, Kirkup (pers. comm.) estimates that the overall effect of such modifications can reduce the pollen production per comparable-sized plant by a factor of about 100.

Similar specialized vent mechanisms have developed in other groups of species that at times have all been included in a large genus *Tapinanthus*. Figure 3D represents *Tapinanthus subulatus*, which should perhaps be referred to a separate genus *Agelanthus*. Irrespective of that segregation, *Agelanthus* and *Phragmanthera* appear to owe their similarity to a remarkable degree of convergence in two quite separate lines of evolution (see next section).

The pollination mechanism in *Globimetula* (Fig. 4A-C) is quite different. Here, the distinct head of the mature flower-bud darkens and when pecked, the petals coil outwards (Fig. 4B) to reveal vents at the base of the staminal column, probing of which causes the stamens to coil inwards explosively bending the style in the same direction. The species of *Tapinanthus* in a strict sense (Fig. 4D-F) also have a bud-head that darkens at maturity. Pecking causes a V-shaped rent and a precise deposition of pollen and inflexion of the style. The genus *Moquiniella*, from the Cape Region of South Africa, is quite unrelated but has a pollination mechanism similar to *Globimetula*, the tip of the bud darkening but not swollen. Examples from *Plicosepalus* and *Vanwykia* of less specialized pollination mechanisms adapted to sunbirds are given by Feehan (1985).

Whereas the dispersal strategy of African Loranthaceae is very constant, the floral diversity is spectacular with levels of specialization that are exceptional in the family. The variation provides the principal criteria for the classification into sections and genera. The divergence of ecogeographic races and species is often marked, among other features, by shifts in the signals given to visiting birds. It is noticeable that the modifications in flowers with swollen bud-heads tend to occur on that structure, with various ridges, wings and crowns, whereas those with vents show more variation related to color-banding, shape, and internal hardening of the corolla-lobes.

Although some work is in progress (Kirkup, pers. comm.), there is insufficient published data to make any generalizations about variation in breeding behavior. The pollination mechanisms and observations elsewhere are indicative of strategies for at least some outbreeding, but the development of colonies frequented by territorial sunbirds probably encourages considerable inbreeding. Populations with asynchrous flowering and different colored flowers are not uncommon.

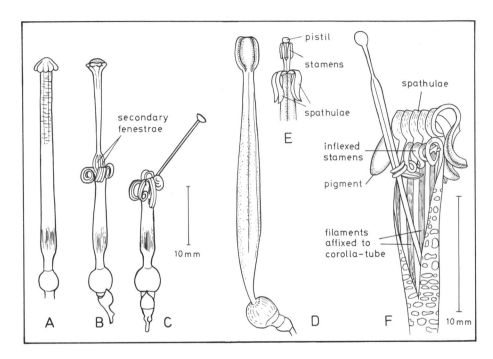

Figure. 4. Pollination mechanism of *Globimetula mweroensis* (A-C). A - Flower before dehiscence. B - Flower after first stage of dehiscence. Note internal column of stamens with secondary *fenestrae* caused by separation of filaments in middle of column. C - Flower after second stage of dehiscence. The stamens have reflexed over the petals. *Tapinanthus oleifolius* (*T. quinquangulus*) (D-E). D - Flower immediately before dehiscence; pigment zone in head stippled. E - Stage 1 in dehiscence. The *spathulae* have reflexed, and the stamens are still grouped around the central pistil. F - *Tapinanthus dependens* (*T. guttatus*). Flower after stage two in dehiscence (after Feehan, 1983).

Biogeography

The family is austral in origin, the most primitive relictual genera scattered in South America, New Zealand and Australia-Papuasia (Barlow, 1983). All the African genera derive from Asia, mostly endemic if the Arabian peninsula is included, but the relatively primitive genera *Helixanthera, Dendrophthoë* and *Taxillus* extend from Asia into Africa. Even in these instances the species form distinctive groups, and there is a case for segregating the two latter into separate genera. Only two close segregates of *Taxillus* extend to Madagascar (Balle, 1964).

Table 1. List of African genera of Loranthaceae, showing numbers of species. *Plicosepalus* (incl. *Tapinostemma*); *Oncocalyx* (incl. *Danserella, Odontella, Tieghemia*); *Dendrophthoë* (incl. *Botryoloranthus, Oedina*).

Tapinanthoid group		Taxilloid group	
Helixanthera	12	*Taxillus*	1
Plicosepalus	11	*Vanwykia*	2
Emelianthe	1	*Septulina*	2
Pedistylis	1	*Dendrophthoë*	3
Actinanthella	2	*Oncella*	4
Oncocalyx	13	*Erianthemum*	16
Spragueanella	2	*Phragmanthera*	33
Oliverella	2		
Berhautia	1	Subtotal	61
Moquiniella	1		
Englerina	22		
Agelanthus	58		
Tapinanthus	32		
Globimetula	13		
Subtotal	171	Total 232	

There seem to be two main groups of genera in Africa (Table 1). The Tapinanthoid group has simple or irregularly branched hairs and cupular bracts, whereas the Taxilloid group has hairs that are stellate and dendritic, *i.e.* with branches in whorls, and unilateral bracts. The basal genera of these two groups are regarded as closely related in Asia, all belonging to the subtribe Elytranthinae, and the characters which distinguish the African lines reoccur rather sporadically within the Asian genera. The characters which have been found useful to classify the Asian genera are mostly of little relevance in Africa where the selective pressures seem to have been significantly different.

The grouping of species still largely follows the system of sections adopted by Sprague (1910) under the omnibus genus *Loranthus*. These sections have been redistributed under a number of genera by subsequent workers, notably Danser (1933), Balle (1956), and Wiens and Tölken (1979). In our present revision of the family throughout Africa (Polhill and Wiens, ined.), some decisions still need to be made on the generic status of the most primitive groups in relation to their Asian relatives and whether or not to separate the fenestrate sections of *Tapinanthus* into *Agelanthus*. The circumscription of species, however, is sufficiently complete to make a general analysis.

About 70% of the species in Africa are confined to one of the main chorological divisions recognized by White (1983). They occur principally in the Guineo-Congolian forests, the Afromontane forests, the Zambezian woodlands, the Somali-Masai bushland, and in the mosaic along the east coast (Table 2, Fig. 5). A good proportion of the transgressors from one major region to another are still ecologically coherent, extending along

---

Table 2. Chorological regions of Africa and the number of species of Loranthaceae in each. See also Figure 5.

I	Guineo-Congolian	23	
II	Zambezian	31	
III	Sudanian	4	
IV	Somali-Masai	20	
VI	Karoo-Namib + XIV	12	
VIII	Afromontane	25	
X	Guinea-Congolia/Zambezia transition	5	
XI	Guinea-Congolia/Sudania transition	10	
XII	Lake Victoria regional mosaic	4	
XIII	Zanzibar-Inhambane regional mosaic	19	
XIV	Kalahari-Highveld transition - see VI		
XV	Tongaland-Pondoland regional mosaic	9	
		162	162
I + X or XI		15	
II + XIII or XV		13	
II + VIII		8	
III + IV, XI or XVI		6	
IV + VIII		2	
VI + XIV		8	
VIII + IV or XI		3	
XII + IV, XI or II		3	
XIII + IV or II in valleys		9	
XV + II or XIV		3	
		70	70
Total			232
Regional endemism			70%

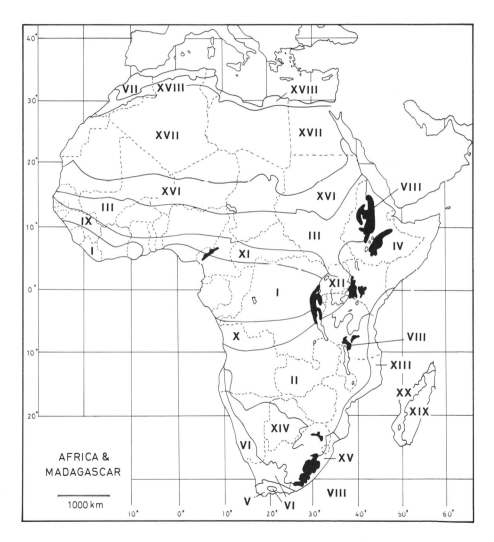

Figure 5. Main phytochoria of Africa and Madagascar (after White 1983). For explanation of relevant numbers see Table 2.

rivers or parasitizing hosts such as *Ficus* that are themselves transgressors.

The general patterns of distribution in tropical Africa can be illustrated by three genera. *Helixanthera* is the most primitive genus of any size in Africa, and its species are scattered round the edge of the continent (Fig. 6). Some of the populations are now very small. Some, such as the three forest species, *H. mannii, H. verruculosa* (ined.) and *H. woodii* (*spp.* 4-6 on Fig. 6), are very distinctive as a group, and similar to each other but widely separated geographically. *Helixanthera kirkii* (*sp.* 7) has populations on the east and west coasts, the largest disjunction within any single African species of Loranthaceae. Its close relatives (*spp.* 8-10) are also widely separated in dry bushland peripheral to the Zambezian woodlands. Only the two distinctive species from the Horn of Africa show any marked morphological adaptation to bird pollination, which is otherwise general on the continent. This sort of distribution pattern may be associated with initial speciation on the older pre-Miocene landforms of Africa (Axelrod and Raven, 1978).

More recent genera have patterns more strongly related to the equatorial forests or the eastern Rift Valleys. The latitudinal pattern may be illustrated by the genus *Globimetula* (Fig. 7). The most generalized species, *G. braunii* (*sp.* 1), is widely distributed in the Guineo-Congolian forests and is replaced by several closely related species in the Zambezian woodlands (*sp.* 3-6). There are then two other series; one in the forests of the Congolian Region (spp. 7-10), with two of the species extending into the peripheral montane forests, one in Cameroon, the other on the flanks of the Western Rift Valley. The last rather distinctive series (*spp.* 11-13) occurs in the Guinean and western Congolian forests, extending slightly into the forests of the transition zone towards the Sudanian Region.

The other characteristic pattern is well shown by *Erianthemum* (Fig. 8). Here the common species, *E. dregei* (*sp.* 1), is widely distributed along the flanks of the eastern Rift Valley from Ethiopia to the eastern Cape, extending to the east coast and westwards to southern Angola. It is notably polymorphic and common in a wide variety of habitats. Its closest relatives are *E. schelei* (*sp.* 2) in montane forests from eastern Tanzania to Malawi and *E. taborense* (*sp.* 3) in the *Brachystegia* woodlands of the Zambezian Region. *Erianthemum taborense* has often been treated as a variety of *E. dregei* but where the ranges approximate, it flowers in the dry season after *E. dregei* has finished, the slight but distinctive features are consistent, it is much less variable, and no obvious intermediates have been found. Species 4-6 are all morphologically isolated, with odd features not found elsewhere in the genus (some putatively primitive), occur outside the main range of the other species, and appear to be relics of early divergence in the genus. The remainder of the genus shows clear morphological modifications for drier habitats with the development of short shoots, few-flowered inflorescences, and enlarged bracts. The first series (*spp.* 8-12) are allopatrically distributed in the Zambezian woodlands, the second series, with further xeromorphic adaptations, in the deciduous bushlands of the Somali-Masai Region and coastal bushlands of East Africa.

Most of the other sizeable genera follow broadly similar patterns, though in the larger ones subsidiary patterns may be increasingly superimposed. In the Tapinanthoid group (Table 1) *Plicosepalus, Emelianthe,* and *Pedistylis* are all eastern genera in semi-arid and coastal regions. *Actinanthella, Oncocalyx, Spragueanella,* and *Oliverella* are all related and eastern. *Berhautia* and *Moquiniella* are isolated monotypic genera in Senegal and the Cape Region, respectively. *Englerina* is equatorial with groups spreading into western Africa, the eastern Afromontane region, thence into the Zambezian woodlands, and with two distinctive groups diversifying in eastern Africa. *Agelanthus* (if segregated from *Tapinanthus*) has several sections - the *Longiflori* with two species, one in the Congolian forests and one in the Sudanian wooded grasslands, section *Agelanthus* both equatorial and eastern, the

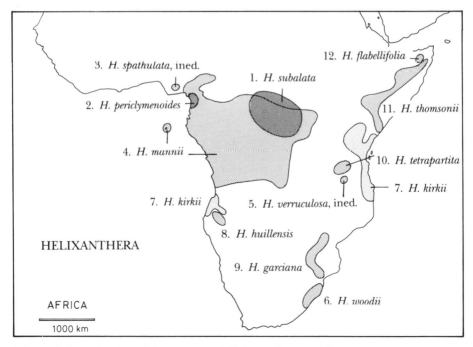

Figure 6. Distribution of the species of *Helixanthera* in Africa.

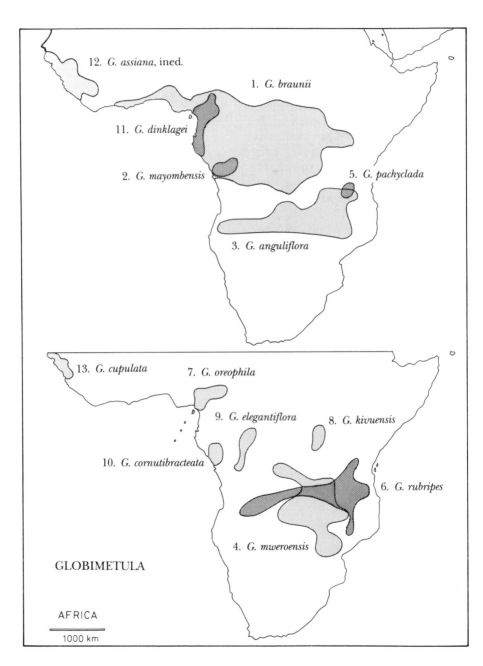

Figure 7. Distribution of the species of *Globimetula* in Africa.

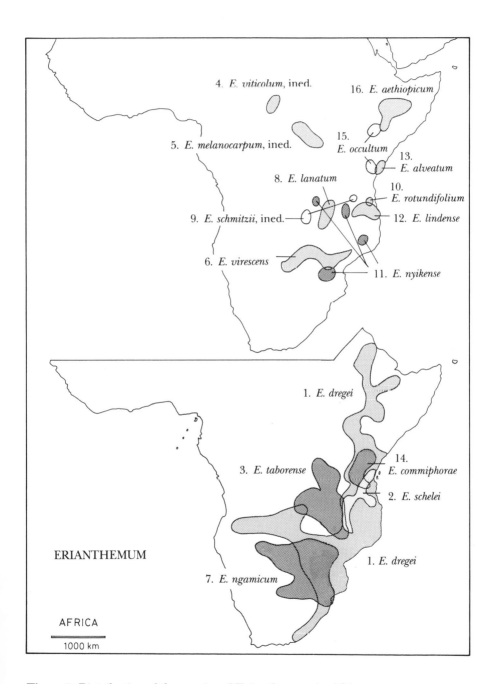

Figure 8. Distribution of the species of *Erianthemum* in Africa.

Tapinanthus in a strict sense is equatorial, with extensions both north other four sections eastern, only the largest section *Erectilobi* extending as far west as Cameroon. *Tapinanthus* in a strict sense equatorial, with extensions both north and south, and a distinctive development in western most Africa, but only marginally entering East Africa.

The Taxilloid group of genera divides into *Taxillus* and its nearest allies and *Dendrophthoë* and its derivatives, both of the mentioned genera having most of their species in Asia. In Africa, *Taxillus* and its nearest allies are eastern and near-coastal from Kenya to Namibia, with *Bakerella* and *Socotrina* in Madagascar. *Dendrophthoë* has three species in the eastern forests from Tanzania to Malawi which should perhaps be placed in a separate African genus *Oedina*. *Oncella* is closely related to *Erianthemum* and is restricted to near-coastal regions of eastern Africa. *Phragmanthera* is a much larger genus with several sections. The least specialized section *Lepidotae* is equatorial, the large section *Rufescentes* likewise but mainly peripheral to the Guineo-Congolian forests and extending to Afromontane regions of eastern Africa, thence into drier habitats. Sections *Phragmanthera* and *Eubracteatae* are more restricted, the former in the Karoo-Namib and Kalahari Regions and the latter in the Zambezian Region.

Conclusions

The distillation of many years study of the woody flora of Africa is included by White in his Vegetation Map of Africa (1983). The study of the 500 species of *Crotalaria* (Leguminosae) in Africa (Polhill, 1982) demonstrated that a genus that is largely herbaceous and characteristic of open disturbed sites showed remarkable correlation with his main chorological divisions. In an independent analysis of the data, White (pers. comm.) showed that 82% of the species were endemic or near-endemic to one of the major regions he defined. The added dimension of this study is to show that the patterns remain very much the same even when the selective factors are more biotic than physical. The sections of *Crotalaria* showed similar patterns of distribution to that described for the genera of Loranthaceae. The oldest sections showed a fragmented distribution of species within Africa and strong intercontinental connections, whereas more advanced sections showed either a strong latitudinal or longitudinal bias.

Speciation from the equatorial forests into the woodlands and wooded grasslands north and south is often accompanied by only slight visible modifications. Along the eastern side of Africa, the diversity of land surfaces and climatic regimes is associated with much more marked adaptive shifts. The eastern groups of Loranthaceae show very evident adaptations with increasing penetration into dry areas and notably exploit most habitats with a woody vegetation.

There is a striking contrast between the uniformity of the dispersal arrangements compared to the high levels of specialization and convergence of the pollination mechanisms, the trends in which have departed markedly from those of their Asian relatives. The conditioning of the African avifauna would appear to have had a significantly cumulative effect on the speciation of African Loranthaceae. The degree of interdependence of components of African ecosystems is important when considering measures for plant conservation. As protected areas come increasingly under pressure, the recognition of regional diversity and the space needed to maintain biological systems becomes an essential part of the management.

Acknowledgments

This paper is prepared as part of a collaborative program on the mistletoes of Africa with Professor Delbert Wiens, University of Utah. The field work has been generously supported by the National Science Foundation, USA. I am also grateful to Dr. Françoise Dowsett-Lemaire, Dr. John Feehan, and Mr. Don Kirkup for much of the biological information quoted above.

Literature cited

Axelrod, D. I. and Raven, P. H. (1978). "Late Cretaceous and Tertiary vegetation history of Africa," pp. 77-130 *In* Werger, M. J. A. (ed.), "Biogeography and Ecology of Southern Africa." *Junk, The Hague.*

Balle, S. (1956). "A propos de la morphologie des Loranthus d'Afrique." *Webbia* 11, 541-585.

Balle, S. (1964). "Loranthacées 60." *In* Humbert, H. (ed.), "Flore de Madagascar et des Comores." *Museum National d'Histoire Naturelle, Paris.*

Barlow, B. A. (1983). "Biogeography of Loranthaceae and Viscaceae," pp. 19-46 *In* M. Calder and P. Bernhardt (eds.), "The Biology of Mistletoes." *Academic Press, Sydney.*

Danser, B. H. (1933). "A new system for the genera of Loranthaceae Loranthoideae, with a nomenclator for the Old World species of this subfamily." *Verh. Kon. Ned. Akad. Wetensch., Afd. Natuurh., Tweede Sect.* 29(6), 1-128.

Docters van Leeuwen, W. M. (1954). "On the biology of some Javanese

Loranthaceae and the role birds play in their life-history."
Beaufortia **4,** 105-208.

Dowsett-Lemaire, F. (1982). "Tinkerbirds and mistletoes on the Nyika Plateau, south-central Africa." *The Golden Bough* **1,** 3.

Dowsett-Lemaire, F. (1988). "Fruit choice and seed dissemination by birds and mammals in the evergreen forests of upland Malawi." *Revue d'Ecologie (Terre et Vie)* **43,** 251-285.

Feehan, J. (1985). "Explosive flower opening in ornithophily: a study of pollination mechanisms in some Central African Loranthaceae." *J. Linn. Soc., Bot.* **90,** 129-144.

Gill, F. B. and Wolf, L. L. (1975). "Foraging strategies and energetics of East African sunbirds at mistletoe flowers." *Amer. Naturalist* **109,** 491-510.

Godschalk, S. K. B. (1983). "Mistletoe Dispersal by Birds in South Africa," pp. 117-128 *In* Calder, M. and Bernhardt, P (eds.), "The Biology of Mistletoes." *Academic Press, Sydney.*

Liddy, J. (1983). "Dispersal of Australian Mistletoes: The Cowiebank Study," pp. 101-116 *In* Calder, M. and Bernhardt, P. (eds.), "The Biology of Mistletoes." *Academic Press, Sydney.*

Polhill, R. M. (1982). "Crotalaria in Africa and Madagascar." *A. A. Balkema, Rotterdam.*

Polhill, R. M. and Wiens, D. (ined.). "The Mistletoes of Africa."

Sprague, T. A. (1910). "Loranthaceae." *In* Thistleton-Dyer, W. T. (ed.), "Flora of Tropical Africa" **6(1),** 225-411. *Lovell Reeve, London.*

White, F. (1983). "The Vegetation of Africa." *UNESCO, Paris.*

Wiens, D. and Calvin, C. L. (1987). "Epiparasitism in mistletoes." *The Golden Bough* **9,** 2-4.

Wiens, D. and Tölken, H. R. (1979). "Loranthaceae." *In* Leistner, O. A. (ed.), "Flora of Southern Africa" **10(1),** 1-41. *Botanical Research Institute, Pretoria.*

Diversity

Species richness in tropical forests

P. S. ASHTON

Arnold Arboretum, Harvard University, USA

To the systematic or evolutionary biologist, it is the extraordinary richness of species which distinguishes tropical forests above all other terrestrial ecosystems, and which makes them so fascinating. It seems curious then that, though this richness was extolled by Humboldt, Darwin and the 19th century explorer botanists, the floristic structure of tropical forests has only recently become of major interest to theoretical biologists. Following Clements' (1936) hypothesis of the climatic climax in vegetational composition, through the dispute between proponents of the Gleasonian (1926) view of ecological independence among species and Tansley's (1920) concept of the plant community type as a quasiorganism, the nature of rain forest floristic structure was all but ignored. Paul Richards (1952) was the first to systematically describe the distinct rain forests on limiting soils which had been noticed by the early naturalists such as Beccari (1904) and Spruce (1908), but the intrinsic structure of the great mixed species pan-climax (in the sense of Clements, quoted in Richards, 1952) remained enigmatic. Indeed, in an era dominated by Gause's (1932) hypothesis of competitive exclusion and the concept of niche specificity, experienced tropical botanists still view the species composition of mixed rain forests as entirely unpredictable (Corner, 1954; van Steenis, 1969). Species richness was considered by them to be attributable to the constancy of the rain forest climate over geological time and to a low rate of extinction. This interpretation must assume that pair-wise interspecific competition in a benign environment is relaxed or absent.

Such a view poses a severe challenge to the evolutionary biologist. If the environment, biological as well as physical, remains effectively constant and generalized over evolutionary time so that the species are effectively complementary and their survival a matter of chance, then what constrains genetic diversification within panmictic populations and what, other than founder effects, promotes speciation should they become subdivided ?

Clearly, we have two kinds of questions to answer. An ecological one: How deterministic is the floristic structure of tropical forests ?, and an evolutionary one: How specialized are the species components, and is the pattern of genetic variation within their populations compatible with an

TROPICAL FORESTS
ISBN 0–12–353550–6

environment of stringent or of lax natural selection, or, alternatively, high or low specificity of interspecific competition ?

Floristic structure

Our understanding of the floristic structure of tropical forests and its maintenance has grown rapidly in the twenty years which have followed the milestone publication in–1969 of the first issue of the Biological Journal of the Linnean Society of London, which was devoted to speciation in the tropics. Though there remains a very great deal to learn there is now, I think, a broad understanding of the relative role of chance and predictability in the composition of tropical forests.

Interpretation of the floristic structure of tropical forests, its variation, and the mechanisms which sustain it is a complex matter. Confusion can only be averted if there is clear recognition of the possibility that determinism of floristic structure can be both present and absent in the same forest, by being so at different scales, either of space or of time. Overall species composition in relation to soil or physiography may be predictable, for instance, but the rank order of species abundance within one uniform habitat, or the spatial relationships between individuals in different species populations, may not. Similarly, it may be that the sequence of species in gap succession is predictable, but the eventual replacement of individuals within the canopy of the mature phase is not.

Another potential scale of variation, too often ignored, is that of the taxon. Selection can occur on variation in a character state at the level of a family as well as a species (Ashton, 1988). Increased understanding of the cladistic relationships among rain forest species is critical to our understanding of the forest. Chemosystematics in particular can assist in our understanding of the taxonomic levels at which insect herbivores, seed predators, pathogenic microorganisms, and also allelopaths can influence population densities, while the chemical arbiters of ectotrophic mycorrhizal symbioses may determine the specificity of modes of soil resource use.

Findings from work over the last 30 years in the mixed lowland forests of northwestern Borneo may be used to evaluate the several current hypotheses which seek to predict and explain spatial patterns of species richness in tropical rain forest. Northwest Borneo is almost certainly the center of species diversity in tropical Asia, and is exceptionally rich in endemics. One reason for this is the high edaphic heterogeneity of the geologically young terrain. Also, earlier speculation that the region may represent a Pleistocene refugium (Ashton, 1972) has recently begun to gain some respectability from paleontological evidence (Morley and Flenley,

1987). The region is currently one of low wind speeds and high rainfall, distributed more or less evenly but without any month in which mean evapotranspiration exceeds precipitation. There is no detectable correlation between the regional rainfall patterns that exist and species composition, including tree species richness. Nevertheless, as shall be seen, the different forests vary greatly in species composition, species richness, and the degree of dominance of individual species. Also, the different landforms and soils differ greatly in their stability, so that there is great site related variation in the frequency and spatial scale of canopy disturbance.

Maintenance of species richness

The various hypotheses for the maintenance of species richness in tropical forests principally either seek to explain habitat-related patterns between communities or the maintenance of species richness within one community. As Hubbell and Foster (1983) have explained, they can also be divided into equilibrium, and non-equilibrium hypotheses. The former predicts that floristic composition, and species rank order of abundance, always proceeds towards an asymptote, and will even be reconstituted following forest destruction if a seed source is available and the habitat remains unaltered. The latter holds that floristic structure is a consequence solely of past and present opportunities for immigration, and accidents of extinction, and that there are no processes which favor predictability or uniformity of species composition in the forest succession following catastrophe.

Among the first category, Connell (1971) has argued that species richness will be determined by the extent to which the forest canopy experiences catastrophic disturbance. Forests which experience frequent and/or drastic disturbance, he claims, will generally be comparatively poor in species because the components of the mature phase cannot survive. Forests which rarely experience disturbance will similarly be relatively poor, because habitats for pioneers will be far scattered and possibly also small in area, increasing rates of extinction among them. Species richness will be greatest, therefore, at intermediate levels of disturbance, when all dynamic phases of the forest are well represented by area. It should be noted that this hypothesis implies, at least when intermediate levels of disturbance are obtained, that the species richness of the various dynamic phases are similar.

This interpretation is consistent with a general equilibrium hypothesis for species richness in plant communities elaborated by Grubb (1977, and elsewhere). Grubb emphasizes the great potential habitat diversity at and immediately above and below the soil surface into which plants can adapt and specialize through competition with other plants and also with animals. He too sees disturbance, whether on the scale of a windstorm or the mound of a burrowing animal, as a major contributor to species enrichment through its influence on the diversity of microhabitats for establishment and regeneration. On the other hand though, Grubb lays stress on the vast array of characters of the plant itself, including those of flower and seed biology which affect seed viability, seed dispersal and dormancy, germination, establishment, and onward growth which, independently or together, can influence success in competition during establishment and early regeneration. Ultimately, he argues, these allow for accumulation of species richness within the plant community through specialization of the regenerative niche in evolution. Grubb's hypothesis implies then that the influence of interspecific competition, on both species survival and evolutionary changes in gene frequencies within populations, is concentrated during the phases of reproduction, establishment, and early growth.

In the view of Tilman (1982, 1986), species richness is primarily determined by the abundance of physical environmental resources, and is enhanced by small scale heterogeneity in those resources which are limiting to plant survival and growth. If one resource is severely limiting, its potential heterogeneity is likewise *per se* limited also. Relatively few species will be adapted to survive, therefore, when one or more resources essential to plant growth are limiting. With increase in a limiting resource opportunities increase for spatial heterogeneity in its distribution. This provides opportunities for more species to occupy the habitat as a whole, while opportunities for specialization in the exploitation of the resource also increases as its availability increases. These opportunities for specialists to coexist in a common environment greatly increase when the availability of two or more limiting resources, such as different soil nutrients, covaries. There is a point, however, when one or several covarying resources increase beyond the levels at which they are limiting to the plant species present. Once that happens, the spatial heterogeneity in resources availability ceases to affect plant distribution and community species richness. Tilman (1982) predicts that one or a limited number of species are then likely to competitively exclude others from the one ubiquitous resource, light. Species richness will therefore, once again, decline, and the decline will be inversely correlated with increased species dominance. In short, species richness is correlated with soils' nutrient status where the nutrient status is low, but where soil resources are not limiting species richness is depressed by an increase in species dominance. The resulting model is a peaked species richness distribution in relation to soil nutrients with the peak skewed towards the less fertile end of the range, and the correlation between resources and species richness increasingly lost once resource levels exceed the level at which the peak is reached.

The field conditions required for the model of Tilman to prevail are incompatible with those required for Connell's (1971) model of inter-mediate disturbance. Both are equilibrium models in the sense of Hubbell and Foster (1983), but Tilman's is dependent on a steady-state ecosystem, where unpredictable perturbations to species numbers are small and infrequent enough to allow an intricate jig-saw of edaphic specialists to co-occur. Such a community structure is more likely, therefore, to occur in the mature than the establishment or building phase of a forest. Abundant successional patches will obscure it, and disturbance could then become the dominant correlate of species richness instead. The growing view that occasional catastrophe is a major factor in the structure and dynamics of tropical forests is similarly incompatible with the Tilman model.

The Borneo test

Examination of the applicability of these various models can reveal much concerning the overall structure and dynamics of mixed rain forest communities. A data set is now available from Borneo which does permit a partial test of those hypotheses (Ashton *et al.*, submitted; Ashton and Hall submitted); partial because it represents a sample of trees exceeding 9.7 cm dbh only, and because recent canopy gaps were subjectively excluded from samples except in certain sets of permanent plots, now 20 years old, within which they have gradually increased. There are 205, 0.4 and 0.6 hectare. samples from mixed lowland forests distributed over 14 sites on yellow to red ultisols, inceptisols, and entisols. They were subdivided into 0.2 hectare units for analysis of floristic variation. In addition, because stand density varied greatly and independently of species composition, samples of 1000 individual trees, derived from groups of floristically similar plots, were used to compare species richness on different sites. Several soil samples, from two standard depths, were taken within each sample.

Analyses of these data confirmed Tilman's prediction to a remarkable extent. First, the pattern of variation in tree species com-position, analysed by numerical classification and ordination, showed that it was correlated with concentrated HCl extractable phosphorus and magnesium where the values are below c. 200 ppm phosphorus and c. 1200 ppm magnesium. Above that threshold, these correlations were much looser or absent and were overtaken by a correlation with topography. Second, below the same nutrient threshold there was a positive linear correlation between species richness and soil magnesium. Maximum species richness was attained at about the same threshold, above which species richness became negatively correlated with both nutrients. There was a high, negative correlation above the nutrient threshold between species richness and the collective dominance of the five most abundant species, measured as relative density, whereas these two were uncorrelated below the threshold.

Further analysis and observations suggest the mechanisms by which this striking pattern of species richness, at areal scales of 0.2-2 ha., is attained.

Field observations suggest that the critical threshold in soil nutrients coincides with the threshold below which the rate of litter accumulation exceeds its rate of breakdown. Surface raw humus, with associated dense root mat, can reach 10 cm depth at the Mixed forest-Heath forest, ultisol-spodosol, transition and is ubiquitous on poorly draining low nutrient yellow-red soils. Species richness in the major ectotrophic mycorrhizal family Dipterocarpaceae, and others including Myrtaceae, is highly positively and linearly correlated with soil magnesium and to a lesser extent with phosphorus below the critical nutrient threshold. Magnesium is thought to support phosphorus uptake by mycorrhiza . Smits (1983) has evidence that some dipterocarp mycorrhiza are host specific, and it seems a real possibility that the narrow and often distinct soil ranges exhibited in series of related and co-occurring dipterocarp species (see Ashton, 1969) are due to these tolerance ranges of their specific mycorrhizal systems, in keeping with Tilman's expectations. Regional endemism in these families is also heavily concentrated on these soils. In contrast, the trends of species richness in families such as Meliaceae and Rubiaceae do not follow those of Dipterocarpaceae, Myrtaceae, or that shown less precisely by the tree flora as a whole.

The species which become dominant in the forest stands, once the initial nutrient threshold is exceeded, and which apparently depress species richness through competitive exclusion, include members of both canopy and sub-canopy. Here, decline in species richness is also correlated with increase in representation of species with large leaves in both the main canopy and sub-canopy, though not in the emergent stratum where predominant leaf size is correlated instead with emergent canopy stature. Among nutrients, potassium has the highest of the negative correlations with species richness on fertile soils and is uncorrelated with species richness in low fertility soils. Soil potassium is derived from clay, and soil clay content is likely to strongly correlate with potassium. Water retaining capacity and oxygenation, hence, rooting depth in these upland soils is largely dependent on clay content. We interpret these correlations on fertile soils as a whole to indicate that dominance increasingly depresses species richness as soil water retaining capacity becomes increasingly favorable.

As this occurs, trees can afford to carry an increasing heat load in their leaf canopy. This enables certain species with large, densely arranged leaves to shade out most other competitors.

The pattern of species richness on fertile soils is not only compatible once again with Tilman's predictions but also with Grubb's hypothesis, because increasing canopy density leads to a decline in the heterogeneity of the microclimate at the forest floor, hence a narrowing of options for specialization in the "regeneration niche."

What then, can be said concerning Connell's intermediate disturbance hypothesis ? The samples whose species richness we have been comparing were sited to exclude pioneer vegetation. Nevertheless, the trend of species richness on fertile soils can be said to be compatible with the hypothesis, because dense canopy closure is dependent on minimal disturbance. Our permanent plots, after twenty years, provide another means to test Connell's hypothesis. They have accumulated gaps during the twenty years of observation, and evidence of their state of dynamic equilibrium is also at hand.

One permanent site represents the low end of the soil fertility range, and is also coastal and prone to drought; another is on deep soils somewhat below the nutrient threshold, and a third is on highly fertile, basalt derived soils.

In spite of the recorded increase in the number of gaps at all these sites, there was no significant increase in the total number of species present on the low nutrient and the intermediate site but a 24% increase (44 added to an original of 184) at the fertile site. This is substantially because few pioneer species reach the 9.7 cm dbh minimum size recorded in the study, and these contributors to the building phase are strongly concentrated on fertile soils.

Nevertheless, our current research is revealing that the pattern of gap formation varies greatly with soils. On friable soils, whether they are the least fertile, poorly draining, leached yellow, podsolic sands or the well-structured fertile clays overlying basalt, large gaps are less frequent than on poorly structured, less fertile clays overlying shale and the soils overlying interbedded clays and sandstones where species richness is highest. The latter land slips, though non-randomly distributed, play an important part in forest succession. Also, group mortality of canopy trees, associated with uprooting and usually due to windthrow, is significantly more common on clay soils of both intermediate and high fertility than on leached sands: There, most trees die before falling and established regeneration of mature phase species dominates succession. Field observation is therefore compatible with the view that patterns in the scale and frequency of canopy disturbance may also contribute to the enhancement of species richness on intermediate sites and its depression on both the most and least fertile.

Unlike the other two permanent sites which, on the time and aerial scales measured, appear to be close to dynamic equilibrium, the forest in the site on xeric, less fertile soils experienced a 16% increase in standing volume during the twenty years of measurement. This increase was concentrated among the largest size classes, while the smallest experienced a decline. Both suggest that the forest canopy is increasing in density following some earlier catastrophe, presumably a severe drought. In further support of this inference, canopy species richness was relatively lower at this site than at the other two and in particular at the intermediate site. This is consistent with observations following the 1982-83 drought in eastern Borneo, when up to 30% of emergent species died in forests not subject to fire and where rates of mortality were highly species specific.

Conclusions

We tentatively conclude therefore that variation in the frequency and scale of canopy disturbance almost certainly has a leading influence on species richness on clay soils. It may also play some part on the more poorly draining, less fertile sands, though here nutrient status would appear to be the main determinant. Occasional, intense drought or other drastic catastrophes are correlated with a depression of species richness, at least in the emergent stratum, which is also in keeping with Connell's prediction.

Perhaps the most interesting finding from this study has been that the pattern of species richness, and the dynamic patterns for the most part as well, implies that these diverse forests of northwest Borneo are mostly at or close to dynamic equilibrium. The stability of climate that this implies may also explain the exceptional tree species diversity of the region.

To what extent can our results be generalized ? That historical biogeography is a major influence on the richness of whole floras is indisputable. It is particularly intriguing, therefore, that our richest sites are closely comparable to the richest sampled by Gentry (1982) in the Peruvian Amazon. Unfortunately, comparable soils and rainfall data from the two regions do not exist.

I understand that the soils at Gentry's sites lack the surface raw humus which is an apparently critical adjunct to high species richness in Borneo. This could imply either that the climate at Gentry's site is seasonally dry or, I think more likely, that soil nutrient status is above the critical threshold. Whichever is correct, Gentry's results are not obviously consistent with what would be predicted from ours. A simple explanation is that his samples are, for a similar site, richer for historical biogeographic reasons. If his rich samples came from a region with an annual dry season, even if short, patterns in Asian forests would anticipate a dramatically lower species richness than in aseasonal regions, irrespective of total annual rainfall. There does, however, exist the possibility that Gentry's forests have experienced more frequent, small to medium scale canopy disturbance over an evolutionary time scale and therefore have a richer pioneer flora. His data could therefore be more consistent with Connell's than Tilman's predictions. This would be in keeping with Hartshorn's (1978) estimate from La Selva forest, Costa Rica, which does experience frequent small scale canopy disturbance, that 80% of the tree species there require treefall gaps to regenerate. We do not know what this percentage would be in our Asian forests, but suspect that it would be lower.

Outside the tropics and outside forest vegetation, Grubb (1977) has stressed the overriding importance both of continuous grazing and of the patchy disturbance caused by burrowing animals on species richness in grasslands, though soil nutrients also play a major role as well.

The possibility that Connell's model may apply in some rain forests and Tilman's in others is compatible with our results. In both cases, we predict that low variance in the annual march of those climatic factors, particularly rainfall distribution, which influence the species richness of tropical rain forest on a regional scale is obligatory. In Asia, increasing seasonality is correlated with increasing between-year variance in the length of the dry and wet seasons, but this need not be the case elsewhere. Rather, it may be more appropriate to ask whether the climatic predictability which seems to be a necessary condition for Tilman's model is obtained anywhere other than in certain tropical forest regions. The forests of the temperate, moist, equable region of the western coasts of the Americas and Europe, which are poorer in canopy species and more uniform than forests at similar latitudes in lower rainfall regions in eastern North America and Asia, may provide one example.

At low latitudes along these same coasts, the species richness of Mediterranean shrub communities may also conform with Tilman's predictions. There, soil water is too restricted for closed woodland vegetation but not, it seems, for semi-closed sclerophyll-microphyll shrublands, at least on friable, and usually relatively infertile, soils. Though winter rain falls variably as heavy storms, topographic factors ensure a highly predictable seasonality of cool, virtually frost free winters, and warm but not baking dry summers due to climatic inversion. As expected, the shrub vegetation is overshaded by species-poor evergreen woodlands on mesic bottomland.

So far, I have argued solely for equilibrium explanations for the patterns of variation in species richness, and by implication their maintenance, in tropical mixed forests. It is more difficult to champion non-equilibrium models, based as they are on negative evidence. Hubbell and Foster (1983) have rightly sought evidence for the presence and even prevalence of non-equilibrium processes by searching for significant patterns of positive or negative association between species populations, and between age classes within species populations, within single community types. It would seem highly improbable that mixtures of up to one thousand species of sessile organisms sharing a single life form and on a uniform site, such as exist in the tree flora of the Pasoh research forest, Peninsular Malaysia, could be maintained either through differentiation of regeneration niches or by means of differential resource exploitation. In such instances, the hypothesis of Janzen (1970) that host specific seed predation in an equable climate leads to low maximum tree population densities, thus enabling many otherwise ecologically complementary species to coexist, provides one attractive equilibrium explanation. Hubbell (in press) has recently found, in the great majority of the c. 300 species populations which he has under observation at Barro Colorado Island, that juvenile performance is indeed positively correlated with distance from the nearest reproductive adult up to a distance of about 15 m. This represents the first persuasive evidence at the community level, that density dependent factors are not only operating but general in mixed tropical forests, although the causes in this case are unknown.

Nevertheless, the Barro Colorado Forest is not outstandingly rich, and it is hard to imagine that density-dependent processes will operate among the many species of exceptionally low population density which characterize the richest forests.

Comparison of the rich, mixed forests of northwestern Borneo with their relatively poor analogs in the isolated and restricted wet rain forests of southern Sri Lanka, indicates that the difference in species richness between them is almost entirely attributable to differences in the numbers of the species occurring at lowest densities. In Borneo, a majority of the most abundant species are consistently so throughout their geographical range and in their preferred habitat, which implies that their presence and performance is mediated by equilibrium factors. On the other hand, not only are differences in species richness on different soils attributable to differences in the number of rare species, but these mainly belong to a relatively few large genera whose co-occurring species are a characteristic element in the floristic structure of the richest mixed rain forests.

The degree of niche specificity among rain forest tree species can be inferred from the pattern of genetic variation within them. High niche specificity, if biotically mediated implies that the competitive environment for species is highly restrictive but continually changing over evolutionary time. Survival of species would be dependent on maintenance of genetic variability within populations, and the existence of populations with low genetic variability, therefore, provides inferential evidence for lower niche specificity. The great majority of evidence both from Central America and the Far East so far suggests that genetic variability within populations of rain forest trees is generally quite high, and similar to that in populations of temperate dicotyledonous plants (Loveless and Hamrick, 1984). In one exception though, Gan *et al.* (1977) provided preliminary evidence of very low variability within a sibling seedling population of the common Malaysian dipterocarp emergent, *Shorea ovalis* (Korth.) Bl. This species is an allotetraploid and is highly apomictic through adventive embryony (Kaur *et al.*, 1978). To the extent that apomixis can be used as indirect evidence for reduced population genetic variability, it is interesting that its presence has been demonstrated or inferred in a majority of the largest of those genera whose members are the principal cause of variation in tree species richness between sites (Ashton *et al.*, submitted). Tentative though this evidence is, it is compatible with the fact that the presence or absence of the rare species in the richest forests, where rare species comprise the majority and where local site heterogeneity is high and areas of uniform habitat therefore small and fragmented, must be relatively unpredictable. For this reason, the biotic influences on natural selection among these many rare species, being dependent on predictable and relatively high population densities, must here be relaxed.

I conclude from this evidence, admittedly preliminary, that niche specificity may be relatively low among some, at least, of the rare species which comprise the vast majority in the mixed rain forest flora. In other words, it is among these that non-equilibrium phenomena should be expected. In effect, I would refine Steenis' (1969) view by questioning what, other than chance, can cause the extinction of an apomictic rain forest tree species whose population density is too low to be affected by density controlling factors.

Maintenance of high species richness among some tropical epiphytic communities may have a similar explanation. As with rain forest trees, their richness reaches its peak in the most equable climates as Gentry (1982) has stressed in the case of the montane forests of the northern Andes, home of by far the richest epiphytic flora in the world. Here, the substrate is unpredictable and constantly rejuvenating on a small scale, thereby permitting coexistence of large numbers of ecologically complementary species in very low overall population densities, according to the intermediate disturbance model. Interestingly, the breeding systems of epiphytes appear to be very different from those of tropical trees, and the genecological significance of this represents an exciting field, so far virtually unexplored, for future study.

Literature cited

Ashton, P. S. (1969). "Speciation among tropical trees: Some deductions in the light of recent evidence." *Biol. J. Linn. Soc.* **1**, 155-196.

Ashton, P. S. (1972). "The quaternary geomorphological history of western Malesia and lowland forest phytogeography." pp. 35-62 *In* Ashton, P. and Ashton, M. (eds.), "Transactions of the second Aberdeen-Hull symposium on Malesian ecology: The quaternary era in Malesia." *Hull Geog. Dept. Misc. Ser.* **13**.

Ashton, P. S. (1988). "Dipterocarp biology as a window to the understanding of tropical forest structure." *Ann. Rev. Ecol. Syst.* **19**, 347-370.

Ashton, P. S. and colleagues (submitted). "Comparative ecological studies in the mixed Dipterocarp forests of northwestern Borneo. I: Patterns of floristic variation. III: Patterns of species richness."

Ashton, P. S., Hall, P. and colleagues (submitted). "Ibid II: Variation in structure and dynamics."

Beccari, O. (1904). "Wanderings in the great forests of Borneo." [English translation by E. H. Giglioli]. *Constable, London.*

Clements, F. E. (1936). "Nature and structure of the climax." *J. Ecol.* **24**, 252-284.

Connell, J. H. (1971). "On the role of natural enemies in preventing competitive exclusion in some marine animals and rainforest trees." pp. 298-312 *In* Boer, P. J. and Gradwell, G. R. (eds.), "Dynamics of numbers in populations." *Advanced Study Institute, Wageningen.*

Corner, E. J. H. (1954). "The evolution of tropical forest." *In* Huxley, J. S., Hardy, A. C. and Ford, E. B. (eds.), "Evolution as a process." *Allan and Unwin, London.*

Gan, Y. Y., Robertson, F. W., Ashton, P. S., Soepadmo, E. and Lee, D. W. (1977). "Genetic variation in wild populations of rain-forest trees." *Nature* **269**, 5626: 323-325.

Gause, G. F. (1934). "The struggle for existence." *Williams and Wilkins, Baltimore.*

Gentry, A. H. (1982). "Patterns of neotropical plant species diversity." *Evol. Biol.* **5**, 1-84.

Gleason, H. A. (1926). "The individualistic concept of the plant association." *Bull. Torrey Bot. Club* **53**, 7-26.

Grubb, P. J. (1977). "The maintenance of species-richness in plant communities: The importance of the regeneration niche." *Biol. Rev.* **53**, 107-145.

Hartshorn, G. S. (1978). "Tree falls and tropical forest dynamics." pp. 617-638 *In* Tomlinson, P. B. and Zimmermann M. H. (eds.) "Tropical trees as living systems." *Cambridge Univ. Press, Cambridge.*

Hubbell, S. P. and Foster, R. B. (1983). "Diversity of canopy trees in a neotropical rain forest and implications for conservation." pp. 25-42 *In* Sutton, S. L., Whitmore, T. O. and Chadwick, A. C. (eds.), "Tropical rain forest: Ecology and management." *Blackwell, Oxford.*

Janzen, D. H. (1970). "Herbivores and the number of tree species in tropical forests." *Amer. Naturalist* **104**, 501-528.

Kaur, A., Ha, C. O., Jong, K., Sands, V. E., Chan, H. T., Soepadmo, E. and Ashton, P. S. (1978). "Apomixis may be widespread among trees of the climax rainforest." *Nature* **271**, 5644: 440-442.

Loveless, M. D. and Hamrick, J. L. (1984). "Ecological determinants of genetic structure in plant populations." *Ann. Rev. Ecol. Syst.* **15**, 65-95.

Morley, R. J. and Flenley, J. R. (1987). "Late Cainozoic vegetational and environmental changes in the Malay archipelago." pp. 50-59 *In* Whitmore, T. C. (ed.), "Biogeographical evolution of the Malay archipelago." *Clarendon, Oxford.*

Richards, P. W. (1952). "The tropical rain forest." *Cambridge Univ. Press. Cambridge.*

Smits, W. Th. M. (1983). "Dipterocarps and mycorrhiza. An ecological adaptation and a factor in forest regeneration." *Fl. Males. Bull.* **36**, 3926-3937.

Spruce, R. (1908). "Notes of a botanist on the Amazon and Andes." 2 vols. Wallace, A. R. (ed.). *Macmillan and Co., London.*

Steenis, C. G. G. J. van (1969). "Plant speciation in Malesia, with special reference to the theory of nonadaptive saltatory evolution." *Biol. J. Linn. Soc.* **1**, 97-133.

Tansley, A. G. (1920). "The classification of vegetation and the concept of development." *J. Ecol.* **8**, 118-149.

Tilman, D. (1982). "Resource competition and community structure." *Princeton Univ. Press, Princeton, New Jersey.*

Tilman, D. (1986). "Evolution and differentiation in terrestrial plant communities: The importance of the soil resource: light gradient." pp. 359-380 *In* Diamond, J. and Case, T. J. (eds.), "Community ecology." *Harper and Row, New York.*

Sri Lankan forests: Diversity and genetic resources

D. B. SUMITHRAARACHCHI

Royal Botanic Gardens, Peradeniya, Sri Lanka

Sri Lanka is a tropical island of continental origin which shares tectonic plates with peninsular India. Both formed part of southern Gondwanaland, with separation taking place during the Miocene. The island covers 65 610 square kilometers and is made up of three peneplanes: (i) the low country with elevations to 30 m above sea-level, (ii) the mid-country reaching to 500 m, and (iii) the uplands above 500 m. From a central mountain massif of rivers flow in all directions, especially from the prominent peaks Pidurutalagala (2524 m), Adams Peak (2238 m), Namunukula (2036 m), and the Knuckles Range (1836 m) (Fig. 1A). An arid zone in the northwest and southeast receives less than 1250 mm rain per year, the dry zone covering most of the island receives between 1250 and 1875 mm per year, and the wet zone in the southwest more than 1875 mm (Fig. 1B). Sri Lanka receives two monsoons; the northeast monsoon between October and February (Fig. 1C), and the southwest monsoon from May to September (Fig. 1D). These monsoons affect differently the three climatic zones because the central mountain massif acts as a barrier.

The diverse climate, soil types, and altitude have provided the basis for a high vegetation and plant diversity (Holmes, 1958). The flora includes 192 families of flowering plants with 1290 genera and 3268 species of which 853(25%) are endemic to the island. High endemism is also seen in the pteridophyte flora where 57 of 314(18%) species are endemic. The lichen family Thelotremataceae has 39 endemic species out of 110(35%). The natural forest cover of Sri Lanka has dwindled to 9% of the surface area. The diverse nature of the vegetation was always recognized by the state, and many national parks and protected areas were declared after 1938. Up to 1986 about 13% of the land area, or 878 986 hectars, have been protected, but a good proportion of the protected areas are grasslands and secondary forests.

There are eight kinds of natural vegetation formations: (i) lowland wet evergreen rain forests at elevations to 1000 m; (ii) lower montane or sub-montane forests at elevations of 1000-1500 m; (iii) upper montane or cloud forests at elevations above 1500 m; (iv) intermediate zone semi-evergreen forests; (v) mixed deciduous or monsoon forests; (vi) thorn

TROPICAL FORESTS
ISBN 0–12–353550–6

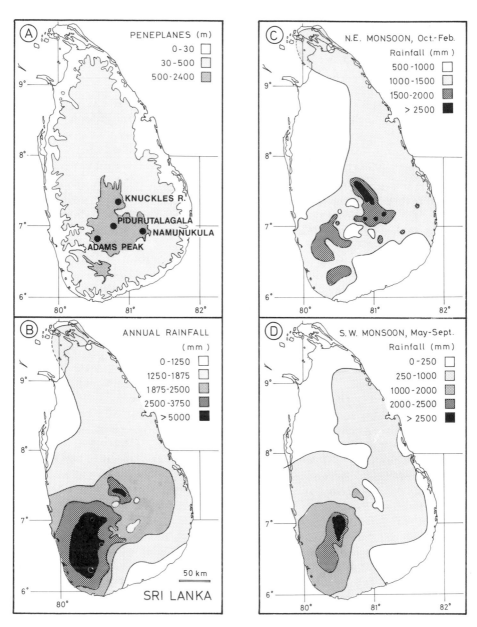

Figure 1. A - Topography of Sri Lanka showing three peneplanes. B - Distribution of rainfall over Sri Lanka. C - Rainfall pattern from the northeast monsoon. D - Rainfall pattern from the southwest monsoon. (Redrawn from Ceylon Meterological Office Yearly Report, 1958)

scrub formations; (vii) mangrove forests; (viii) fire-induced savannas (see also Champion, 1936).

Lowland evergreen forests

These forests are situated in the southwestern parts of the island, which receive a rainfall of 2500-5000 mm per year concentrated especially in May through September and October through February. This area has supported large tropical evergreen forests from sea-level to about 1000 m over the last 200 years, but now much has been cleared for plantations and settlements. The wet zone of Sri Lanka covers about 14 850 square kilometers or 23% of the total land area. In Sri Lanka species of Dipterocarpaceae dominate the lowland rain forests, which are therefore called Mixed Dipterocarp Forests. The trees are 30-45 m tall with emergents reaching to 60 m. About 326 out of 853 endemic plant species of Sri Lanka are found in these forests. Timber plants include *Dipterocarpus zeylanicus, D. hispidus, Shorea trapezifolia,* and *Cullenia zeylanica.* Medicinal and spice plants include *Coscinium fenestratum, Acronychia pedunculata, Vateria copallifera, Myristica dactyloides, Mesua nagasarium,* and *Elletaria ensal.* Fruit trees and food plants include *Podadenia sapida, Caryota urens,* and *Elettaria ensal.* Ornamental plants include *Horsfieldia iryaghedhi, Areca concinna, Loxococcus rupicola,* and *Filicium decipiens.*

Montane forests

Montane forests occur at elevations above 1000 m. The lower montane forests occuring to 1500 m can be recognized by the presence of *Calophyllum, Syzygium, Turpinia, Myristica, Cullenia, Litsea* etc., while *Gordonia, Michelia, Elaeocarpus, Rhododendron, Calophyllum, Syzygium, Cinnamomum, Vaccinium,* and *Ilex* characterize the upper montane forests above 1500 m. The trees are usually up to 15 m tall. The average temperature is 15°C while minimum temperatures reach 5°C during the first week of February, or occasionally below 0°C in the Horton Plains area. Air frost may occur in tea estates and grasslands in the Nuwara Eliya and Horton Plains during early February. The day temperature during this period may reach 25°C. Annual precipitation varies from 2000 mm in the eastern to 5000 mm in the western parts of the upper montane zone. In addition to heavy rainfall, heavy mist often occur in the upper montane areas. Strong winds occur from June through early August. Because of these conditions and the shallow soil, the forests may be dwarfed and only reaching 1-1.5 m tall along exposed ridges. They are then referred to as elfin forests. The trees have twisted trunks with flat, slanting crowns possessing thick, brittle, and small leaves. The leaves of understory plants are mesophyllous as in *Actinodaphne speciosa* and species of *Strobilanthes*. The forest is basically a two story structure consisting of a canopy tree layer with occasional emergents, and a shrub layer in which many species of *Strobilanthes, Impatiens, Indocalamus, Coleus,* and *Hedyotis,* are gregarious. An inconspicuous herb layer sometimes grows in exposed patches within the forest while herbs are abundant along well exposed forest edges where it is disturbed. Montane forests support climbers and scandent shrubs such as *Rubia cordifolia, Elaeagnus latifolia, Toddalia asiatica,* and *Zanthoxylum tetraspermum.*

The montane forests are second to the lowland forests in numbers of endemic species, but unlike the lowland forests, where many trees are endemic, the montane forests are rich in endemic herbs. At present, over 26 682 hectars of montane forests are legally protected, but large plantations of *Eucalyptus* and *Pinus caribaea* along with natural forest elimination for housing and vegetable cultivation also occur in these areas.

Timber plants from the montane forests include *Michelia nilagirica, Calophyllum walkeri, Canthium montanum,* and *Vaccinium symplocifolium.* Medicinal and spice plants include *Acronychia pedunculata, Rubia cordifolia, Symplocos spicata, Valeriana moonii, Gaultheria rudis, Semecarpus coriaceus, Cinnamomum ovalifolium,* and *Spillanthus acmilla.* Fruit trees and food plants include *Rhodomyrtus tomentosa.* Ornamental plants include *Hypericum mysorensis, Rhododendron arboreum, Disporum leschenaultianum, Anaphalis spp., Medinilla fuchsioides, Osbeckia spp., Gordonia zeylanica, Calophyllum walkeri,* and *Impatiens spp.* The presence of Ericaceae, Ranunculaceae and Valerianaceae in the upper montane forest shows its affinity to the temperate flora. The abundance of epiphytic mosses and lichens such as *Usnea barbata* indicates a high atmospheric humidity.

Monsoon forests

These forests, commonly known as the dry zone forests, are found in the northern, eastern and southern provinces. The dry zone covers 77% of the land mass and extends from sea-level to 1000 m elevation. These forests differ from the lowland evergreen and montane forests in many aspects. The canopy of the monsoon forests are open and up to 18 m tall, consisting mostly of deciduous trees. The monsoon forests can be roughly divided into two types: (i) the *Manilkara* community and; (ii) the mixed community.

The *Manilkara* community is two-storied and occurs in the driest areas where the forest is open. The emergent trees usually attain heights of 18 to 20 m. The second stratum in these forests include mostly thorny species and species with latex. Tree species include *Manilkara hexandra, Chloroxylon swietenia,* and *Drypetes sepiaria.* Shrubs include *Ziziphus mauritiana, Carissa spinarum, Dimorphocalyx glabellus,* and *Randia dumetorum.*

The mixed community stratification is different from that of the lowland evergreen forests. It has an emergent layer, about 20 m in height, with wide spacings and a lower layer which can be considered as having a continuous canopy about 18 m in height. The tall trees are deciduous and open. A third understory layer, up to about five m, consists of young trees. A fourth layer is composed of shrubs, and a fifth layer of herbs. A difference between this type of forest and the lowland evergreen forest is that the latter is composed of many large individuals, whereas the dry monsoon forest is composed of a few large individuals and many small ones. Emergent tree species include *Manilkara hexandra, Diospyros ebenum, Berrya cordifolia, Syzygium cumini, Pterospermum canescens, Dialium ovoideum, Mischodon zeylanicus,* and *Chloroxylon swietenia.* Canopy tree species include *Euphorbia longana, Vitex pinnata, Schleichera olerosa,* and *Drypetes sepiaria.* Understory trees and shrubs include *Tarenna asiatica, Polyalthia korinti, Diospyros ebenum, Memecylon umbellatum, Dimorphocalyx glabellus, Croton spp.,* and *Phyllanthus polyphyllus.* Climbers include *Derris uliginosa, Ventilago maderaspatana,* and *Morinda umbellata.* Timber trees include *Chloroxylon swietenia, Azadirachta indica, Manilkara hexandra, Vitex pinnata, Adina cordifolia, Chukrasia tabularis, Diospyros oocarpa, Diospyros ebenum, Drypetes sepiaria,* and *Syzygium cumini.* Fruit trees and food plants include *Euphorbia longana, Dialium ovalifolia,* and *Drypetes sepiaria.* Medicinal and spice plants include *Azadirachta indica, Vitex negundo, Croton aromaticus, C. laccifer, Euphorbia antiquorum, E. tirucalli, Cissus quadrangularis, Cassia fistula, Micromelum zeylanicum,* and *Murraya koenigii.* Ornamental plants include *Cassia fistula, C. roxburghii, Gloriosa superba, Bauhinia spp.,* and *Memecylon umbellatum.*

Conclusions

Sri Lanka has been isolated from India since the Miocene. It has a varied topography and a wide range of climates and soil types. Of the species, 25% are endemic to the island, and endemic elements are found in each of the natural vegetation types. At present extensive areas are being developed for irrigation and cultivation in the dry zone. If diversity is to be kept for future generations not only the rain and montane forests should be preserved. The first priority in conservation should, therefore, be to declare national parks in the northeastern part of the island in order to prevent deciduous forests from becoming annihilated. Extensive shifting cultivation and logging have cleared several mountains. The water catchment areas of the big reservoirs are now nearly devoid of natural forest which results in silting of reservoirs and in lower precipitation. Replanting schemes with local genetic resources from the montane forests should be initiated. In this manner the erosion of the genetic resources could come to an end. A steady supply of water for the reservoirs would be a useful side effect.

Literature cited

Champion, H. G. (1936) "A preliminary survey of the forest types of India and Burma. *Indian For. Rec. n. ser.* **1**, 1-286.

Holmes, C. H. (1958). "The broad patterns of climate and vegetation distribution in Ceylon. *In* Proceedings of the symposium on humid tropics vegetation, Kandy. *UNESCO, Paris.*

The forests of the Paraguayan Chaco

R. SPICHIGER AND L. RAMELLA

Conservatoire et Jardin Botaniques, Geneva, Switzerland

The Boreal Chaco is a part of the Gran Chaco and covers a vast alluvial plain of about one million square kilometers which measures 1500 x 700 km. This plain rises gradually 350 m from southeast to northwest. The Gran Chaco includes part of Paraguay, Argentina, Bolivia, and, to a lesser extent Brazil. The Paraguayan Chaco covers one half of the Republic of Paraguay or about 250 000 square kilometers located west of the Paraguay river. Its geographical limits are shown in Figure 1A-B. The few hills and mountains which occur in the Chaco are concentrated in the northern part and along the western bank of the Paraguay River.

The Paraguayan Chaco has been inaccessible for a long time because of its climate and the extent and hostility of its forests. Exploration, collecting and scientific studies in the countries comprising the Chaco were of unequal standard. Therefore, floristic information on the Chaco is incomplete. There are some publications on botany and ethnobotany (Arenas, 1981), as well as on general ecology and environment (Gorham, 1973), and Hueck (1978) gave a general account on the Chaco forests.

In Bolivia, the first botanical surveys and collections in the Chaco were made by Herzog in 1906-1907 and 1910 during two journeys to Santa Cruz (Herzog, 1910, 1923). These studies, together with Cardenas' research (1940, 1951), provide an excellent general overview of the vegetation. Biogeographical and geological maps of the Chaco were compiled by Brockmann (1978) and Pareja *et al.* (1978). In 1934, during the Chaco war, collections were made by Cardenas near Roboré in the State of Santa Cruz and by Rojas in 1935 near Carandaity in the State of Chuquisaca. In recent times, collections were made by Fernandez Casas, Krapovickas, Schinini and Desloover.

The Argentinian Chaco is much better known and was described in many general publications on phytogeography (Lorentz, 1876; Hauman, 1931; Frenguelli, 1941; Castellanos and Perez Moreau, 1945; Hauman *et al.*, 1947; Cabrera, 1953, 1976; Morello and Saravia Toledo, 1959; Ragonese and Castiglioni, 1970). Recently, Morello and Adamoli (1968, 1974) published two papers on the vegetation of the Argentinean Chaco. Plant collections from this region are more extensive because of the activities of the botanical institutions of Argentina.

TROPICAL FORESTS
ISBN 0–12–353550–6

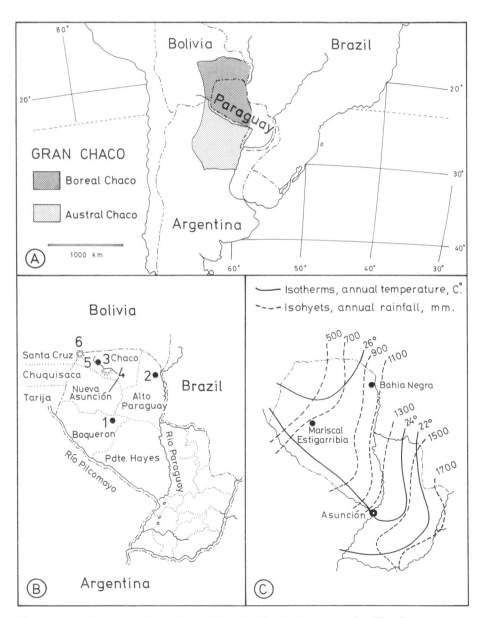

Figure 1. A - Location of the Gran Chaco in South America. B - The Paraguayan Chaco (1 - Filadelfia, 2 - Bahia Negra, 3 - Mayor Pablo Lagerenza, 4 - Cerro León, 5 - Rio Lagerenza, 6 - Cerro Cabrera). C - Temperature and rainfall in Paraguay.

In the Paraguayan Chaco, only the area around the Pilcomayo river had been explored at the beginning of this century. Hassler (1908) published a list of the plants collected in this area by Morong, Britton, and Kerr between 1892 and 1893, and by Rojas in 1906. Fiebrig and Rojas (1933) published the first phytogeographical account of the Boreal Chaco based on Hassler's research (1908) and on their own expeditions. These data concern the area of the lower Pilcomayo River and the banks of the Paraguay River up to Bahia Negra, covering a width of 100-150 km along the banks. In the last twenty years, collections have been made by Bordas, Schinini, Pedersen, Krapovickas, Fernandez Casas, Molero, Bernardi, Spichiger, Loizeau, Billiet, Desloover, Arenas, Hahn, and Ramella. Esser (1982) gives a general and more recent view of the vegetation of the Paraguayan Chaco.

Climate and geology

The mean annual temperature ranges from 18° C in the southern part of the Gran Chaco to 25° C in the northern part. Absolute maximum and minimum are 48° C and -5° C, respectively. Frequent frosts from May until September are caused by intrusion of cold air masses from the South Atlantic. Annual rainfall decreases from 1000 mm near the Paraguay River to 500 mm or less in the center of the Boreal Chaco and then increases towards the west to 800 mm near the foot of the Andes (Fig. 1C). Except for the central part of the Boreal Chaco where the mean annual rainfall is lowest, the irregular distribution of rainfall is largely responsible for the arid character of the environment. Local drought in the Chaco occurs because wet Atlantic air loses most of its humidity over Brazil and eastern Paraguay.

The Chaco is located on a Paleozoic belt at the margin of the Brazilian shield. It subsided during the Mesozoic and Cenozoic eras and accumulated more than 3000 m of marine and terrestrial sediments. From the end of the Cenozoic era (Pleistocene), large quantities of Andean sediments were transported by streams into this area. These streams were progressively filled with deposits and today only a fossil net of dry riverbeds remains. Aeolian sediments were added to the fluvial deposits. Fine components, such as sands, silt, clay, and loess, dominate the Chaquean soils. The color is light brown in the upper horizon and dark brown in the B-horizon. The *pH* is alkaline, neutral, or even slightly acid. The frequent presence of salt is characteristic of these soils and partly explained by accumulation in the capillary system of the soil caused by the strong evaporation under the semi-arid climatic conditions. Several authors believe that salt perhaps was transported in solution from the Andes as well.

Vegetation

Southeast Paraguayan Chaco. This region belongs to the Department of Presidente Hayes. It includes the moister part of the Chaco and its vegetation is much influenced by the seasonal changes of the Pilcomayo and Paraguay rivers. The local rainfall is higher than that further to the west. The vegetation is a mosaic of forests, palm-savannas, and savanna grassland. Forests colonize the banks of the streams and form gallery forests and also the topographically highest terrain where the soils are well drained. Gallery forests consist of hygrophilous species, most of which extend their range into eastern Paraguay. Upland forests, however, are the typical formations of the Chaco. The extremely xeromorphic forest, or *quebrachal* (Fig. 2A), contains the *quebrachos, Aspidosperma quebracho-blanco, Schinopsis quebracho-colorado, Ruprechtia triflora, Capparis spp., Stetsonia coryne, Cereus spp., Bulnesia sarmientoi, Chorisia insignis, Bromelia serra,* and *B. hieronymi* as dominant species. Palm-savannas, called palmares, colonize lower and moister parts of the land. Savanna grassland colonize the lowest and wettest places.

The palm-savannas which dominate the landscape of the forest/savanna mosaic in the southeast, are progressively reduced towards the northwest and give away to the *quebrachal*. With further decrease in air and soil moisture, palm-savannas disappear completely. At the border of the Department of Boquerón in the central Chaco, the vegetation is a continuous, extremely xeromorphic forest with scattered open formations. These gaps support impoverished forms of the *quebrachal*, called *peladares* (Fig. 2B). The shrub cover of the *peladares* is discountinuous with mineral soil exposed between. *Trithrinax biflabellata, Ziziphus mistol, Maytenus sp.* and *Bulnesia sarmientoi* are characteristic species. The extreme form of these *peladares*, a monospecific *Stetsonia* or *Cereus*-formation, occurs in several places.

Figure 2. A - Extremely xeromorphic *Quebrachal* forest. [A.qb=*Aspidosperma quebracho-blanco.* B.s.=*Bulnesia sarmientoi.* Br.h.=*Bromelia hieronymi.* Br.s.=*Bromelia serra.* C.i.=*Chorisia insignis.* Ca.=*Capparis spp..* R.t.=*Ruprechtia triflora.* S.c.=*Stetsonia coryne.* S.qc.=*Schinopsis quebracho-colorado.* C.=*Cereus sp.*] B - *Peladar* which is an impoverished form of the *quebrachal.* C - Vegetation sequence near Río Pilcomayo influenced by summer floods and dry winter

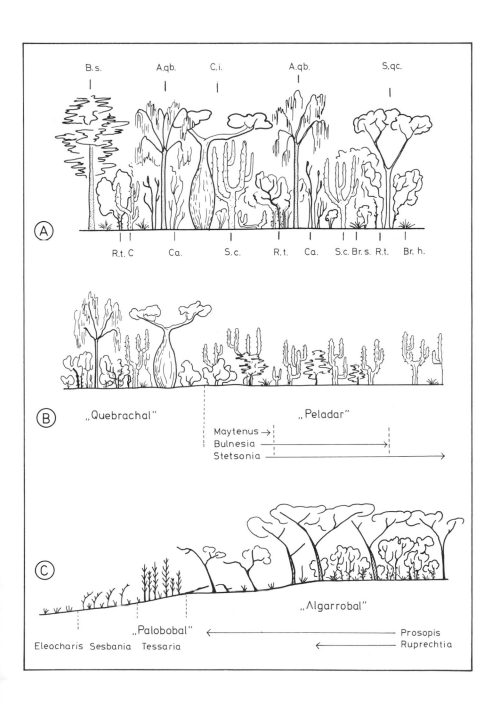

The *peladar*es are generally situated on elevated ground and, consequently, higher above the groundwater level. The presence of salt could play an important role in the differentiation of the *peladares*.

During winter, the beds of several rivers are completely dry and then usually colonized by herbaceous species of Amaranthaceae, Compositae and Gramineae. Some of the extremely xeromorphic forest islands are surrounded by a belt of homogeneous *Prosopis*-shrubs. These relatively hygrophilous trees tend to invade the adjacent savanna and apparently anticipate the advance of the *quebrachal*.

The vegetation of the areas close to the Pilcamayo River experience long summer-floods. In winter, the land is dry. The following sequence can be observed (Fig. 2C) (i) sparse grassland with *Eleocharis* in the parts which experience prolonged inundations; (ii) savanna grassland with scattered *Sesbania*; (iii) homogeneous *Tessaria integrifolia* shrubland at the mean level of the summer-floods; (iv) deciduous *Prosopis* forests *(algarrobal)* very damaged at their outer fringes as a result of the anaerobic conditions periodically brought by the changing flow of Pilcomayo River; (v) *Quebrachal*.

Gallery forests are frequent in the moister areas of southeastern Chaco. They can tolerate four months of continuous flooding. The tree stratum reaches more than 20 m in height. *Calycophyllum multiflorum*, *Pisonia zapallo* and *Chlorophora tinctoria* are very abundant (Fig. 3A) and are typical of all moister forests of the Chaco.

Another type of vegetation, characteristic of the colonization in the Pilcomayo River backwaters, is the *espartillar* (Fig. 3B). This formation is a savanna grassland of *Elionurus muticus* with scattered trees of *Schinopsis balansae* and *Astronium urundeuva* growing on sandy soils. The *espartillar* depressions are colonized by a *matorral* of *Acacia aroma* called *tuscal*.

Figure 3. A - Gallery forest. B - *Espartillar* savanna grassland with scattered trees and *Acacia aroma*. C - Transect showing the progressively impoverished vegetation from a gallery forest to a *peladar*. [A.qb. = *Aspidosperma quebracho-blanco*. B.s.=*Bulnesia sarmientoi*. C.i.=*Chorisia insignis*. R.t.=*Ruprechtia triflora*. S.c.=*Stetsonia coryne*.]

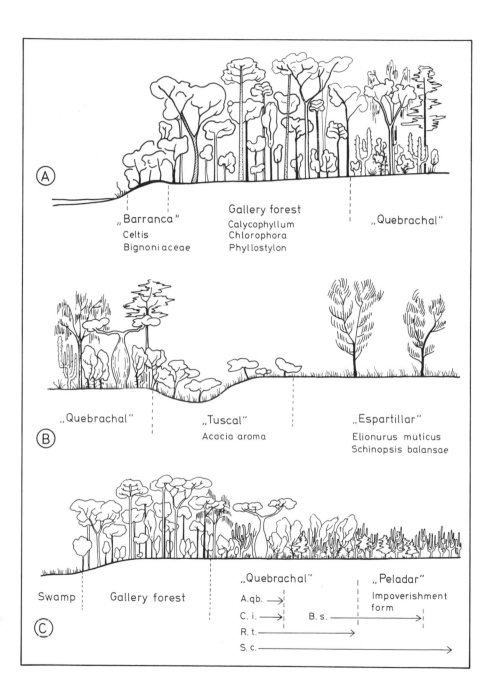

A

"Barranca"
Celtis
Bignoniaceae

Gallery forest
Calycophyllum
Chlorophora
Phyllostylon

"Quebrachal"

B

"Quebrachal"

"Tuscal"
Acacia aroma

"Espartillar"
Elionurus muticus
Schinopsis balansae

C

Swamp   Gallery forest

"Quebrachal"

A.qb. ⟶

C. i. ⟶

R. t. ⟶

S. c. ⟶

B. s. ⟶

"Peladar"
Impoverishment
form

Central and Northwest Paraguayan Chaco. These regions belong to the Departments of Boquerón and Nueva Asunción. Harsh environmental conditions, rainfall of less than 500 mm per year, and extreme temperatures occur in the central Chaco. These areas are generally covered with the typical formation of the Paraguayan Chaco. The extremely xeromorphic *quebrachal* forest are characterized by *Aspidosperma quebracho-blanco, Schinopsis quebracho-colorado, Ruprechtia triflora,* and Cactaceae. In the region of Filadelfia in Boquerón, occupation by the Mennonite community has created large agricultural monocultures. West of Boquerón, at the border with Bolivia, there is an area with sparse vegetation of Cactaceae and *Bulnesia* which is an impoverished form of the *quebrachal* and correlated with the more rigorous climatic conditions and, apparently, the soil salinity. Figure 3C shows the progressive impoverishment of the vegetation from gallery forest to *peladar.*

North Paraguayan Chaco. The northern parts of the Department of Chaco presents some important distinctions. The most important hills of the Chaco, the Cerro León complex with an altitude of about 720 m above sea-level, are found here. The Lagerenza River, called Timane on old Bolivian maps, is a superficial stream generated by summer rains and originating from a small hill complex near Lagerenza. It disappears near the Cerro León.

The *quebrachal* (Fig. 2A), is the type of vegetation that still dominates the lower and drier parts. This forest is generally taller and denser than in the Departments of Boquerón and Nueva Asunción.

The Río Lagerenza (Fig. 4A) is unusual among Chaco streams in that it rarely dries up completely. A gallery forest with hygrophilous species like *Cathormion polyanthum, Geoffroea striata, Coccoloba sp.,* and *Celtis sp.* colonizes the banks of the river. The first temporarily flooded zone is a savanna grassland with scattered shrubs of *Cathormion polyanthum* and *Copernicia australis.* The inner zone is separated from an outer flooded zone by a fringe of hygrophilous species on an intervening bank located on higher ground called *barranca.*

The outer flooded zone is a xeromorphic, *peladar*-like vegetation growing on bare, stony and clayish soil. Between the outer flooded zone and the *quebrachal* another *barranca,* with gallery forest species occurs.

Cerro León (Fig. 4B) is the highest hill of the Paraguayan Chaco. It now comprises part of *Parque Defensores del Chaco* but has been very scantily explored botanically. The first collections were made during the last decades by Bordas, Schinini, and Hahn. It consists of an northwest-southeast orientated series of geological folds. The complete formation covers an area of about 30 x 40 kilometers. Cerro León is surrounded by

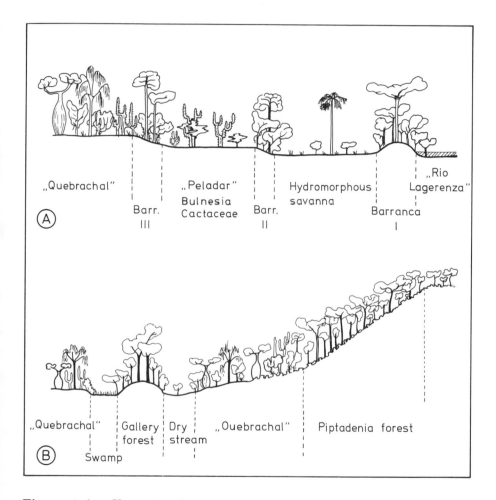

Figure 4. A - Vegetational sequence in the Río Lagerenza complex. B - Vegetational sequence in the Cerro León complex.

swamps and gallery forests with *Pisonia zapallo, Chlorophora tinctoria,* typical chaquean *quebrachal*. At the foot of the mountain is a mosaic of *Bumelia obtusifolia, Astronium urundeuva, Schinopsis balansae, Aporosella chacoensis,* and *Cathormion polyanthum.* These swamps are replenished by the streams draining Cerro León during summer. Like the

Lagerenza River, these temporary torrents disappear near the Cerro León.
The slopes of the hills are colonized by a deciduous forest of
Anadenanthera colubrina (= *Piptadenia macrocarpa*), *Pterogyne
nitens, Aspidosperma pyriformis, Cordia sp., Astronium urundeuva,
Erythroxylum sp.* and many species of Euphorbiaceae. This forest seems
to have considerable floristic resemblance to the western forests along the
foot of the Andes and to forests of eastern Paraguay. The top of the Cerro
León bears either the same *Anadenanthera* forest or a savanna shrubland
with scattered *Tabebuia caraiba, Jacaranda mimosifolia*, and a shrub of
Meliaceae.

Conclusion

The Chaco is poorly studied compared to eastern Paraguay. Many areas
have never been explored by botanists. The Chaco is an area covered by
relatively homogeneous vegetation. However, there are important
differences between the zonal Chaco forest which we call *quebrachal*, and
the forests of the hills, swamps, dunes, and gallery forests. This diversity
of ecosystems includes rare components of the Chaco which are often
sensitive to exploitation by man. A better knowledge of the Chaco forest
and especially of the regional variation in composition and physiognomy
is needed. Further projects should include extensive collecting as well as
edaphic and climatic studies.

Because of beginning deforestation and extensive use of the natural
resources in this region environmental problems may soon arise. Four
major threats may be mentioned. The agricultural colonization of the
central Chaco has resulted in the disappearance of its forest. The
construction of the paved Trans-Chaco road improves the accessibility to
the area and the value of the lands along the road. The exploitation of some
very valuable tree-species such as the "Quebracho colorado" (*Schinopsis
balansae*) and the "Palo Santo" (*Bulnesia sarmientoi*), will cause the
disappearance of these species in the near future. The extraction of rocks
required for the construction of the Trans-Chaco road, threatens the Cerro
León forests.

Acknowledgments

We thank the National Geographic Society, General Ismael Otazu,
Colonel Victorino Amarilla, Dr. L. Berganza, his wife, and his staff, Dr.
I. Basualdo, Lic. F. Mereles, Lic. N. Soria, Don Teodoro Brusquetti and
family, our colleagues Lic. P. Arenas and Ing. L. Spinzi for their
participation and collaboration.

Literature cited

Arenas, P. (1981). "Etnobotánica Lengua - Maskoy." *F.E.E.C., Buenos Aires.*

Brockmann, C. E. (1978). "Mapa de cobertura y uso actual de la tierra en Bolivia. Memoria Explicativa." *Geobol., La Paz.*

Cabrera, A. L. (1953). "Esquema fitogeográfico de la República Argentina." *Rev. Mus. La Plata, Sec. Bot.* **8,** 87-168.

Cabrera, A. L. (1976). "Regions fitogeográficas Argentinas" pp. 1-85 *In* Kugler, W. F. (ed.), "Enciclopedia Argentina de agricultura y jardinería." **2**(1). *Editorial Acme, Buenos Aires.*

Cardenas, M. (1940). "Formaciones vegetales del Chaco." *Rev. Cienc. Biol.*

Cardenas, M. (1951). "Un viaje botánico a la Provincia de Chiquitos del Oriente boliviano." *Rev. Agric. (Cochabamba)* **6.**

Castellanos, A. and Perez Moreau, R. A. (1945). "Los tipos de vegetación de la República Argentina." *Univ. Nac. Tucuman, Tucuman.*

Esser, G. (1982). "Vegetationsgliederung und Kakteenvegetation von Paraguay." *Trop. Subtrop. Pflanzenwelt* **38,** 5-113.

Fiebrig, C. and Rojas, T. (1933). "Ensayo fitogeográfico sobre el Chaco boreal." *Rev. Jard. Bot. Mus. Hist. Nat. Paraguay* **3,** 3-87.

Frenguelli, J. (1941). "Rasgos principales de fitogeografía Argentina." *Rev. Mus. La Plata, Sec. Bot.* **3,** 65-181.

Gorham, J. R. (ed.), (1973). "Paraguay: ecological essays." *Acad. Arts and Sci. Americas, Miami.*

Hassler, E. (1908). "Contribuciones a la flora del Chaco Argentino-Paraguayo; primera parte: florula Pilcomayensis." *Fac. de Cienc. Med, Buenos Aires.*

Hauman, L. (1931). "Esquisse phytogéographique de l'Argentine subtropicale et ses relations avec la géobotanique sud-américaine." *Bull. Soc. Roy. Bot. Belgique* **64,** 20-80.

Hauman, L., Burkart, A., Parodi, L. R. and Cabrera, A. L. (1947). "La vegetación de la Argentina." *Coni, Buenos Aires.*

Herzog, Th. (1910). "Pflanzenformationen Ostbolivias." *Bot. Jahrb. Syst.* **44,** 346-405.

Herzog, Th. (1923). "Die Pflanzenwelt der bolivischen Anden und ihres östlichen Vorlandes." *In* Engler, A. and Drude, O. (eds.), Die Vegetation der Erde **15,** 84-105. *Engelmann, Leipzig.*

Hueck, K. (1978). "Los bosques de Sudamérica." *Soc. Alem. Coop. Tec. (GTZ), Eschborn.*

Lorentz, D. P. G. (1876). "Cuadro de la vegetación de la República Argentina." *Peuser, Buenos Aires.*

Morello, J. H. and Adamoli, J. (1968). "La vegetación de la República

Argentina: las grandes unidades de vegetación y ambiente del Chaco Argentino; primera parte: objetivos y metodología." *Inst. Bot. Agric., Buenos Aires.*

Morello, J. H. and Adamoli, J. (1974). "La vegetación de la República Argentina: las grandes unidades de vegetación y ambiente del Chaco argentino; segunda parte: vegetación y ambiente de la provincia del Chaco." *Cent. Inv. Rec. Nat., Buenos Aires.*

Morello, J. H. and Saravia Toledo, C. (1959). "El bosque chaqueño." *Rev. Agron. Noroeste Argent.* **3,** 5-81, 209-258.

Pareja, J., Vargas, C., Suárez, R., Ballón, R., Carrasco, R. and Villarroel, C. (1978). "Mapa geológico de Bolivia; memoria explicativa." *Yacim. Petrol. Fisc. Boliv., La Paz.*

Ragonese, A. E. and Castiglioni, J. C. (1970). "La vegetación del parque chaqueño." *Bol. Soc. Argent. Bot.* **11,** 133-160.

Shrublands of the Venezuelan Guayana

O. HUBER

C. V. G. - Electrificación del Caroni C. A. (EDELCA) and Instituto Venezolano de Investigaciones Cientificas, Caracas, Venezuela

Among the many remarkable biological features of the Guayana region, southeastern Venezuela shrublands and related vegetation types play an important role in almost all landscape units. Perhaps, the frutescent life form is the most successful architectural and functional structure in the diverse plant cover of the region, because it presents an unparalleled diversity, both physiognomically and floristically there.

Despite their wide distribution and ecological significance, shrublands of the Venezuelan Guayana have received little attention. Only recently they were shown as an own entity on vegetation maps of the area (MARNR, 1982). The following overview gives the first detailed account of this biome, its spatial distribution, floristic composition, ecological significance, and dynamic relations.

Shrubs and shrublands

The arboreal or herbaceous growth forms are in general readily identified by their evident physiognomy and structure. Shrubs are more problematical in this respect in tropical vegetation types where an overwhelming variety of growth and life forms occupies every habitat.

In the following account, "shrubland" is used for a vegetation type in which a low, usually 0.5-5 m tall, ligneous compartment, formed by shrubs and shrub-like plants constitutes the main functional unit of the ecosystem. This circumscription emphasizes the fact that the principal energetic processes, photosynthetic rates, and biomass production of the ecosystem are performed in the shrub layer, regardless of occasional presence of trees and/or patches of herbaceous vegetation. A shrub as here defined, is a plant, mainly 0.5-5 m tall, with ligneous or sometimes subligneous stems and twigs, of predominantly basal ramification or not ramified at all. Also included as shrub-like are monopodial, mostly

TROPICAL FORESTS
ISBN 0–12–353550–6

unbranched and often monocarpic plants, which have their foliage either distributed along the axis or concentrated towards the apex. These have also been called the caulirosulate life form, rosette trees, or *Schopfbäumchen* (Fig. 3).

Geographic distribution

The Venezuelan portion of the Guayana Shield region covers some 400 000 square kilometers located south and east of the middle and lower Orinoco river (Fig. 1). The main physiographic provinces are the flat, upper Orinoco and lower Caroni lowland plains; the mainly hilly northern and northwestern piedmont and intermediate uplands of 100-1000 m altitude; and the mountainous Guayana Highlands, rising 1000-3000 m above sea-level and including the famous sandstone table mountain *tepuis*. Corresponding to these physiographic units a macro-thermic belt with mean annual temperatures above 24°C, a meso-thermic belt with mean annual temperatures between 12° and 24°C, and a submicro-thermic belt, with mean annual temperatures between 6° and 12°C, occur along an altitudinal gradient from the lowlands to the mountain summits. Most of the area is subject to a high rainfall of 2000-4000 mm per year and a short or no dry season except for the northern plains along the Orinoco river where a distinctly biseasonal climate with 800-1500 mm precipitation per year prevails.

Although shrublands seldom occupy large areas but occur in pockets or islands within the forest biome of the upper Orinoco/Guayana region, a number of areas with high proportions of shrublands can be identified. According to their physiognomic, floristic, and ecological characteristics, these shrublands can be grouped into macro-thermic (lowland), meso-thermic (upland), and submicro-thermic (*tepui* summit) associations. Whereas the limit between macro- and meso-thermic shrubland appears rather fluid in several instances, the differentiation between meso- and submicro-thermic shrubland is usually distinct, especially at the floristic and ecological level. The submicro-thermic shrubland is collectively called Pantepui-shrublands in this paper, because of its strong correlation with tepui summit habitats, a characteristic habitat of the Pantepui floristic province (Huber, 1987). Figure 1 shows the location of the main shrubland areas in the Venezuelan Guayana. Table 1 summarizes the main physical characteristics associated with these shrublands.

Rio Negro shrublands ("Bana"). These well studied shrublands of the upper Rio Negro basin occur in small patches on deep sandy podsols (Klinge *et al.*, 1977; Bongers *et al.*, 1985). They are typically formed of 0.5-3 m tall, dense shrub islands on low hills and ridges, and separated by shallow, temporarily inundated depressions with ephemeral herbaceous vegetation (Fig. 2).   Dominant families are Bombacaceae, Rubiaceae, Humiriaceae, Nyctaginaceae and Leguminosae.

Atabapo-Sipapo shrublands. Numerous peculiar shrub islands of varying density and extension are interspersed in the lowland areas of these two river systems at 50-100 m above sea-level. Dense, 1-5 (-8) m tall, shrubs form small groups separated by areas of bare white sand. This shrubland always occurs on dune-like features which are up to two meters higher than the surrounding forest terrain (Fig.2).   Dominant families are Malpighiaceae, Combretaceae, Humiriaceae, Leguminosae, Sapotaceae and Rubiaceae.

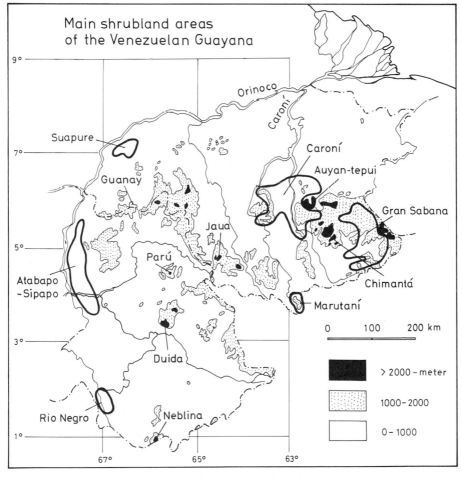

Figure 1. Main shrubland areas of the Venezuelan Guayana.

Table 1. Shrubland types of the Venezuelan Guayana and their main physical characteristics (Klinge *et al.*, 1977; Galán, 1984: Huber, 1986). t°C - mean annual temperature, **prec. mm** - mean annual precipitation in millimeters, **gr** - granitic, **s** - sandstone.

Shrubland type	Altitude m	Substrate	t°C	prec. mm
Rio Negro	100	sand	26	3500
Atabapo-Sipapo	50-100	sand	26	3000
Suapure	300-600	rock (gr)	25-23	2000
Caroní	400-1200	rock (s)	24-20	2500-3800
Gran Sabana	1000-1500	rock (s)	21-18	1600-2100
Marutaní	1000-1400	rock (s)	21-19	2500?
Pantepui	1600-2600	rock (s)/peat	17-11	> 3000

Suapure shrublands. These shrublands occur at 300-600 meters above sea-level on drier hilltops of the piedmont region and grow on plinthic and bauxite substrates derived from the underlying granitic rocks. They are always surrounded by dense sclerophyllous or mesophyllous forests and exhibit a low, 3-6 m tall, stunted physiognomy with few lianas, and a sparse understory. The small crowns of the shrubs are relatively open (Fig. 2). Dominant families are Humiriaceae, Olacaceae, Chrysobalanaceae, Annonaceae, Flacourtiaceae and Leguminosae.

Caroní shrublands. These shrublands occupy one of the largest areas, extending from the western edge of Sierra de Lema to Cerro Guaiquinima in the north, and to the vicinity of Uriman in the south. The altitudinal range is 400-1200 meters above sea-level. These usually rather dense and homogeneous shrublands of 2-7 (-10) m in height grow almost exclusively on sandstone slopes with varying degrees of inclination (Fig. 2). The dominant families are Rubiaceae, Combretaceae, Clusiaceae, Burseraceae, Chrysobalanaceae and Vochysiaceae.

Gran Sabana shrublands. This type of shrubland is frequently found at intermediate altitudes of 1000-1500 meters above sea-level in the eastern Guayana Highlands, growing preferably on a rocky substrate derived from sandstone but, in some instances, also on deep white sands. Density and height vary greatly depending on substrate but in general, the shrublands on sandstone are more evenly spaced, whereas the shrubs on sandy soils show a tendency to cluster into numerous small dense shrub islands separated by areas of bare white sand. Dominant families are Clusiaceae, Loganiaceae, Theaceae, Humiriaceae, Linaceae, Vochysiaceae and Compositae.

Marutaní shrublands. These dense shrublands occupy altitudes of 1000-1400 meters above sea-level in the Marutaní (Pia-zoi) massif. They were only recently discovered and described by Steyermark and Maguire (1984). The 1.5-4 m tall shrubland presents a continuous, rather homogeneous cover dominated by the shrub *Tyleria floribunda* (Ochnaceae). Other common families are Humiriaceae, Theaceae, Cyrillaceae, Malpighiaceae and Compositae.

Auyan-tepui shrublands. The summit of this large, approx. 700 square kilometers table mountain, rises from 1600 to 2400 meters above sea-level and harbours a relatively homogeneous 1-3 m tall shrubland type, growing preferably on deep organic soils (histosols). Low shrubs with rounded or stunted crowns of the families Theaceae, Rubiaceae, Tepuianthaceae, Malpighiaceae and Melastomataceae predominate. Caulirosulate plants are less common.

Chimantá shrublands. It appears that this huge, fragmented mountain massif contains the largest variety of frutescent life forms and vegetation types in the entire Pantepui province. The summit area covers 730 square kilometers at elevations between 1900 and 2600 meters above sea-level. It contains at least three physiognomically well distinct shrubland types with a series of local floristic variants on different summit sections (Huber, in press). Dominant families are Theaceae, Compositae, Ericaceae, Ochnaceae and Melastomataceae.

Jaua shrublands. Although this large tepui massif, which attains elevations between 1800 and 2400 meters above sea-level, is still little explored, the presence of larger shrubland areas has already been noted (Steyermark and Brewer-Carías, 1976). These seem to consist of rather homogeneous communites, formed by typical shrubs of 1-5 (-7) m, most of which belong to the Theaceae, Tepuianthaceae, Rutaceae and Compositae.

Guanay shrubland. The summit area of Cerro Guanay rises from 1600 to 2000 (?) meters above sea-level. It is covered by an almost continuous, dense shrubland type of 0.5-4 (-6) m tall plants growing on very broken rocky substrate. The dominant families are Theaceae, Melastomataceae and Rubiaceae.

Parú shrublands. A great portion of the large dissected internal plateaus of Cerro Parú rises from of 1200 to 1800 meters above sea-level. It is covered by a variety of shrubland types, mainly associated with the very broken rocky terrain. These almost unexplored shrublands appear heterogeneous in their aspect and floristic composition. Dominant families are Malpighiaceae, Tepuianthaceae, Theaceae, Ochnaceae and Aquifoliaceae.

Duida shrublands. Despite the four explorations made on the summit of this famous table mountain, little is known about its vegetation. According to Tate (in Gleason, 1931), different low shrubland types occur on crests and ridges on the southern part of the summit. The dominant families are Ericaceae, Theaceae, Ochnaceae, Compositae and Melastomataceae.

Neblina shrublands. On the summits of this large mountain complex, including also Cerros Avispa and Aracamuni to the north, a variety of shrubland types are found mainly on organic peat soils or on rocky substrates, at elevations between 1600 and 2500 meters above sea-level. Some shrublands of Neblina and Chimantá are physiognomically similar. The dominant families are Theaceae, Compositae, Melastomataceae, Rubiaceae and Rutaceae.

Physiognomy

A variety of different structural types exists among the Guayana shrubland formations. Although a certain basic and more widely occurring physiognomy can be recognized, each of the shrubland areas mentioned exhibits one or more autochthonous shrubland types with a particular morphological and structural constitution.

One important parameter for characterizing the physiognomy of a shrubland type is the height of the shrub layer. Generally, macro-thermic and meso-thermic shrublands are taller than those of the Pantepui province. The lowest type however is represented by the open "Bana" shrubland, with an average height of 0.3-2 m, but several transitional phases show progressive increase in height towards the surrounding low, medium- and tall Amazon *caatinga* forests. It also appears that shrubs growing on rocky substrates usually attain a taller size than those growing on sandy soils. The tallest shrubland types are found in the Caroní region where average heights of 6-8 m are frequent. The general height of lowland shrubland is 1.5-4 m with occasional emergents up to 6-7 m in height. Shrubland formations of open sites in the Pantepui province are usually 0.5-3 m high and very often show a remarkable uniformity of the upper crown level, influenced by the frequent, strong winds. If the shrubland grows in depressions and/or on irregular rocky terrain, the average height is 3-5 m and a few emergents reach up to 7 m.

Another important parameter is stand density. Although this parameter depends on local terrain conditions and is highly variable even within the same shrubland type, some patterns in the spatial distribution of the individual shrubs can be recognized (Fig. 2). On one hand the shrubs are grouped in islands of variable size, separated by more or less extensive open spaces, and showing a clustered aspect from the air. This pattern is characteristic of shrubland growing on sandy soils, such as of the Rio Negro "Bana", the Atabapo-Sipapo shrubland, and some shrubland types in the Gran Sabana region. On the other hand, shrublands growing on rocky slopes usually present a rather uniform distribution of the shrubs, although in some cases a slight tendency towards clustering can be observed, *e.g.* in some Gran Sabana shrublands. Where the individual shrubs are widely spaced, a herbaceous layer usually develops in the space between shrub islands giving rise to a shrub savanna; such transitional types are frequently found toward the margin of a shrubland area, especially in the Caroní region. The densest shrubland communities are found in the Pantepui region, on rocky terraces, and among some *páramo*-like shrubland types, which are virtually impenetrable.

The third important structural parameter is the vertical arrangement and stratification of shrubs within a given shrubland type. In many cases, only one shrub stratum is present, and the lower individuals belong mostly to juvenile regrowth. More commonly, however, two compartments can be recognized. A dominant main stratum of 1-3 m is complemented by a lower stratum of sprawling ground shrubs, the crowns of which occupy the space up to 1 m above the soil surface. *Humiria balsamifera* is a typical example of a sprawling low shrub, occupying the niche at the base of shrubs in the main stratum (Fig. 3). A third layer, formed by emergent shrubs and treelets, can also be present in some cases; their usually rounded and small crowns rise to more than one meter above the main stratum. Other life forms associated with the crown space, such as lianas, epiphytes, or parasites are normally present but rarely with a large number of individuals.

Figure 2. Shrubland types of lowland and upland Venezuelan Guayana (on sand, bauxite, and sandstone).

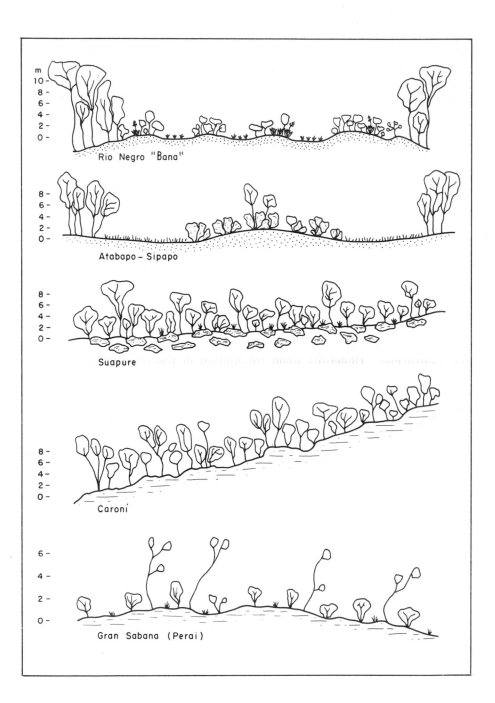

No other vegetation type of Guayana offers a wider spectrum of fruticose growth-forms than the various shrublands. All transitions from dense, umbrella-shaped shrubs to low gnarled, almost leafless shrubs are present (Fig. 3). The most peculiar and specific Guayanan growth-forms are found in Theaceae. Several species of *Bonnetia sensu lato* have thick, fleshy leaves densely arranged in the apical portion of the twigs looking like an artichoke. *Acopanea*, recently discovered on Chimantá, is the most unusual member of the family. The stem is reduced to a few centimeters and large stolons arising from the basal leaf-rosettes, connect hundreds of individuals (Fig. 3). In the Compositae, an entire genus, *Chimantaea*, has developed the caulirosulate habit, forming large *páramo*-like shrubland communities. This is the only example of such a typical tropical, high-altitude, life form in the Guayana mountains.

Floristics

A large number of spermatophyte taxa is involved in the floristic composition of the shrublands in the Guayana region. A detailed numerical analysis of the families, genera, and species has not yet been undertaken, partly because of the floristic complexity of these communities and partly because of the uncertain taxonomic status of several important groups, such as Clusiaceae, Myrtaceae, Leguminosae, and Lauraceae. Hopefully upon completion of the forthcoming *Flora of the Venezuelan Guayana* by J. A. Steyermark and collaborators such studies can be initiated. A clear-cut differentiation exists between the flora of the lowland and upland shrublands and that of the Pantepui province.

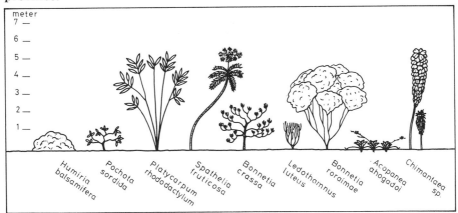

Figure 3. Some of the most common growth forms of shrubs in Venezuelan Guayana.

These differences can already be observed at the family level but become even more evident at the generic and specific level. Typical lowland shrub representatives are Bombacaceae (*Pochota* [syn. *Rhodognaphalopsis*]), Combretaceae (*Terminalia, Ramatuella*), Leguminosae (*Aldina, Macrolobium, Taralea* [syn. *Dipteryx*]), and Burseraceae (*Dacryodes, Protium, Trattinnickia*). *Pochota sordida* is the main shrub in the Rio Negro "Bana", but other species of this genus also occur widely in all other lowland shrubland formations of Guayana. Leguminous genera and species are frequent in the eastern Guayana shrub areas of Caroní and Gran Sabana, whereas Combretaceae play an extremely important role in both the eastern and the western sectors of the Guayana region.

In the Pantepui shrublands, the above mentioned families are either completely lacking (Leguminosae, Burseraceae), or are only scarcely represented (Bombacaceae). Evidently, the high mountain environment of Guayana with its acidic rock types and high rainfall regime, favors other, more specialized families, such as Theaceae *sensu lato,* Ochnaceae, Tepuianthaceae, Ericaceae and Compositae *pro parte.* Widely dispersed tropical families like the Rubiaceae, Melastomataceae, and Clusiaceae have also produced a large number of endemic species and even genera in the Pantepui shrublands.

For the Guyana region it is very important to note that almost monospecific communities in which one woody species tends to form large and dense colonies exist only in the Pantepui province. The phenomenon is particularly frequent in the eastern Pantepui area. On the 42 tepui-summits visited so far, we found at least eight associations absolutely dominated by one or two woody species. In these cases, a characteristic assemblage of accompanying herbaceous species usually occurs together with single individuals of other shrub species; they are, however, all clearly subordinate accessory elements of these plant associations. The most impressive examples of such almost monospecific communities are found in shrublands of *Bonnetia, Neogleasonia*, and *Neblinaria* of the Theaceae (the older nomenclature is maintained to show the three different communities, see Steyermark, 1984). Other examples are represented by *Chimantaea* (Compositae), *Mallophyton* (Melastomataceae), *Adenanthe* (Ochnaceae), and *Ledothamnus* (Ericaceae). Although the same genus may be present on several mountain summits, usually only one or few species of that genus form these communities on each tepui (Table 2.).

Recently, Steyermark (1986) has published an extensive account on the endemism in the Venezuelan Guayana. At least 30 of the approximately 40 endemic genera cited, represent important constituents of shrublands in the Pantepui province. This fact emphasizes the prominent role of these ecosystems in the process of floristic differentiation.

Table 2. Monospecific shrub communities in Pantepui.   *) now both in *Bonnetia* (Steyermark, 1984).

Shrub type	Family	Location
Chimantaea shrub	Compositae	Chimantá
Mallophyton shrub	Melastomataceae	Chimantá
Neogleasonia shrub *)	Theaceae	Chimantá
Adenanthe shrub	Ochnaceae	Chimantá
Adenanthe/Ledothamnus shrub	Ochn./Ericaceae	Chimantá
Bonnetia sessilis shrub	Theaceae	Auyan-tepui
Bonnetia sessilis/tristyla shrub	Theaceae	Yutajé/Guanay
Neblinaria shrub *)	Theaceae	Neblina

Ecology

Guayana shrubland types are edaphic plant communities, because climatic parameters do not seem to act as limiting factors in any of the mentioned shrubland areas. As a matter of fact, within a particular altitudinal belt each shrubland type grows under similar meso-climattic conditions as the adjacent forests and savannas.

In general, Guayana shrublands are associated with either sandy or rocky substrate. In the first case, the sand cover may be deep, as in the "Bana" shrubland and the Atabapo-Sipapo shrub islands, or shallow as in some Gran Sabana shrublands. The rocky substrate may be sandstone as in Caroní, Gran Sabana and most of the Pantepui shrublands, or granite as in Suapure shrublands. In the latter case, however, different communities are found on granite outcrops and on derived aluminum-rich soils, respectively.

The soil moisture conditions in shrublands on both, sand and rock, seem to be highly unfavorable because of low water retention capacity, excessive desiccation, *etc.* Most of the shrubs have developed deep and extensive root systems which are able to reach the water table or the water in rock crevices, thereby minimizing the adverse effects of highly variable water availability. Deciduous species are found only in the shrubland communities on granite rocks in the Suapure region. Here annual rainfall reaches only 2000 mm per year, and it is concentrated in an 8-10 months rainy season alternating with a distinct dry season. All other Guayana shrubland types of the lowlands, intermediate uplands, and tepui summits are evergreen. The foliage is usually coriaceous and grayish-green because of a thick protective cuticle.

In the case of the Rio Negro "Bana", it has been shown (Sobrado and Medina, 1980) that the predominant sclerophylly of shrubs growing on extremely acidic sandy soils is correlated with low availability of nitrogen and phosphorus. Although comparative studies have not yet been made in any of the other lowland shrublands of the Venezuelan Guayana, it seems reasonable to assume that similar conditions of oligotrophism exist here. In the case of the *páramo*-like Pantepui shrubland on Chimantá growing in organic, peat-like soils up to 2 m deep, recent nutrient analyses have also shown extremely low contents of nitrogen and phosphorus in vegetation and soil (Cuevas, 1987). In order to explain the presence of these well-differentiated and dense shrub communities under pronounced oligotrophic conditions, one has to assume extremely efficient and specialized nutrient cycling mechanisms. Considering that the underlying acidic and mineral-poor sandstones of the Roraima group are usually not participating in the soil forming process, the principal input of nutrients into these ecosystems must be by air.

Other peculiar ecological adaptations, such as the vertical leaf arrangement observed by Medina *et al.* (1978) on shrubs of the Rio Negro "Bana", apical clustering of leaves, or the frequent presence of pachycaulous stems and twigs with a leathery, often glossy bark typical of several community-forming *Bonnetia* species, have been observed in many families of almost all Guayana shrublands that range from the lowlands up to the highest tepui summits.

Although fires from adjacent savannas occasionally penetrate into shrublands, especially in the lowland and the Gran Sabana, this factor apparently does not play a major selective role in these ecosystems. Neither can it be asserted that fires are degrading forests toward shrublands, nor that they are responsible for the maintenance of a certain shrubland type. Where frequent burning occurs, a significant floristic change takes place very rapidly, giving rise to clearly recognizable secondary shrublands. The hypothesis of fire origin for certain Pantepui shrublands, recently advanced by Givnish *et al.* (1986), cannot be supported if the distribution of similar shrublands all over the Pantepui summits, where burning occurs extremely sporadically, is taken into account.

Dynamics

Klinge (1978) considers the "Bana" shrubland as the lowest stage of a spatial sere ranging from "tall *caatinga*" forest to "low *caatinga*" forest and then to "Bana." Bongers *et al.* (1985) recognize a further gradient from tall "Bana" to low and finally open "Bana." Both authors base their conclusion on biomass comparison and on certain characteristic species shared by all vegetation types involved. It must be noted, however, that the floristic composition of true "Bana" shrubland differs notably from that of the adjacent *caatinga* forests, especially if one considers the dominant shrub species and the important proportion of terrestrial rosette bromeliads in the former which are completely lacking in the forest.

On the other hand, the shrublands on lateritic bauxite soil in the Suapure region show a clear correlation with a low sclerophyllous forest growing nearby on sites with deeper topsoil. In this case, the shrubland clearly represents an edaphic variant of the low forest, also demonstrated by the high proportion of common species occupying the same niches in both formations (Huber and Guánchez, unpublished data).

One of the most surprising phenomena, has recently been observed in the Caroní shrublands. A few very typical and common shrub species, such as *Terminalia quintalata, Bonnetia sessilis* and *Platycarpum rhododactylum* were found to form extensive and dense forests up to 12 m tall on a rocky substrate covered by a relatively thick leaf litter. This rather homogenous forest type is suffering from a die-back pattern characterized by numerous distinct circular areas with many standing dead trees. Regeneration in these areas is reduced and in some instances, the forest cover is abruptly broken by almost treeless patches covered only by tall herbs, mainly Marantaceae and lianas. No explanation has yet been found for this phenomenon, but local conditions of severe nutrient stress could play an important role.

The other Guayana shrublands and especially the Pantepui shrublands, show certain floristic and ecological connections with the ligneous compartment of contiguous herbaceous ecosystems (shrub savannas, shrub meadows). Nevertheless, the overall ecological, floristic, and physiognomic characteristics of these shrublands are always sufficiently well differentiated from any other vegetation type of that particular landscape, therefore, justifying their recognition as one of the major and most diverse biomes of the Guayana region.

Acknowledgments

I thank Ministerio del Ambiente y de los Recursos Naturales Renovables (MARNR), Consejo Nacional de Investigaciones Científicas y Tecnológicas (CONICIT, Grants DDCT-ECO-4, S1-1343), C.V.G. - Electrificación del Caroní C.A. (EDELCA), C.V.G. - Técnica Minera C.A. (TECMIN), C.V.G. - Bauxita Venezolana C.A. (BAUXIVEN), The New York Botanical Garden, and Herbario "V.M. Ovalles." Special thanks go to Prof. S. S. Tillett for his kind support and the critical revision of this manuscript. My wife María and Mr. Tomás Rodríguez are sincerely acknowledged for the drawings.

Literature cited

Bongers, F., Engelen, D. and Klinge, H. (1985). "Phytomass structure of natural plant communities on spodosols in southern Venezuela: the Bana woodland." *Vegetatio* **63,** 13-34.

Cuevas, E. (1987). "Perfil nutricional de la vegetación de turberas en el Macizo del Chimantá, Edo. Bolívar, Venezuela. Resultados preliminares." *Acta Ci. Venez.* **38,** 366-375.

Galán, C. (1984). "Memoria explicativa del mapa de zonas bioclimáticas de la cuenca del Río Caroní." *CVG-EDELCA, Caracas.*

Givnish, T. J., McDiarmid, R. W. and Buck, W. R. (1986). "Fire adaptation in Neblinaria celiae (Theaceae), a high-elevation rosette shrub endemic to a wet equatorial tepui." *Oecologia (Berlin)* **70,** 481-485.

Gleason, H. A. (1931). "Botanical results of the Tyler-Duida expedition." *Bull. Torrey Bot. Club* **58,** 287-298.

Huber, O. (1986). "La vegetación de la cuenca del Río Caroní." *Interciencia* **11,** 301-310.

Huber, O. (1987). "Consideraciones sobre el concepto de Pantepui." *Pantepui* **1,**(2), 2-10.

Huber, O. (in press). "La Vegetación." *In* Huber, O. (ed.) El Macizo del Chimantá: un ensayo ecológico tepuyano." *O. Todtmann Editores, Caracas.*

Klinge, H. (1978). "Studies on the ecology of Amazon Caatinga forest in southern Venezuela - 2." *Acta Ci. Venez.* **29,** 258-262.

Klinge, H., Medina, E. and Herrera, R. (1977). "Studies on the ecology of Amazon Caatinga forest in southern Venezuela - 1." *Acta Ci. Venez.* **28,** 270-276.

MARNR (1982). "Mapa de la vegetación actual de Venezuela," 75 sheets, 1:250 000, *Caracas.*

Medina, E., Sobrado, M. and Herrera, R. (1978). "Significance of leaf orientation for leaf temperature in an Amazonian sclerophyll vegetation." *Rad. and Environm. Biophys.* **15,** 131-140.

Sobrado, M. A. and Medina, E. (1980). "General morphology, anatomical structure, and nutrient content of sclerophyllous leaves of the "Bana" vegetation of Amazonas." *Oecologia (Berlin)* **45,** 341-345.

Steyermark, J. A. (1984). "Flora of the Venezuelan Guayana - 1." *Ann. Missouri Bot. Gard.* **71,** 297-340.

Steyermark, J. A. (1986). "Speciation and endemism in the flora of the Venezuelan tepuis." pp. 317-373 *In* Vuilleumier, F. and Monasterio, M. (eds.), "High altitude tropical biogeography." *Oxford Univ. Press, New York and Oxford.*

Steyermark, J. A. and Brewer-Carías, C. (1976). "La vegetación de la cima del Macizo de Jaua." *Bol. Soc. Venez. Ci. Nat.* **132/133,** 179-405.

Steyermark, J. A. and Maguire, B. (1984). "Informe preliminar sobre la flora de la cumbre del Cerro Marutaní." *Acta Bot. Venez.* 14(3), 91-117.

Diversity of east Ecuadorean lowland forests

H. BALSLEV AND S. S. RENNER

Botanical Institute, Aarhus University , Denmark

One of the most intriguing topics in botany today centers on the diversity of the terrestrial flora. The questions associated with tropical diversity, its origin and maintenance, are in recent years being approached from many different perspectives (see, for example, Ashton, and Gentry, this volume). The primary thrust of this paper is the examination of different kinds of information concerning vascular plant species diversity in eastern Ecuador in an attempt to give a more accurate picture of the diversity among the plants of this country.

Ecuador spans the equator from 1°30′ northern to 5° southern latitude. It is traversed from north to south by the Andean cordillera, which divides the country into three natural regions: the coastal plain, the cordillera, and the eastern lowlands. The proximity of the Niño sea-current to the northern parts of the coastal plain and of the Humboldt sea-current to its southern parts creates a gradient from deserts to pluvial forests there. The vegetation belts along the Andean cordillera grade from tropical forests in the lowlands to *páramo* vegetation near the perpetual ice caps of the highest peaks that reach more than 6000 meters above sea-level. A complex geological history with much volcanic activity and tectonic movements have contributed to the high diversity of soils and habitats in Ecuador. This setting has created excellent opportunities for the diversification of the flora, and despite its restricted size of about 300 000 square kilometers, Ecuador has one of the richest floras among tropical countries.

Oriente - its geography, climate, and vegetation

The eastern part of Ecuador, locally called *Oriente*, is a flat terrain comprising the bulk of the provinces Napo, Pastaza, and Morona-Santiago. It covers an area of circa 130 000 square kilometers below 600 meters elevation or of 136 500 square kilometers, *i.e.*, 5% more, if the Andean slopes up to an altitude of 900 m are included. In our experience,

TROPICAL FORESTS
ISBN 0–12–353550–6

the belt between 600 and 900 meters belt is the natural distributional limit of many species of flowering plants, which tend to be confined either to the montane forests above this belt or to the lowland forests below. To the north, the *Oriente* is limited by the border between Colombia and Ecuador. To the west, the 600-900 meters elevation lines form winding but generally north/south oriented borders. To the south and east the *Oriente* is limited by the *Línea del Protocolo de Río de Janeiro de 1942*, which Peru claims as its border with Ecuador, a claim Ecuador disputes.

The climate is perpetually wet with annual precipitations between 2500 mm and 5000 mm, the parts receiving most rain being those closest to the Andean foothills. The precipitation is distributed throughout the year and no month receives less than 100 mm rain on the average; May through July are the wettest months and December through February are the least wet months. The mean temperature for all months is about 25°C. The soils of the area constitute a mixture of old weathered and more recently deposited fluvial sediments, as described in the contribution of Salo *et al.* (this volume). There are no peat swamps or podsols.

The San Miguel, Putumayo, Coca, Napo, Pastaza, Cangaime, and Santiago are all major rivers draining the eastern slopes of the Andes and flowing through the *Oriente*. The rivers are generally white water rivers with a large load of clay particles which become suspended in the waters when they move down the Andean slopes with high velocity and eroding capacity. The water level in the white water rivers originating in the Andean cordillera does not follow a seasonal pattern. Instead, the water level may be high or low at any time of the year, and changes from one extreme to the other may occur overnight, for example, after a rainstorm in a river´s catchment area. For this reason, the floodplain forests associated with the white water rivers are not subject to any seasonal regime of water level changes. In the lower Amazon basin, floodplain forests inundated by white waters are called *várzea* (Prance, 1979); however, the Ecuadorean white water inundated forests hardly qualify for this designation because of the lack of seasonality in flooding regime.

Between the large white water rivers there is a system of smaller rivers with black, oligotrophic water draining the flat plain of the *Oriente*. In some places, black water lakes have formed. In contrast with the white water rivers, the water level in the black water rivers and the lakes is entirely dependent on the precipitation within the *Oriente* itself. The local seasonal changes in rainfall are just great enough for evapotranspiration sometimes to exceed precipitation so that some lakes, such as, for example, those of the Cuyabeno area, dry up from December through March. Thus, the forests along their margins are exposed to a seasonal change in water level and correspond to the black water flooded *igapó* forest of the lower Amazon (Prance, 1979). Beyond the floodplains, the uplands or *terra firme* are covered with tall, species rich *terra firme* forest.

Inventory data

Our inventory data set concerning species richness from the *Oriente* originates from Añangu on the southern shore of the Río Napo and includes two transects, one in the non-flooded forest and one in the floodplain forest, and a one hectare quadrat plot (Balslev *et al.*, 1987; Korning and Thomsen, 1987). The area within which the three samples were taken is about 3.5 x 0.5 kilometers. In these inventories, only trees above 10 cm dbh were included. The transects were marked at every 20 meters and sampled according to the point-centered quarter method of Cottam and Curtis (1956). A detailed description of the area and sampling methods is given in Balslev *et al.* (1987).

The non-flooded forest transect. This transect is located along a low ridge some 40 to 80 meters above the mean level of Río Napo, but only 300 meters to the south of its shore and is never flooded. It is four kilometers long with 201 sample points and 804 sampled trees. There are 244 tree species from 53 different families. We calculated a density of 728 individuals per hectare, with the first 728 trees along the transect including 228 species. The 10 most important families, their numbers of species, and their percentage of the total sample of species are shown in Table 1. A few common species make up the majority of individuals. For example, the 24 most common species (*i.e.*, the top 10% when species are ranked by frequency) account for 50% of the individuals. The "species/area curve" does not level out in the sample.

The flood plain transect. This transect is located on the floodplain about 200 meters away from the Río Napo and is flooded when the river rises. The water level within the forest reaches 1-1.5 meters when the river is at its highest. The water is a mixture of white water from the Río Napo and black water from the small tributary Río Añangu. This transect is 2.1 kilometers long with 105 sample points and 420 sampled trees. It includes 149 species from 44 different families. We calculated a density of 417 individuals/hectare, and the first 417 trees along the transect included 146 species. The 10 most important families, their numbers of species, and their percentages of the total sample of species are shown in Table 1. As in the non-flooded forest, the trees on the floodplain are not equally divided among the species; the 18 most common species account for 50% of the individuals.

Table 1. The most important families in inventories of trees above 10 cm dbh at Ãnangu, eastern Ecuador (Balslev *et al.*, *1987;* Korning and Thomsen, 1987). For each family the number of species is given and the percentage this represents of the total number of species in the sample (=relative diversity

Non-flooded transect			*Floodplain transect*			*One hectare plot*		
	Species	%		Species	%		Species	%
Moraceae*	26	11	Moraceae*	18	12	Moraceae*	18	11
Lauraceae	20	8	Rubiaceae	12	8	Legumin.**	17	11
Legumin.**	20	8	Legumin.**	12	8	Lauraceae	12	7
Meliaceae	15	6	Lauraceae	7	5	Arecaceae	9	6
Rubiaceae	15	6	Meliaceae	6	4	Myristicaceae	7	5
Burseraceae	10	4	Sterculiaceae	6	4	Annonaceae	6	4
Euphorbiaceae	10	4	Arecaceae	5	3	Burseraceae	6	4
Arecaceae	8	3	Chrysobalan.	5	3	Lecythidaceae	5	3
Myristicaceae	8	3	Flacourtiaceae	5	3	Melastomatac.	5	3
Annonaceae	7	3	Annonaceae	4	3	Sapotaceae	5	3
Total		56			55			60

*Moraceae *sensu stricto, i.e.* excluding Cecropiaceae. **Leguminosae *sensu lato, i.e.* including Caesalpiniaceae, Fabaceae, and Mimosaceae.

The one hectare study plot in non-flooded forest. This plot was established by Korning and Thomsen in 1985 for their M.Sc. thesis work (Korning and Thomsen, 1987; Korning, 1987; Thomsen, 1987). It is located close to the non-flooded transect on the top of a ridge. The one hectare square plot has 734 trees over 10 cm dbh, and these belong to 153 species. The 10 most important families, their numbers of species, and their percentage of the total sample of species are shown in Table 1. The fact that few species account for the majority of individuals is even more pronounced in this square plot than in the transect. Here, the nine most common species (*i.e.*, the top 6% of the species ranked by frequency) account for 50% of the individuals.

The three samples at Añangu together include 1958 trees over 10 cm dbh belonging to 394 species. The square plot and the transect in the unflooded forest share 30% of their species; the square plot and the floodplain transect share 8% of their species; and the non-flooded and floodplain transects share 19% of their species. These species numbers and percentages of shared species clearly show that the area around Añangu must contain many more species of trees than the 394 encountered so far.

Distribution data

The second kind of information bearing on species diversity in the eastern lowlands comes from an analysis of the total distribution of 536 species of vascular plants occurring in Ecuador. These species were selected at random from available monographs and belong to a variety of growth and life forms such as trees, shrubs, herbs, saprophytes, parasites, and epiphytes, to 31 different families, and to a variety of different life zones in the country (Balslev, 1988).

Of the 536 species, 42% occur in the lowlands below 900 meters elevation, 49% occur at mid-elevations between 900 meters and 3000 meters, and 9% occur at high elevations above 3000 meters. Of the Lowland species 44% occur only west of the Andes, and 36% only east of the Andes, whereas 20% are trans-Andean. Of the total sample of 536 species of Ecuadorean plants, these occurring in *Oriente* account for 24% of which 23 % are widespread species and the remaining 1 % is endemic to the eastern lowlands of Ecuador below 900 meters elevation. There are no physiographical barriers separating the eastern Ecuadorean lowlands from the rest of the Amazon basin, which may account for the low endemicity.

Checklist data

Our final set of data derives from an enumeration of all vascular plants collected in eastern Ecuador below an elevation of 600 meters (Renner *et al.*, in prep.). It was already pointed out that a considerable number of lowland species reaches their upper altitudinal limit around 600 meters. The checklist includes information on altitudinal range for species growing near the 600 meters cut-off. It also includes information on a species´ distribution within Oriente and its growth and life form. The list is based on all Ecuadorean material in the herbaria of Aarhus (AAU), Göteborg (GB), Stockholm (S), and the Universidad Católica in Quito (QCA) and for selected families also on holdings of the Field Museum (F), the Missouri Botanical Garden (MO), the New York Botanical Garden (NY), and the United States National Herbarium (US).

Of the 145 families of flowering plants in the checklist, 30 (20%) have already been treated in the *Flora of Ecuador* series; another 20 family enumerations (14%) have been checked by the specialist working on the respective family for *Flora of Ecuador* or *Flora of Peru*. Of the remaining 96 families, 66 (45%) have less than 11 species in *Oriente*. Among the 30 families with more than 11 species, the most problematic and the most poorly collected ones are the Orchidaceae, Leguminosae, and Myrtaceae, followed by the Araceae, Solanaceae, and Monimiaceae. Even though the relevant literature (*e.g.*, Flora of Peru, Flora Neotropica) was consulted, 11% of the 3000 species in the checklist remain unnamed.

The largest and relatively most explored province, Napo, contains 73% of the species. Only 39% of the species have been collected in the two southern provinces, Pastaza and Morona-Santiago, and 28% of the species are restricted to them. A history of the botanical exploration of the *Oriente* (Renner, in prep.) shows the very pronounced concentration of collecting in Napo to the neglect of Pastaza and particularly Morona-Santiago.

Therefore, our impression that the southern part of the eastern Ecuadorean lowlands close to the Peruvian Amazon houses a flora distinct from the northern part close to the Colombian border must await support from future intensive collecting in the southern provinces.

The most common growth form which accounts for 40% of the species is the tree. Herbs, shrubs, and epiphytes each account for 20, 16, and 11% of the species in the checklist, and the remaining 13% are divided among climbers, hemiepiphytes, vines, lianas, parasites, and stranglers.

The 10 most important families in the checklist are shown in Table 2, which also gives the current number of species recorded in each family and their percentage of the total number of species in the checklist. The Orchidaceae, Melastomataceae, Araceae, and Solanaceae consist mainly

---

Table 2. The most important families in the Checklist of flowering plants in Oriente (Renner *et al.*, in prep.). For each family the number of species is given and the percentage it represents of the total number of species in the sample (=relative diversity).

	Species	%
Orchidaceae	200	7
Rubiaceae	179	6
Melastomataceae	168	6
Leguminosae (incl. Caesalpiniaceae, Fabaceae, and Mimosaceae)	152	5
Moraceae (excl. Cecropiaceae)	83	3
Arecaeae	80	3
Annonaceae	73	3
Solanaceae	72	3
Lauraceae	69	2
Myrtaceae	66	2
Piperaceae	66	2
Euphorbiaceae	63	2
Arecaceae	60	2
Total		46

of epiphytes, climbers, shrubs, and herbs and, therefore, were not important in the inventories at Añangu. Otherwise, the most important families in the checklist are also among the most important ones in the inventories. These families - listed in Table 1 - mostly coincide with those Gentry (1988) sees as the contributors of about half the species richness in any 0.1 ha sample of neotropical lowland forest. An exception are the Burseraceae which contribute 4% of all species at Añangu, but are not mentioned as important by Gentry. Burseraceae are also very important in central Amazonian inventories (Rankin-de-Merona *et al.*, 1989).

Conclusion

The checklist shows that 3000 species of flowering plants are currently known from the eastern lowlands of Ecuador. The trees - of any size - account for 1200 or 40% of these. However, in a recent progress report on their work towards a guide to the trees of the eastern Ecuadorean lowlands, Neill and Palacios (1987) estimated a final number of 3000 trees, which would mean that the number of currently known tree species would have to rise two and a half times. At Añangu, a small area of 3.5 x 0.5 kilometers already has 394 tree species (over 10 cm dbh) in a sample of 1958 individuals. Clearly, the inventory method is extremely effective in sampling tree species richness, and continued inventory work in carefully chosen areas (*i.e.*, away from the few centers of previous collecting activities) is the surest way to increase the number of large tree species known from the *Oriente*.

If the sample of species whose distribution in Ecuador we analyzed is indeed representative of the Ecuadorean flora, as we believe it is, about 1/4th (24%) of all Ecuadorean species of vascular plants pertain to the *Oriente* below 900 meters altitude. Based on information in the checklist about the altitudinal ranges of numerous species, we estimate that at least 20% come from below 600 meters. Benjamin Øllgaard (pers. comm.) estimates that there are about 300 species of pteridophytes in the eastern lowlands, which together with the 3000 species of flowering plants already included in the checklist would give a total of 3300 vascular plants for *Oriente* below 600 meters. Based on this figure and our estimate that the *Oriente* contributes 20% to the vascular plants of Ecuador, we extrapolate a minimum of 16 500 species of vascular plants in Ecuador. For our method of estimating species richness to yield a figure coinciding with earlier estimates of 20 000 vascular plants in Ecuador (Gentry, 1978; Harling, 1986), the number of species in the eastern lowlands would have to be circa 4000. Keeping in mind what was said above about the richness in tree species revealed by the inventories and about large, poorly known families, such as orchids, legumes, and aroids, this increase is entirely likely.

Acknowledgments

We are grateful for comments on this manuscript by P. Ashton, B. Boom, and U. Molau. We are particularly indebted to David Neill of the Missouri Botanical Garden for sharing information on all tree species he and his collaborators have recently collected in the eastern lowlands and to Calaway Dodson for a list of almost 200 species of orchids his team has collected near Misahualli and Tena.

Literature cited

Balslev, H. (1988). "Distribution patterns of Ecuadorean plant species." *Taxon* **37,** 567-577.

Balslev, H., Luteyn, J. L., Øllgaard, B. and Holm-Nielsen, L. B. (1987). "Composition and structure of adjacent unflooded and floodplain forest in Amazonian Ecuador." *Opera Bot.* **92,** 37-57.

Cottam, G. and Curtis, J. T. (1956). "The use of distance measures in phytosociological sampling." *Ecology* **37,** 451-460.

Gentry, A. H. (1978). "Floristic knowledge and needs in Pacific tropical America." *Brittonia* **30,** 134-153.

Gentry, A. H. (1988). "Changes in plant community diversity and floristic composition on environmental and geographical gradients." *Ann. Missouri Bot. Gard.* **75,** 1-34.

Harling, G. (1986). "Flora of Ecuador - its present status." pp. 9-10 *In* Øllgaard, B. and Molau, U. (eds.), "Current Scandinavian Botanical Research in Ecuador." *Rep. Bot. Inst., Univ. Aarhus* **15**.

Korning, J. (1987). "Studies of Amazonian tree and understory vegetation and associated soils in Añangu, east Ecuador. Part. II, An analysis of vegetation structure and discussion of sample methods and the relation between vegetation and soil." *Unpublished cand. scient. thesis, Aarhus Univ.* pp. 1-40.

Korning, J. and Thomsen, K. (1987). "Studies of Amazonian tree and understory vegetation and associated soils in Añangu, east Ecuador. Part. I, a. General background, b. Analysis of physical and chemical properties of three soil profiles." *Unpublished cand. scient. thesis, Aarhus Univ.* pp. 1-43.

Neill, D. and Palacios, W. (1987). "Proyecto Arboles de la Amazonia Ecuatoriana - Informe interno de los avances del proyecto." *USAID and MAG, Quito,* pp. 1-24.

Prance, G. T. (1979). "Notes on the vegetation of Amazonia III. The terminology of Amazonian forest types subject to inundation." *Brittonia* 31, 26-38.

Rankin-de-Merona, J. M., Hutchings H., R. W. and Lovejoy, T. E. (1989). "Tree mortality and recruitment over a five year period in undisturbed upland rain forest of the central Amazon." *In* Gentry, A. (ed.), "Four Neotropical Forests." *Yale Univ. Press, New Haven.*

Thomsen, K. (1987). "Studies of Amazonian tree and understory vegetation and associated soils in Añangu, east Ecuador. Part. III. An analysis of variation and affinities in species composition among forest types within a limited locality of terra firme and relation of these to soil conditions." *Unpublished cand. scient. thesis, Aarhus Univ.,* pp. 1-74.

Speciation and diversity of Ericaceae in Neotropical montane vegetation

J. L. LUTEYN

New York Botanical Garden, U.S.A.

The Ericaceae, including Monotropaceae and Pyrolaceae, are a large, cosmopolitan family of 110-120 genera and perhaps 4000 species which inhabit the temperate regions of the world including cooler montane areas in tropical latitudes. In the Neotropics there are 48 genera and nearly 800 species (Appendix 1). Although Ericaceae are prominent features of tropical montane regions throughout the world and often dominate a major vegetation type - "Befaria belt" and "ericaceous belt" - they are, paradoxically, little known ecologically or biologically. Virtually no investigations have been made of tropical Ericaceae with regard to soils, breeding systems, physiology, dispersal mechanisms, presence of mycorrhiza, ecotypic variation, predation, chromosomes, chemistry, *etc.*

Morphology

There is a diverse array of morphological features in Ericaceae with ample variation important for taxonomic consideration. For example, stamens vary considerably and yet are generically constant. Therefore, at the generic level, taxonomists have historically used stamen type along with a combination of other major floral and vegetative features, such as ovary position (superior vs. inferior), fruit type (loculicidal or septicidal capsule vs. berry), corolla type and shape (choripetalous vs. sympetalous, terete vs. winged), *etc.* At the specific level, traditional characters such as size and shape of leaves, floral bracts, *etc.*, inflorescence type and length, and pubescence patterns are very useful (Smith, 1932). A previously unnoticed character, the types of glands of the inflorescence, especially the calyx, were used extensively at the sectional and series levels in *Cavendishia* (Luteyn, 1976, 1983) and seem to be unique to that genus. Furthermore, my observations during extensive field work have revealed another series of characters that may be taxonomically useful, which have not previously been used. Many of these are probably strongly associated with pollination ecology and include floral color, especially the

TROPICAL FORESTS
ISBN 0–12–353550–6

contrasting colors of the corolla, calyx, pedicel, and floral bracts, as well as corolla shape - both characters that do not preserve well after drying (*cf.* Luteyn, 1976, 1983, 1984, 1986, and 1987 for examples of how these characters have been taxonomically useful).

Geographical distribution

Within the Neotropics, the Ericaceae are concentrated in northwestern South America, primarily Colombia, Ecuador, Peru, and Venezuela. From there the numbers of both genera and species taper off in every direction (Fig. 1A). The distribution of the family is concentrated in five biogeographical regions shown in Figure 1B, the structural units of which are defined by distinct geologic histories.

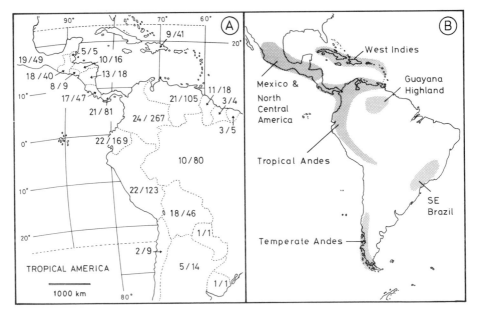

Figure 1. A - Distribution of Latin American Ericaceae. Numbers of genera and species are indicated for each country. B - Centers of diversity of the Ericaceae in Latin America.

Figure 2 shows the numbers of genera and species and the extremely high percentages of endemic species of Ericaceae found within each of the five biogeographic regions, along with an indication of the numbers of species common to the regions and in overall relation to North America and the temperate Andes. Figure 3 shows the distribution of *Cavendishia*, the largest Andean genus, giving the numbers of species found in each degree-square of its range. The important thing to note in Figs. 1, 2, and 3 is the pronounced endemism of the genera and species in each of the five biogeographic regions. For example, 750 out of 756 species or 99% of neotropical Ericaceae are found nowhere else in the world, 94% of the 540 species in the tropical Andean biogeographic region are endemic to that area, and 75% of the 69 species of *Cavendishia* in Colombia are endemic to that country. An example of high diversity in a very local area comes from the wet western slopes of the Western Cordillera of Colombia.

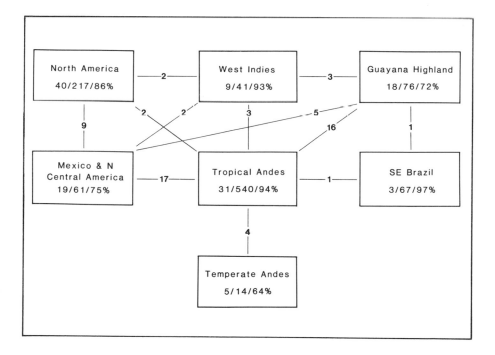

Figure 2. Biogeographical relationships of the New World Ericaceae. Numbers refer to the number of genera/number of species/percentage of species endemic to the region; numbers along the connecting lines refer to the number of species common to the linked regions.

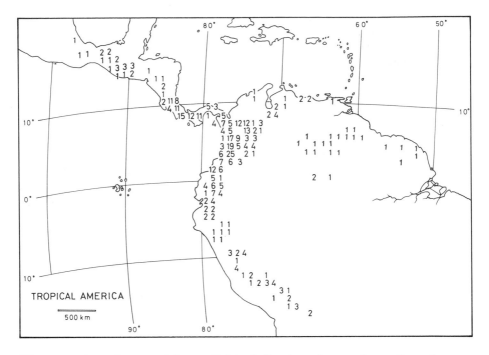

Figure 3. Distribution of *Cavendishia* indicating the numbers of species per degree-square.

There, after only seven days along a five-kilometer stretch of road through a disturbed premontane forest at 1300-1800 m above sea-level, I collected nine genera and 41 species of Ericaceae which is equal to 15% of the species in Colombia. No fewer than 12 species were new to science. Of the 41 species, 16 belonged to *Cavendishia* and only five had been recorded from that particular degree-square in my recent monograph (Luteyn, 1983).

At both the generic and specific level I am now beginning to recognize discernible floristic assemblages in the Neotropics. Some of these correspond to the biogeographic regions mentioned above: The Mexican element is characterized by *Comarostaphylis*, *Arbutus xalapensis*, and *Arctostaphylos pungens*; the West Indian element by *Lyonia* with 25 endemic species, *Gonocalyx*, and *Vaccinium racemosum*; the Guayana Highlands element by the endemic genera *Ledothamnus*, *Tepuia*, *Mycerinus*, and *Notopora*, as well as *Pernettya marginata*, *Vaccinium puberulum*, *V. euryanthum*, and *Thibaudia nutans*; and the southeastern Brazilian element by *Agarista* with 25 species, and *Gaylussacia* with 36 endemic species.

The tropical Andean biogeographic region has been further subdivided into finer structural units by Simpson (1975) and modified by Berry (1982). In this region, the Ericaceae demonstrate a *páramo* element above 3000 m elevation characterized by the genus *Plutarchia* in Colombia, and *Gaultheria* spp., *Vaccinium floribundum*, *Pernettya prostrata*, and *Disterigma empetrifolium* throughout the range; a southern Peru-northern Bolivia element characterized by *Demosthenesia*, *Siphonandra*, and *Pellegrinnia*; and a lowland Pacific "Chocó" element below 1000 m elevation, first noted by Smith (1946), and characterized by *Anthopterus, Killipiella, Macleania pentaptera*, 16 species of *Cavendishia*, and 10 species of *Psammisia*.

Within the tropical Andean region, moisture gradients definitely affect ericad distribution: There are always more species of Ericaceae on the wet side of the mountain than on the dry. For example, along the western Pacific-facing slopes there are few species in Costa Rica and Panama where the slopes are dry, many species in Colombia and northern Ecuador where they are wet, and virtually no species from extreme southern Ecuador south to Chile where they are extremely dry. The reverse is true on the eastern Amazon-facing slopes where from Venezuela to Colombia there are only moderate numbers of species, probably because of the drying effect of the *llanos*, while from Ecuador to northern Bolivia where it is wet there are many species.

Although the Ericaceae show regional development in each of the three cordilleras of Colombia and in the Amotape-Huancabamba transition zone of southern Ecuador and northern Peru, I have not yet analyzed all the data for those areas. *Satyria panurensis* and *Psammisia guianensis* have a distinctive distribution in an arc around the Amazon Basin at the Andes/Amazonia ecotone and frequently occur in white-sand savannahs.

Notable disjunct distributions include the West Indian genus *Gonocalyx* with one species in Colombia and two in Costa Rica; the Andean genus *Befaria* with one species in Cuba and one in southern Florida and Georgia, USA.; and the circumboreal *Arctostaphylos uva-ursi* with subspecies *cratericola* in Guatemala. The world-wide distribution of *Monotropa uniflora* includes the Neotropics where it follows the natural distribution of *Pinus* through the Mexican-Central American highlands into Belize and then continues south to Colombia in *Quercus* forests, and pine plantations.

General ecology

In the Neotropics, Ericaceae predominate in moist, semi-disturbed habitats of cool montane regions. They are most common in the areas called "cloud forest", *yungas, ceja*, or Andean forest-wet forests of middle elevations between 1000 and 2800 m elevation with ample light and abundant precipitation in the form of clouds, fog, mist, or rain. In this habitat they may be terrestrial or epiphytic, prostrate, pendent to erect shrubs, or rarely small trees. Some species occur above the treeline at elevations between 3000 and 4000 m in the *páramo*-areas which are always cold and wet, often receive frost in the early morning, and are dominated by grasses, sedges, and species of the composite genus *Espeletia*. A few species are found in mangrove swamps as epiphytes (Smith, 1946). Because of their heliophilic ecology, they are frequently found as pioneers near the craters of volcanoes or in recent landslide areas. Man has had great influence on the distribution of Ericaceae in many instances by creating disturbed areas, especially along the forest edge and steep slopes of clearings after logging or road building.

Pollination of Ericaceae in temperate and subtropical latitudes is primarily by bees, but in the Neotropics bird-pollination is the rule with hummingbirds acting as the main vector (Fig. 4). Dispersal of the small, light seeds in genera such as *Befaria, Lyonia*, and *Agarista*, which have capsular fruits, is by wind; however, most of the neotropical genera are berry-fruited and birds or small mammals act as agents of dispersal.

Adaptations to montane habitats

The high number of species of Ericaceae in the neotropical mountains, especially the wet tropical Andes and their prominence - and sometimes dominance in certain communities - acknowledge the extensive adaptive radiation and success of the family in this region. Let us now examine some of the adaptations which are responsible for their success in montane habitats.

The family has a physiological tolerance to high light intensity, high constant moisture, cool temperatures, and acid soils - conditions found primarily in the wet tropical Andean biogeographic region. In low-light conditions within forest, the taxa do not flower; however, where there are gaps or forest- or stream-edges they will flower, or they may become scandent as in *Psammisia falcata* and *Thibaudia rigidifolia* to reach the canopy. The preference for consistently high moisture may be readily observed when one crosses to the drier or more seasonal side of a mountain: The Ericaceae almost immediately disappear. There are genera occurring under drier conditions (*e.g.*, genera of the Arbuteae such as *Arbutus, Arctostaphylos*, and *Comarostaphylis*), but these are less specialized within the family and occur outside the wet tropical Andes.

Figure 4. *Lampornis calolaema* (female) feeding at *Cavendishia capitulata* at Monteverde, Costa Rica. Photo by Dr. William H. Busby.

Within Andean genera there are some adaptations to the moderate water stress conditions found in epiphytic habits or in pastures, clearings, and ridge tops. These include coriaceous leaves, succulent leaves and flowers, protective floral bracts, *etc.*, and certain genera such as *Befaria*, *Gaylussacia*, and some *Gaultheria* species seem to prefer somewhat more xeric sites. In the a seasonal habitats there are noticeable periods of flushing as evidenced by the red, newly unfolding leaves, but growth is mostly continuous throughout the year and little periodicity is seen which is also true to some extent for flowering. Cool temperatures are preferred as evidenced by the occurrence of only a few species below 1000 m elevation and above 3000 m. In temperate regions, it is well known that Ericaceae abound in acidic sandy or boggy habitats. This preference carries into the Neotropics where soils of *pH* 2.8 have been measured for some *Lyonia* species (Judd, 1981). Very few species occur on serpentine, and those reported from limestone areas are actually rooted in humus-filled crevices. The acidic humus which accumulates in tree crotches provides an excellent habitat for epiphytic ericads. It has also been well established that temperate Ericaceae have mycorrhizal associates. Although there are very little data for neotropical Ericaceae (St. John, 1980), it may be assumed that under poor soil conditions a mycorrhizal association forms to facilitate nutrient uptake.

The family has evolved the ability to exploit various habitats and can adapt to a changing environment. This is apparent from observing the numerous life forms, which include trees, shrubs, subshrubs, herbs - green autotrophic and achlorophyllous mycotrophic - lianoids, erect or pendent epiphytes, facultative epiphytes, cushion plants, or prostrate mat-forming chamaephytes. For example, *Disterigma empetrifolium*, a normally erect, somewhat wiry small shrub in *subpáramo*, may become a stoloniferous subshrub in páramo seepage areas or a cushion plant in highly exposed páramo; and the genus *Cavendishia* displays species which may be tiny pendent epiphytes (*C. barnebyi*), to wiry (*C. uniflora*) or scandent (*C. bracteata*) subshrubs, to erect shrubs (the majority of species) which are often facultatively epiphytic, to small trees up to seven meters tall (*C. pubescens*).

In very humid forests, some members of the Ericaceae may produce rhizomes (*Themistoclesia*) or roots along their stems (*Cavendishia, Disterigma*). In relatively drier sites such as *páramos* or wind-swept ridges, some species of *Ceratostema* and *Macleania* often bear tuber-like, swollen hypocotyls seemingly for water storage. A few genera such as *Vaccinium* and *Comarostaphylis* produce burls for regeneration after fire or injury in various habitats.

The family has switched from entomophily to ornithophily and from superior to inferior ovaries. Genera and species with superior ovaries predominate at higher latitudes and outside the tropical Andean biogeographical region, whereas the reverse is true at low latitudes, especially in the tropical Andes. For example, the percentages of genera with superior ovaries are: North America 95; Mexico 68; Guatamala 61; Costa rica, Panama and Venezuela 29; Colombia 21; Ecuador and Peru 18; Bolivia 22; Argentina 60; and Chile 100. Ericaceae with superior ovaries are nearly always entomophilous and often have small, white, urceolate flowers. One exception to this is *Befaria* which has large flowers with separate petals. *Befaria aestuans* usually has white to pink, spreading petals, a slight perfumy scent, and is entomophilous; whereas *B. resinosa* has evolved red, tubular flowers by means of long, imbricate petals, and is scentless and hummingbird visited. Ericaceae with inferior ovaries, on the other hand, are nearly always ornithophilous and often have long, red, tubular flowers plus many of the characteristics of the syndrome of ornithophily described by Faegri and Pijl (1966) and Stiles (1981). An exception to this is *Disterigma alaternoides*, a species with small, white flowers and concentrated (mean=27.8%) but little nectar, which is bee-pollinated (Snow and Snow, 1980). The preponderance of inferior-ovaried, bird-pollinated Ericaceae in the low latitudes of the tropical Andean biogeographic region is no accident since the tropical Andes are also the current center of hummingbird species diversity. Snow and Snow (1980) list the Ericaceae as one of the prime families in the Andean flora for providing nectar.

Although Ericaceae pollination syndromes point towards out-crossing, Melampy (1987) has demonstrated that *Befaria resinosa* is self-compatible. The following observations would also lead me to hypothesize that many species, especially the "weedy" ones, are facultatively self-compatible even if one cannot discount that pollinators do a good job! Ericaceae are strongly protandrous with pollen already mature and sometimes shed within the bud; the stigma appears receptive quite soon after flowers open as evidenced by a stigmatic exudate; slight tapping of the pendent or horizontally-oriented flowers often produce a shower of pollen onto the receptive stigma; and several species with high flower densities such as *Cavendishia bracteata, Macleania rupestris,* and *Thibaudia floribunda* always produce abundant fruit.

They are often good colonizers because: Some are probably self-compatible; fruits produce many small seeds and are bird-dispersed sometimes over long distances; seeds apparently tolerate wide-ranging microenvironmental factors and fluctuations such as variable, diurnal temperatures, high insolation, extremes of dryness and moisture. In some genera, germination is very successful as evidenced by the common sight of hundreds of tiny seedlings along steep roadcut slopes below adult plants. In other genera and species, however, rapid germination may indicate low seed tolerance. For example, germination sometimes begins while seeds are still within the fruit. Green embryos seem to be characteristic to *Sphyrospermum, Themistoclesia,* and *Diogenesia.* I have also observed radicle elongation within mature fruits of *Sphyrospermum distichum* and germination with seedlings of several centimeters size rupturing fruits which are still attached to inflorescences, for example in *Cavendishia complectens subsp. striata* var. *cylindrica, Satyria pilosa,* and *Psammisia cf. coarctata.*

The above mentioned characteristics along with the fact that adult plants can often grow under somewhat seasonal conditions have provided the Ericaceae the evolutionary potential for facultative epiphytism, and many species have opted for the epiphytic habit. I have calculated that at least 30 genera and 340 species of neotropical Ericaceae equaling 45% of the species may be epiphytes and this is probably a low estimate. The Ericaceae is the sixth largest epiphytic family of vascular plants in the Neotropics following the Orchidaceae, Bromeliaceae, Araceae, Piperaceae, and Gesneriaceae (Gentry and Dodson, 1987). Within the epiphytic habit, they may be pendulous or erect.

Geological perspective and the role of Ericaceae

As the Andes mountains began to rise during the Miocene and temperate montane, forested habitats became available in the Pliocene (Raven and Axelrod, 1974; Berry, 1982), the Ericaceae were already adapting to a rapidly changing environment. The final uplift of the Andes occurred toward the end of the Pliocene followed by Pleistocene glaciation events which shaped the distributions of vegetation zones. The new, geologically and ecologically diverse areas favored species with colonizing abilities. During the last 5-6 million years tropical Andean Ericaceae diverged with an explosive radiation due to their ability to exploit the various new habitats, aided especially by facultative epiphytism and possible self-compatibility, and their co-evolution with hummingbirds.

Disturbance has been good for neotropical Ericaceae, and I suggest that they have evolved and continue to do so in disturbed sites as disturbed-habitat opportunists. Therefore, their primary role is as pioneers or colonizers in an unstable montane environment.

Acknowledgments

The National Science Foundation supported my field work, and Drs. María L. Lebrón-Luteyn and L. Andersson provided helpful comments.

Literature cited

Berry, P. (1982). "The systematics and evolution of Fuchsia sect. Fuchsia (Onagraceae)." *Ann. Missouri Bot. Gard.* **69,** 1-198.

Faegri, K. and Pijl, L. van der (1966). "The Principles of Pollination Ecology." *Pergamon Press, Oxford.*

Gentry, A. H. and Dodson, C. H. (198 7). "Diversity and biogeography of neotropical vascular epiphytes." *Ann. Missouri Bot. Gard.* **74,** 205-233.

Judd, W. S. (1981). "A monograph of Lyonia (Ericaceae)." *J. Arnold Arb.* **62,** 63-436.

Luteyn, J. L. (1976). "A revision of the Mexican-Central American species of Cavendishia (Vacciniaceae)." *Mem. New York Bot. Gard.* **28,** 1-138.

Luteyn, J. L. (1983). "Ericaceae - Part I. Cavendishia." *Flora Neotropica Monogr.* **35,** 1-290.

Luteyn, J. L. (1984). "Revision of Semiramisia (Ericaceae: Vaccinieae)". *Syst. Bot.* **9,** 359-367.

Luteyn, J. L. (1986). "New species of Ceratostema (Ericaceae: Vaccinieae) from the northern Andes." *J. Arnold Arb.* **67,** 485-492.

Luteyn, J. L. (1987). "New species and notes on neotropical Ericaceae." *Opera Bot.* **92,** 109-130.

Melampy, M. N. (1987). "Flowering phenology, pollen flow and fruit production in the Andean shrub Befaria resinosa." *Oecologia* **73,** 293-300.

Raven, P. H. and Axelrod, D. I. (1974). "Angiosperm biogeography and past continental movements." *Ann. Missouri Bot. Gard.* **61,** 539-673.

Simpson, B. B. (1975). "Pleistocene changes in the flora of the high tropical Andes." *Paleobiology* **1,** 273-294.

Smith, A. C. (1932). "The American species of Thibaudieae." *Contr. U. S. Natl. Herb.* **28(2),** 311-547.

Smith, A. C. (1946). "Studies of South American plants, XI. Noteworthy species of Hippocrateaceae and Vacciniaceae." *J. Arnold Arb.* **27,** 86-120.

Snow, D. W. and Snow, B. K. (1980). "Relationships between hummingbirds and flowers in the Andes of Colombia." *Bull. British Mus. Nat. Hist. (Zool.)* **38(2),** 105-139.

St. John, T. (1980). "Una lista de espécies de plantas tropicais brasileiras naturalmente infectadas com micorriza vesicular-arbuscular." *Acta Amazonica* **10(1),** 229-234.

Stiles, F. G. (1981). "Geographical aspects of bird-flower coevolution, with particular reference to Central America". *Ann. Missouri Bot. Gard.* **68,** 323-351.

Appendix 1. Genera of Neotropical Ericaceae.

Genus and no. of Neotropical species		Distribution (no. of species)
Agarista D. Don ex G. Don	28	Florida (1); Mexico-Bolivia and E to Venezuela and SE Brazil (28); Africa, Madagascar, Réunion, and Mauritius (1).
Anthopteropsis A. C. Smith	1	C Panama (1, endemic).
Anthopterus Hook.	6	SE Panama-Ecuador (5); NE Peru (1, disjunct).
Arbutus L.	3	Europe and the Middle East (2); Canary Islands (1); W Hemisphere (5) incl. Mexico-Nicaragua (3).
Arctostaphylos Adans.	2	W USA (ca. 70); circumpolar and boreal (1); Mexico-Guatemala (2).
Befaria Mutis ex L. f.	14	SE USA (1); Cuba (1); C Mexico-N Bolivia, E to Guyana (13).
Calopteryx A. C. Smith	2	W-C Colombia (1); E Ecuador (1).
Cavendishia Lindley	110	S Mexico-N Bolivia, E to Amapá, Brazil.
Ceratostema Juss.	21	S Colombia-N Peru incl. Guayana Highlands (1, disjunct).
Chimaphila Pursh	2	Boreal N America, Europe, and Asia (5); Neotropics S to Panama (2, disjunct).
Comarostaphylis Zucc.	9	S California-W Panama.
Demosthenesia A. C. Smith	10	S Peru-N Bolivia.
Didonica Luteyn & Wilbur	2	E Costa Rica-C Panama.
Diogenesia Sleumer	14	Colombia to N Bolivia and E to W Venezuela.
Disterigma (Klotzsch) Niedenzu	30	Guatemala-N Bolivia, E to Guyana.

Appendix 1. (cont.)

Gaultheria L.	ca. 40	N America (5); E Asia and Japan (32); Australia, Tasmania, and New Zealand (10); Malesia (24); Neotropics (40).
Gaylussacia H.B.K.	41	N America (7); Venezuela-N Argentina (4); SE Brazil, mostly Minas Gerais (36).
Gonocalyx Planchon & Linden	8	West Indies (Hispañiola, Puerto Rico, Dominica); Colombia (1, disjunct); Costa Rica (2, disjunct).
Kalmia L.	1	N America (6); Cuba (1).
Killipiella A. C. Smith	2	Colombia-N Ecuador.
Lateropora A. C. Smith	3	W - C Panama (endemic).
Ledothamnus N. E. Br.	9	Guayana Highlands (endemic).
Lyonia Nutt.	26	E North America (5); SE Mexico (1); Greater Antilles (24); S and SE Asia (5).
Macleania Hook.	43	S Mexico-Peru, and E to W Venezuela.
Monotropa L.	2	Boreal Europe, Asia, and N America (2); Neotropics S to N Colombia (disjunct populations).
Mycerinus A. C. Smith	3	Guayana Highlands (endemic).
Notopora Hook. f.	5	Guayana Highlands (endemic).
Oreanthes Benth.	4	S Ecuador (endemic).
Orthaea Klotzsch	31	S Mexico-N Bolivia and E to Guyana and Trinidad.
Orthilia Raf.	1	Circumboreal; W N America; Mexico-Guatemala (disjunct).
Pellegrinnia Sleumer	4	S Peru (endemic).
Pernettya Gaudich.	4	Tasmania and New Zealand (5); temperate S America (5); C Mexico-Bolivia (4).

Appendix 1. (cont.)

Pieris D. Don	1	E and SE USA (2); Cuba (1); E Asia (4).
Plutarchia A. C. Smith	10	Colombia (endemic).
Polyclita A. C. Smith	1	N Bolivia (endemic).
Psammisia Klotzsch	66	Costa Rica-Bolivia and E to Guyana and Trinidad.
Pterospora Nutt.	1	W N America; NE USA and adjacent Canada and Mexico (disjunct populations).
Pyrola L.	1	Boreal, temperate, and Arctic N Hemisphere (20-30); Mexico-Guatemala (disjunct populations).
Rusbya Britton	1	N Bolivia (endemic).
Satyria Klotzsch	23	S Mexico-N Bolivia and E to French Guiana.
Semiramisia Klotzsch	4	C Colombia-N Peru; coastal range of Venezuela (1, disjunct).
Siphonandra Klotzsch	1	S Peru-N Bolivia.
Sphyrospermum P. & E.	25	Haiti, S Mexico-N Bolivia and E to French Guiana and Trinidad.
Tepuia Camp	8	Guayana Highlands (endemic).
Themistoclesia Klotzsch	32	Costa Rica-Peru and E to Venezuela.
Thibaudia R. & P. ex J. St-Hil.	62	Costa Rica-N Bolivia and E to Surinam.
Utleya Wilbur & Luteyn	1	C Costa Rica (endemic).
Vaccinium L.	39	North America (65); Europe (6); E Asia and Japan (98); the Pacific (11); SE Africa and Madagascar (5); Malesia (240); Neotropics (39).

Diversity and distribution of Guianan Passifloraceae

C. FEUILLET

ORSTOM, French Guiana

Since Spanish priests interpreted the floral structure of the passion flower as an image of the instruments of the death of Jesus, botanists and plant lovers have been fascinated with the genus *Passiflora*. The showy flowers, edible fruits, medicinal properties, and taxonomic problems generated numerous papers, but still very little is known about distribution patterns of *Passiflora* in the New World.

For the Flora of the Guianas project, data were collected in the field and in herbaria, and the literature about distribution of Guianan species of *Passiflora* was studied. The Guianan region is on the northern coast of South America limited by the Orinoco and Amazon rivers and the Atlantic Ocean. It consists of five administrative areas, Guyana, Surinam, and French Guiana surrounded by the Venezuelan Guayana and the Brazilian Guiana. The species of *Passiflora* in the Guianas are mostly herbaceous or woody climbers, but two or three species in Guyana may be shrubs.

Distribution patterns

Specimens belonging to 78 taxa, 12 of which are new species, have been collected in the Guianan region. This is a large number considering that the total number of taxa of *Passiflora* in South America is less than 500 (Killip, 1938). In species richness, the Guianan region is comparable to Ecuador (Holm-Nielsen *et al.*, 1988) and Mexico (MacDougal, pers. comm.), which are countries with a richer flora and a larger diversity of soils, climates, and elevations.

Appendix 1 gives a list of the taxa and their occurrence in the five parts of the Guianan region. There are 30 taxa which occur outside of the Guianas and two cultivated species (*P. edulis* and *P. quadrangularis*) among the 78 taxa. The remaining 46 species are, as far as we know, endemic to this region.

The distribution patterns of the 30 taxa that also occur beyond the Guianan region can be divided into six types (Fig. 1): (i) *Passiflora*

TROPICAL FORESTS
ISBN 0–12–353550–6

foetida var. *moritziana* occurs on the sandy beaches of Venezuela, in sandy areas at the foot of the Venezuelan Andes, and between Surinam and French Guiana at the mouth of the Maroni river (type 1); (ii) *Passiflora costata* and *P. riparia* have a circum-Amazonian distribution from Peru to the Guianas (type 2); (iii) three species are more widespread from central America to central Brazil and Uruguay (type 3); (iv) four species were collected only in a part of Amazonia and in the northeast of Brazil (type 4); (v) nine species are characteristic of the Guianas and the Amazon basin (type 5), and finally; (vi) eleven taxa are distributed in tropical South America and the West Indies (type 6).

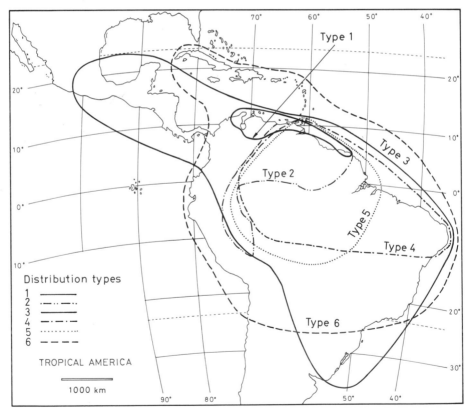

Figure 1. Distribution types of 30 Guianan *Passiflora* species with distributions that reach beyond the Guianas.

As for *Passiflora*, the Guianan flora in general is heterogenous. The two richest areas are Guyana and French Guiana, with only a few species occuring in Surinam. The number of endemics in each territory shows two additional striking facts (Appendix 1). The first is that there are six endemics in Guyana, 18 in French Guiana, and none in Surinam. The second is that among the 46 taxa endemic to the Guianas only five occur in both Guyana and French Guiana.

Clearly, there are two floras of *Passiflora* species in the Guianas. One center of endemism is in the western part of Guyana in the forests east and north of the Rupununi savannas. The other center is in French Guiana.

Distribution in French Guiana

The five distribution types within French Guiana are of unequal importance (Fig. 2): (i) a group of seven taxa occur more or less all over French Guiana (*P. coccinea, P. glandulosa, etc.*); (ii) the western part of the country has three characteristic species (*P. acuminata, P. costata* and *P. misera*); (iii) eleven taxa are restricted to the northern half of the country (*P. auriculata, P. citrifolia, etc.*); (iv) two species were collected only in the northeastern part of the country (*P. candida* and *P. sp. nov. 1*); and (v) twelve species occur exclusively in a zone where the average rainfall is above 3500 mm per year and in the cloud forest where the humidity is locally high (*P. fanchonae, P. rufostipulata, P. crenata, P. plumosa. etc.*).

Among the 37 taxa which are native to French Guiana two are from an unknown locality, and 33 have been collected at least once in the northern half of the country. From the latter group, 29 are known in the northeastern corner of French Guiana, which is also the richest area for endemism, particularly the mountain ranges, Montagne Cacao, Montagnes de Kaw, Montagne des Nouragues and Montagne Tortue (C, K, N and T respectively in Figure 2), which have elevations between 300 and 500 m above sea-level and about 4000 mm of annual precipitation on the summits.

Density of species

Inventory data for lianas were collected in two places, both located in the high rainfall zone. At the base of the southern cliff of the inselberg in the Montagne des Nouragues range, a clearing of 5000 square meters was inventoried one year after cutting (without burning the logs or compacting

Figure 2. Distribution of
Passiflora species in
French Guiana.
C - Montagne Cacao,
K - Montagne de Kaw,
N- Montagne de Nougues,
T- Montagne Tortue.

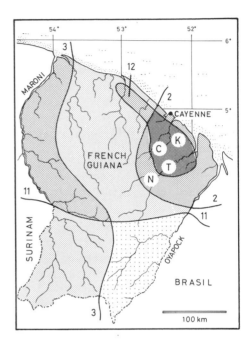

the soil). Eleven species of *Passiflora* were collected in this plot in which only 13 species are known from the surrounding forest. At Montagnes de Kaw, five plots of 20 x 20 m were inventoried three years after clearing (without burning but the soil was severely compacted). They contained nine species (5-8 per plot), and the first square of 5 x 5 m was exceptional with six species (*P. glandulosa*, *P. garckei*, *P. rufostipulata*, *P. candida*, *P. sp. nov. 1* and *P. sp. nov. 2*). The Montagnes de Kaw as a whole contains 18 species.

Conclusion

The dominant affinities of the Guianan *Passiflora* species are with those in the West Indies and Amazonia respectively. Of the *Passiflora* taxa collected in the Guianan region, 28% occur also in these two regions. The remaining Guianan *Passiflora* species have their center of distribution in the Guianas, but they belong to three distinct distribution types. One group of species is Guyanan, a second group is French Guianan and the third is rather widespread in tropical America.

In general, the Guianan species of *Passiflora* are common along roadsides in their young stages, but 90% reach the canopy when they are adult. They have often narrow distribution areas. The species seem to depend on the forest dynamics and may under natural conditions be an integrated element in the succession following landslides, wind fallen trees and other natural light gaps. In French Guiana, it seems likely that speciation was very active around four mountains in the wet northeastern part of the country. In Guyana, diversity and endemism are most important on the western slopes of the Kanuku mountains and from there around the Rupununi savannas and northward to the Ireng River area.

With about the same number of *Passiflora* species, Guyana (34) and French Guiana (37) are comparable, but the level of endemism is three times higher in French Guiana.

Literature cited

Holm-Nielsen, L. B., Jørgensen, P. M. and Lawesson J. E. (1988). "Passifloraceae." *In* Harling, G. and Andersson, L. (eds.), *Flora of Ecuador* **31,** 1-130.

Killip, E. P. (1938). "The American species of Passifloraceae." *Publ. Field Mus., Nat Hist., Bot. Ser.* **19,** 1-613.

Appendix 1. List of of *Passiflora* species in the Guianan region: Ven - Venezuelan Guayana; Guy - Guyana; Sur - Surinam; FG - French Guiana; Bra - Brazilian Guiana; oth - other parts of the Neotropics; cul - cultivated in the Guianas.

	Ven	Guy	Sur	FG	Bra	oth	cul
P. *acuminata* DC.				*	*		
P. *amicorum* Wurd.	*						
P. *araujoi* Sacco					*	*	
P. *auriculata* Kunth	*	*	*	*	*	*	
P. *bomareifolia* Steyerm. & Mag.	*						
P. *candida* (P. & E.) Masters				*	*		
P. *capparidifolia* Killip		*			*		
P. *cardonae* Killip	*	*					
P. *ceratocarpa* Silveira		*					
P. *cirrhiflora* Juss.		*	*	*			
P. *citrifolia* (Juss.) Masters				*			
P. *coccinea* Aublet	*	*	*	*	*	*	
P. *coriacea* Juss.		*				*	
P. *costata* Masters		*	*	*	*	*	
P. *crenata* Feuillet & Cremers				*			
P. *deficiens* Masters		*					
P. *edulis* Sims							*
P. *ernesti* Harms					*	*	
P. *fanchonae* Feuillet				*			
P. *filipes* Benth.	*					*	
P. *foetida* L.							
var. *foetida*		*	*	*	*	*	
var. *hispida* (DC.) Killip		*	*	*	*	*	
var. *moritziana* (Pl.) Killip			*	*		*	
var. *orinocensis* Killip	*					*	
var. *strigosa* S. Moore	*					*	
P. *fuchsiiflora* Hemsley		*	*	*	*		
P. *garckei* Masters	*	*	*	*	*		
P. *glandulosa* Cav.	*	*	*	*	*	*	
P. *gleasoni* Killip		*					
P. *guazumaefolia* Juss.	*					*	
P. *holtii* Killip		*			*		
P. *involucrata* (Masters) Gentry					*	*	
P. *laurifolia* L.		*	*	*			

Appendix 1. (cont.)

	Ven	Guy	Sur	FG	Bra	oth	cul
P. *leptopoda* Harms		*	*		*		
P. *longiracemosa* Ducke		*			*		
P. *maguirei* Killip		*			*		
P. *maliformis* L.	*					*	
P. *misera* Kunth		*	*	*	*	*	
P. *nitida* Kunth	*	*	*	*	*	*	
P. *nuriensis* Steyerm.	*						
P. *ovata* Martin ex DC.				*			
P. *pachyantha* Killip		*					
P. *pedata* L.		*	*		*	*	
P. *phaeocaula* Killip					*		
P. *picturata* Ker			*			*	
P. *plumosa* Feuillet & Cremers				*			
P. *pulchella* Kunth	*					*	
P. *quadrangularis* L.							*
P. *quadriglandulosa* Rodschied		*			*	*	
P. *quelchii* N. E. Br.		*					
P. *retipetala* Masters		*	*		*		
P. *riparia* Masters		*			*	*	
P. *rubra* L.				*		*	
P. *rufostipulata* Feuillet				*			
P. *sclerophylla* Harms	*	*					
P. *securiclata* Masters	*	*			*		
P. *serratodigitata* L.		*	*	*		*	
P. *spinosa* (P. & E.) Masters					*	*	
P. *stipulata* Aublet				*			
P. *stoupyana* DC.				*			
P. *tholozanii* Sacco					*		
P. *tuberosa* Jacq.		*				*	
P. *variolata* P. & E.	*			*	*	*	
P. *vespertilio* L.		*	*	*	*	*	
P. *vitifolia* Kunth	*					*	
P. *sp. nov. 1*				*			
P. *sp. nov. 2*				*			
P. *sp. nov. 3*	*						
P. *sp. nov. 4*				*			
P. *sp. nov. 5*		*					
P. *sp. nov. 6*				*			
P. *sp. nov. 7*				*			
P. *sp. nov. 8*				*			
P. *sp. nov. 9*				*			
P. *sp. nov. 10*				*			

Appendix 1. (cont.)	Ven	Guy	Sur	FG	Bra	oth	cul
P. sp. nov. 11				*			
P. sp. nov. 12				*			
Total:	20	34	19	37	30	30	2
Number of endemic taxa:	4	6	0	18	2		
Number of Guianan taxa:	8	19	6	24	13		

Diversity of Lecythidaceae in the Guianas

S. A. MORI

The New York Botanical Garden, USA

The Lecythidaceae is a pantropical family with 20 genera and 271 species (Kowal *et al.*, 1977) of which 11 genera and 199 species are currently recognized in the Neotropics (Mori and Prance, in press). The Guianas (Guyana, Surinam, French Guiana) harbor 50 species or 25% of known Neotropical Lecythidaceae. Of these, 10 are thought to be endemic to the Guianas, but this total will surely diminish with more thorough botanical exploration of neighboring countries.

Twenty-seven of the species of Lecythidaceae found in the Guianas have all, or at least the major part of their distributions, located to the north and east of the Orinoco, Negro, and Amazon rivers. Although the distributions of some of these species are slightly larger than the Guayana floristic province that overlies the Guayana crystalline shield (Maguire, 1970), I consider them to be taxa of the Guayana floristic province which have migrated into the more recent habitats of the Amazon Basin. *Couratari guianensis*, *Couroupita guianensis*, *Eschweilera coriacea*, and *Lecythis corrugata* are widespread species that are found in the Guianas as well as west of the Andes. In addition, three species pairs indicate the transandean relationships of Lecythidaceae: 1) *Gustavia hexapetala* (east)/ *G. dubia* (west), 2) *Couratari stellata* (east)/ *C. scottmorii* (west), and 3) *Lecythis zabucaja* (east)/ *L. ampla* (west). Only the widespread species, *Gustavia augusta*, is found in the coastal forests of eastern extra-Amazonian Brazil as well as in the Guianas (Mori and Prance, 1987a).

The Guayana floristic province is disproportionately richer in species of zygomorphic-flowered Lecythidaceae. Only three of the 50 (6%) Guianan Lecythidaceae possess actinomorphic flowers whereas 24% of all Neotropical have actinomorphic flowers. The actinomorphic *Allantoma* and *Grias*, and the slightly actinomorphic *Cariniana*, are absent from the Guianas. In contrast, the zygomorphic-flowered *Corythophora*, *Couratari*, *Eschweilera*, and *Lecythis* are most diverse in central Amazonia and the Guayana floristic province (Mori and Prance, in press).

The proposed national park of 133 600 hectares surrounding Saül, French Guiana was selected for study by Mori and collaborators (1987)

TROPICAL FORESTS
ISBN 0–12–353550–6

because it has 27 species of Lecythidaceae and because its forests are little disturbed. The purpose of the study was to gather ecological information about Lecythidaceae which, in turn, could be used to help solve taxonomic problems. The nomenclature for Lecythidaceae in this paper follows Prance and Mori (1979) and Mori and Prance (in press).

High species diversity of Lecythidaceae at Saül

Lecythidaceae are represented in the lowland moist forests surrounding Saül by *Corythophora* (2 *spp.*), *Couratari* (4 *spp.*), *Eschweilera* (12 *spp.*), *Gustavia* (2 *spp.*), and *Lecythis* (7 *spp.*). In addition to being diverse, Lecythidaceae are ecologically important in the forests of Saül. Its Family Importance Value of 23.2 is exceeded only by Burseraceae (34.9) and Sapotaceae (31.5) and also by Leguminosae (49.9) if considered collectively. *Couratari stellata* and *Eschweilera coriacea* with Importance Values of 6.9 and 6.6, are the fifth and sixth most important species in this forest (Mori and Boom, 1987).

How do so many species of Lecythidaceae coexist in the forest surrounding Saül ? To help answer this question, I will discuss abiotic and biotic factors that have led to speciation in Lecythidaceae.

Abiotic factors

In the Guianas, Lecythidaceae are mostly restricted to lowland moist forests. Only *Lecythis alutacea* reaches elevations above 1000 meters. *Lecythis brancoensis* and *L. schomburgkii*, both of which inhabit the savannas of southwestern Guyana, are the only non-forest Lecythidaceae in the Guianas.

Therefore, altitude and differentiation into the savanna habitat have had little impact on speciation of Lecythidaceae in the Guianas. Within lowland moist forests, Lecythidaceae are found in two principal habitats, periodically inundated forests *(várzea)* and non-inundated forests *(terra firme)*. In the Amazon Basin, Spruce (1908) first pointed out that these two habitats have played an important role in plant speciation. Ducke (1948) demonstrated that different adaptations for seed dispersal have evolved in closely related species found in *várzea* and *terra firme* in response to availability of water and animals as dispersal agents, respectively. Although Lecythidaceae are more speciose in *terra firme,* distinct Amazonian species have evolved in response to the selective pressures exerted by the *várzea* habitat (Mitchell and Mori, 1987; Pires and Prance, 1977). The *várzea* habitat is not nearly as extensive in the Guianas as it is in the Amazon, and therefore there are few Guianan

Lecythidaceae restricted to this habitat. However, an undescribed species of *Lecythis*, with pneumatophores and corky seed coats, is found along larger rivers, and *Couratari gloriosa*, appears to be restricted to the margins of smaller streams. Nevertheless, within *terra firme Corythophora rimosa subsp. rubra*, *Eschweilera micrantha*, and *Lecythis confertiflora* prefer ridge tops, whereas *Couratari stellata*, *Eschweilera coriacea*, and *L. corrugata* are more frequent in lower areas (Mitchell and Mori, 1987).

Speciation of Lecythidaceae in response to the white sand habitats that underlie such specialized vegetation types as wallaba forest in Guyana and *caatinga* in central Amazonia has not occurred. The most common Lecythidaceae in wallaba forest is *Lecythis corrugata subsp. corrugata* which is also found in other lowland moist forests, and there are no species of Lecythidaceae known to be restricted to caatinga. Nevertheless, further study of central Amazonia, where there are numerous endemic species of Lecythidaceae, may reveal the presence of edaphic endemics.

Light intensity is significant in the evolution of tropical trees. Denslow (1980) has shown that tropical trees are (i) large-gap specialists, (ii) small-gap specialists, or (iii) understory specialists. She suggests that these specializations are important in promoting tropical tree diversity. Species colonizing large gaps usually possess abundant but small seeds which may remain dormant in the soil for a long time, extended annual fruiting seasons, and rapid growth rates. Small-gap or understory specialists have fewer and larger seeds, seeds with little or no dormancy, and slower growth rates. There is also evidence that different zones within a gap (*e.g.*, the root, trunk, and crown zones) are colonized by different species of trees (Brandani *et al.*, 1988). Adaptation of trees for regeneration in primary forest or for colonizing different size gaps has promoted differentiation of taxa in tropical forests.

Species of Lecythidaceae do not colonize large gaps. Nearly all Neotropical species of this family are inhabitants of primary forests with little tendency for invasion of deforested areas. Exceptions are *Lecythis minor* of the drier areas of northern Colombia and perhaps some species of *Gustavia*, such as *G. augusta*, *G. poeppigiana*, and *G. superba*, which are found in disturbed and undisturbed habitats alike. When primary forest is cut but not burned, Lecythidaceae will often regenerate from stumps. However, when the forest is both cut and burned, species of Lecythidaceae generally will not survive (Prance, 1975). Although species of Lecythidaceae are not adapted to colonization of large gaps, some Lecythidaceae (*e.g.*, species of *Couratari;* Schulz, 1960) may depend on small gaps for their establishment. This is apparently true for many species of tropical trees (Denslow, 1980; Foster and Brokaw, 1982) and may apply to other Lecythidaceae, but studies have not yet been done to show this.

Moreover, the striking differences in dispersal strategy as well as seedling morphology of Lecythidaceae discussed below may be related to the colonization of gaps of different sizes (Denslow, 1980). Guianan species of Lecythidaceae are always trees which, because of interspecific differences in stature, are adapted to different vertical light regimes of the forest. In the vicinity of Saül, some species of Lecythidaceae are understory trees up to 20 meters (*e.g.*, *Gustavia augusta*, *G. hexapetala* Fig. 1, *Eschweilera grandiflora*, *E. pedicellata*, *E. parviflora*), others are canopy trees of 20-35 meters (*e.g.*, *Corythophora amapaensis*, *C. rimosa subsp. rubra*, *Eschweilera apiculata*, *E. collina*, *E. coriacea*, *E. decolorans*, *E. laevicarpa*, *E. micrantha* Fig. 1, *E. sagotiana*, *E. squamata*, *Lecythis confertiflora*, *L. corrugata*, *L. idatimon*, *L. persistens*, and *L. poiteaui*), and others are emergent trees greater than 35 meters at maturity (*e.g.*, *Couratari gloriosa*, *C. guianensis*, *C. multiflora*, *C. stellata* Fig. 2, and *Lecythis zabucaja*). Terborgh (1985) has suggested that the positions of the strata of tropical forests are determined by the angle of light hitting each subsequent stratum as well as by the shapes of the tree crowns in each stratum. This, he adds, contributes a vertical component to tropical forest diversity, a hypothesis which is supported by the adaptation of species of Lecythidaceae into three distinct strata in the forests surrounding Saül. Further research is needed in order to determine what other factors, such as differences in capacity to lift water, are involved in promoting different heights among species of Lecythidaceae.

Biotic factors

The pollinators of Neotropical Lecythidaceae are mostly bees. The only known exception is that of bat pollination in *Lecythis poiteaui*. This species has night-blooming flowers with a musty odor, inflorescences at the periphery of the canopy, green to white, enrolled petals, and more numerous stamens than any other species of the genus (Mori *et al.*, 1978). These are classical features of bat pollination and, surely, must have evolved in response to selective pressures exerted by these important pollinators of tropical plants. Two morphologically similar species, *L. barnebyi* and *L. brancoensis*, may also be bat pollinated, but this has yet to be confirmed with field work. The flowers of Neotropical Lecythidaceae have responded to the selective pressures of bees in a number of ways (Mori and Boeke, 1987). However, I want to stress the importance that evolution of the androecium has had in diversification of Lecythidaceae. There are two main kinds of androecia in Neotropical Lecythidaceae. The first type, found in the Guianas only in *Gustavia*, is actinomorphic, and the second

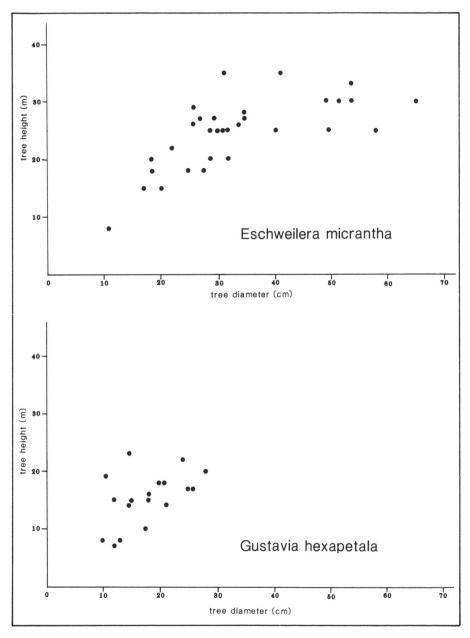

Figure 1. Sizes of *Gustavia hexapetala* (understory species) and *Eschweilera micrantha* (canopy species) as found in the forests surrounding Saül, French Guiana.

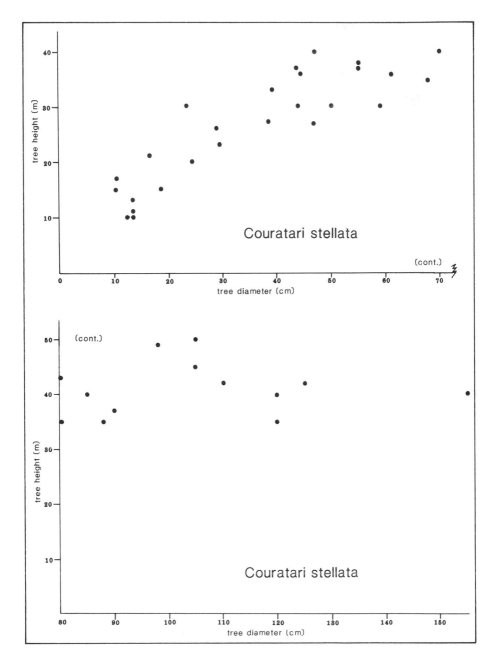

Figure 2. Size of *Couratari stellata* (emergent species) as found in the forests surrounding Saül, French Guiana.

type, represented in the Guianas by species of *Bertholletia, Corythophora, Couratari, Couroupita, Eschweilera*, and *Lecythis*, is zygomorphic (Fig. 3). The complexity of zygomorphic-flowered Lecythidaceae has been described and illustrated (Prance, 1976; Mori *et al.* 1978; Mori and Boeke, 1987; Mori and Prance, 1987b) and will not be discussed further here. The reward presented to pollinators can be (i) undifferentiated, fertile pollen, (ii) differentiated, fodder pollen, and (iii) nectar. In all actinomorphic-flowered species, fertile pollen is the only type produced and serves as the pollinator reward as well as effects fertilization of subsequent flowers visited. In zygomorphic-flowered species, fodder pollen or nectar serves as the pollinator reward and separate fertile pollen effects fertilization. The location of the fodder pollen in the androecium varies in different species. For example *Lecythis persistens, L. poiteaui, L. zabucaja* and *Corythophora rimosa* present fodder pollen in the staminodia of the androecial hood, whereas in *Corythophora amapaensis* and *Lecythis corrugata subsp. corrugata*, the fodder pollen is located in staminodia within the staminal ring itself (Fig. 3). Fertile pollen is always produced within the anthers of the staminal ring. The position of fodder pollen determines the place on the pollinators body upon which fertile pollen is deposited. In general, zygomorphic-flowered species are visited by fewer and more specific kinds of pollinators. This is especially true for those zygomorphic, nectar-producing species which always have their nectar hidden at the apex of a long coil. Moreover, the androecium of these species is closed because the androecial hood is tightly appressed to the staminal ring (Fig. 3). Consequently, only strong bees with long tongues are able to open the flowers and extract nectar. In contrast, the actinomorphic-flowered species, which offer only fertile pollen as a reward, have open flowers which are easily accessible to all species of bees.

Euglossine bees are important in the pollination of zygomorphic-flowered species of Lecythidaceae which offer nectar as a pollinator reward. The geographic distributions of these kinds of Lecythidaceae and of euglossine bees are restricted to the Neotropics where their ranges are nearly identical. Euglossine bees must have played an important role in the evolution of this unique kind of flower. Euglossines appear to be less important in the pollination of actinomorphic-flowered, fertile pollen reward species of Lecythidaceae (Mori and Boeke, 1987).

The Lecythidaceae of Saül bloom mostly in the dry season. Of the 27 species present there, only *Couratari multiflora* and *Gustavia hexapetala* produce significant numbers of flowers in the wet season (Mori and Prance, 1987c). Species with very similar flowers, such as *Eschweilera decolorans / E. coriacea* and *Lecythis confertiflora / L. idatimon* tend to have different flowering times (Fig. 4). In contrast, closely related species blooming at the same time, such as *Corythophora amapaensis* and *C. rimosa subsp. rubra,* have different floral structures and are pollinated

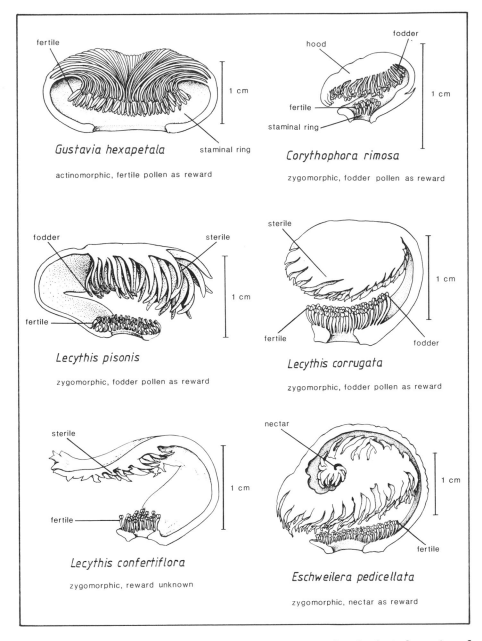

Figure 3. Androecial structure and pollinator rewards of selected species of Neotropical Lecythidaceae (redrawn with permission from Mem. New York Bot. Gard. 44).

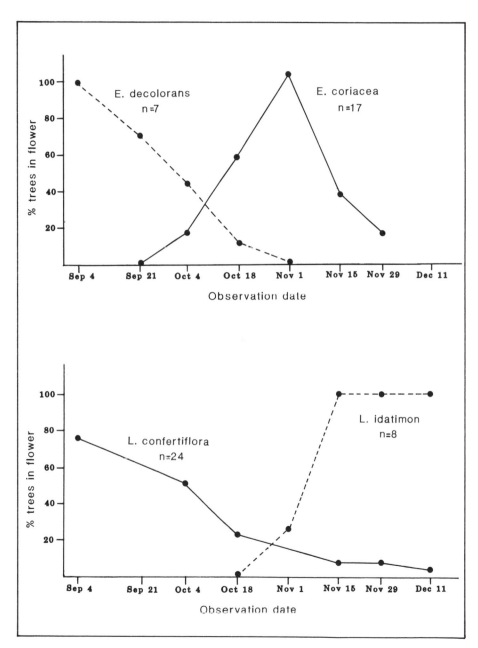

Figure 4. Differences in flowering times of selected species of Lecythidaceae as found in the forests surrounding Saül, French Guiana.

by different species of bees (Mori and Prance, 1987c).

Although further study is needed, phenological separation of Lecythidaceae appears to be a mechanism by which species diversity is maintained. Indeed, the recognition of different phenological patterns is useful in helping solve species problems. For example, at the start of the Saül study, I considered *Lecythis confertiflora* as a synonym of *L. idatimon*. However, a phenological study of this "species" at Saül revealed that many individuals bloomed earlier than other individuals. Subsequent study of these populations revealed that the early bloomer was a larger tree with smaller leaves and differently shaped fruits than the later bloomer (Fig. 4). These phenological and morphological differences were confirmed in Amapá, Brazil, and I now recognize the early bloomer as *L. confertiflora* and the later bloomer as *L. idatimon*.

The considerable variation in fruit and seed morphology of Neotropical Lecythidaceae has been described and illustrated elsewhere (Prance and Mori, 1979, 1983). The fruits are either dehiscent or indehiscent and the pericarp, which is usually hard and woody, may also be soft and edible in *Grias* and *Gustavia*. In addition, the seeds may lack or possess an aril which, in turn, is modified in a number of different ways. Finally, the seedlings may have undifferentiated cotyledons or cotyledons which are plano-convex or foliaceous. These differences, I maintain, are adaptations for seed dispersal and seedling establishment.

In order to illustrate the adaptive significance of fruit and seed structures in Lecythidaceae, I will describe those of *Lecythis pisonis*, *Bertholettia excelsa*, and *Couratari stellata*. The large, woody fruits of *L. pisonis* dehisce via an operculum at maturity. At this time, the fruits remain in the canopy and the seeds are attached to the fruit wall by a funicle which, in turn, is surrounded by a white, fleshy aril (Fig. 5). The seeds of *L. pisonis* possess undifferentiated embryos. In contrast, the fruits of *B. excelsa* are functionally indehiscent and fall to the ground at maturity. Although there is a vestigial operculum, its diameter is smaller than those of the seeds, and therefore the latter are retained within the fruit at maturity. The seeds of *B. excelsa* lack an aril but possess the same undifferentiated embryo found in *L. pisonis* (Fig. 5). *Couratari stellata* has dehiscent fruits which remain on the tree at maturity. However, the aril is flattened and membranous and forms a wing which surrounds the seed. In addition, the embryo possesses two, well-developed, foliaceous cotyledons (Fig.5).

The significance of these morphological differences is apparent when they are examined in light of the dispersers of the above species. *Lecythis pisonis* is dispersed by bats (Greenhall, 1965; Prance and Mori, 1983) which remove the seeds shortly after the fruits dehisce in order to eat the aril. The seeds are dropped, either in flight or under roosts, after the aril has been consumed. Dehiscence, retention of the fruits in the tree at

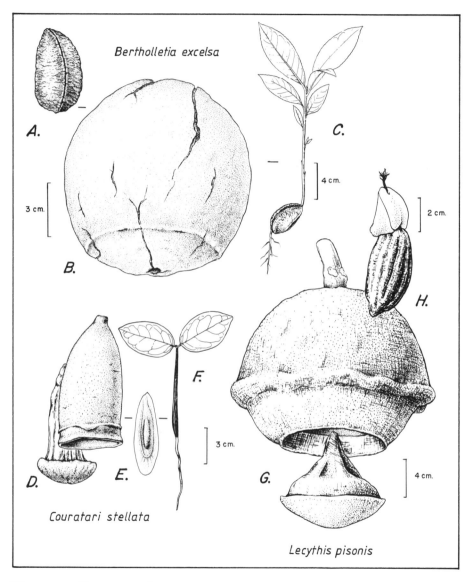

Figure 5. Fruits, seeds, and seedlings of selected species of Neotropical Lecythidaceae. A-C. *Bertholletia excelsa*. D-F. *Couratari stellata*. G-H. *Lecythis pisonis* A - Seed with very hard testa and no aril. B - Secondarily indehiscent fruit. C - Seedling without developed cotyledons. D - Fruit base and operculum. E - Winged seed. F - Seedling with two foliaceous cotyledons. G - Fruit base and operculum. H - Seed with cord-like funicle surrounded by fleshy aril (G-H redrawn with permission from Brittonia 33, the remaining with permission from Fl. Neotrop. Monogr., Mori and Prance, in press).

maturity, and the well-developed aril are adaptations for bat dispersal. In contrast, *Bertholletia excelsa* is dispersed by agoutis, a rodent of the genus *Dasyprocta*, which gnaw through the woody pericarp to remove the seeds for consumption. The agoutis eat some seeds and bury others for future consumption. Those seeds which are forgotten germinate within a year later, resulting in seedling establishment away from the mother tree (Huber, 1910; Prance and Mori, 1983). Functional indehiscence, fall of mature fruits, and loss of the aril are adaptations for agouti dispersal. Finally, the seeds of *C. stellata* drop from the fruits at maturity and are carried away by the wind. Dehiscence and winged seeds are clearly adaptations for wind dispersal (Prance and Mori, 1983). It is possible that differences in embryo structure are related to the gap colonizing strategies of Lecythidaceae. The large, undifferentiated, non-photosynthetic embryos of *L. pisonis* and *B. excelsa* provide a greater energy store which, in turn, may allow these species to remain in the understorey for a longer time waiting for proper conditions for growth into saplings. On the other hand, the foliaceous cotyledons of *C. stellata* may provide a mechanism by which this species is adapted for immediate colonization of small gaps (Denslow, 1980; Schulz, 1960).

Conclusions

Lecythidaceae provide excellent opportunities for study of the evolution of diversity in lowland Neotropical trees. For example, as many as 27 species of this family have been recorded from the 133 600 hectare proposed national park surrounding Saül, French Guiana. Moreover, the family has unique floral and fruit features which are clearly influenced by the animals that pollinate their flowers and the animals, wind, and water that disperse their seeds. In addition, preliminary study suggests that competition among species of Lecythidaceae for abiotic resources, such as light and soil moisture, is also important in promoting species diversity. Continued studies of Lecythidaceae in limited areas of high species diversity will do much to enhance our understanding of the evolution of species of tropical trees. Detailed studies of Lecythidaceae are especially needed from central and western Amazonia and the Chocó of Colombia. It is also important that intact areas of high species diversity be identified and preserved as soon as possible. This is especially critical for Lecythidaceae because of the inability of nearly all of the species to colonize the large gaps caused by the destructive activities of man. Carefully selected biological reserves in the Neotropical lowlands will do much to ensure the preservation of tree diversity. For example, the establishment and protection of the 133 600 hectare proposed national park surrounding Saül, French Guiana would insure the survival of 54% of all of the species of Lecythidaceae known to occur in the Guianas.

Acknowledgments

I thank G. T. Prance for his long collaboration in our studies of Lecythidaceae, Bobbi Angell for preparing the illustrations, my wife, Carol Gracie, for arranging the illustrations and reviewing the manuscript, and Brian Boom, John Pipoly, and John Mitchell for help in gathering field data. Financial suport was provided by the National Science Foundation (BMS 75-03724, DEB-8020920) and the Fund for Neotropical Plant Research of The New York Botanical Garden.

Literature cited

Brandani, A., Hartshorn, G. S. and Orians, G. H. (1988). "Internal heterogeneity of gaps and species richness in Costa Rican tropical wet forest." *J. Trop. Ecol.* **4**, 99-119.

Denslow, J. S. (1980). "Gap partitioning among tropical rainforest trees." *Biotropica* **12**, 47-55.

Ducke, A. (1948). "Arvores Amazônicas e sua propagação." *Bol. Mus. Paraense Hist. Nat.* **10**, 81-92.

Foster, R. B. and Brokaw, N. V. L. (1982). "Structure and history of vegetation of Barro Colorado Island." pp. 67-81 *In* Leigh, E. G., Rand, A. S. and Windsor, D. M. (eds.), "The ecology of a tropical forest: seasonal rhythms and long-term changes." *Smithsonian Institution Press, Washington, D.C.*

Greenhall, A. M. (1965). "Sapucaia nut dispersal by greater spear-nosed bats in trinidad." *Caribbean J. Sci.* **5**, 167-171.

Huber, J. (1910). "Mattas e madeiras amazônicas." *Bol. Mus. Paraense Hist. Nat.* **6**, 91-225.

Kowal, R., Mori, S. A. and Kallunki, J. A. (1977). "Chromosome counts of Panamanian Lecythidaceae." *Brittonia* **29**, 399-401.

Maguire, B. M. (1970). "On the flora of the Guayana highland." *Biotropica* **2**, 85-100.

Mitchell, J. D. and Mori, S. A. (1987). "Ecology." pp. 113-123 *In* Mori, S. A. and collaborators, *Mem. New York Bot Gard.* **44.**

Mori, S. A. and Boeke, J. D. (1987). "Pollination." pp. 137-155 *In* Mori, S. A. and collaborators, *Mem. New York Bot. Gard.* **44.**

Mori, S. A. and Boom, B. M. (1987). "The forest." pp. 9-29 *In* Mori, S. A. and collaborators, *Mem. New York Bot. Gard.* **44.**

Mori, S. A. and collaborators. (1987). "The Lecythidaceae of a lowland

Neotropical forest: La Fumée Mountain, French Guiana." *Mem. New York Bot Gard.* **44,** 1-190.

Mori, S. A. and Prance, G. T. (1987a). "Phytogeography." pp. 55-71 *In* Mori, S. A. and collaborators, "The Lecythidaceae of a lowland Neotropical forest: La Fumée Mountain, French Guiana." *Mem. New. York Bot. Gard.* **44.**

Mori, S. A. and Prance, G. T. (1987b). "Species diversity, phenology, plant-animal interactions, and their correlation with climate, as illustrated by the Brazil nut family (Lecythidaceae)." pp. 69-89 *In* Dickinson, R. E. (ed.), "The geophysiology of Amazonia." *John Wiley and Sons, New York.*

Mori, S. A. and Prance, G. T. (1987c). "Phenology."pp. 124-136 *In* Mori S. A. and collaborators, "The Lecythidaceae of a lowland Neotropical forest: La Fumée Mountain, French Guiana." *Mem. New York Bot. Gard.* **44.**

Mori, S. A. and Prance, G. T. (In press). "Lecythidaceae -- Part II. The zygomorphic-flowered New World genera (Couroupita, Corytho-phora, Bertholletia, Couratari, Eschweilera, and Lecythis)." *Fl. Neotrop. Monogr.*

Mori, S. A., Prance, G. T. and Bolten, A. B. (1978). "Additional notes on the floral biology of Neotropical Lecythidaceae." *Brittonia* **30,** 113-130.

Pires, J. M. and Prance, G. T. (1977). "The Amazon forest: a natural heritage to be preserved." pp. 158-194 *In* Prance, G. T. and Elias, T. S. (eds.), "Extinction is forever." *New York Botanical Garden.*

Prance, G. T. (1975). "The history of the INPA capoeira based on ecological studies of Lecythidaceae." *Acta Amazônica* **5,** 261-263.

Prance, G. T. (1976). "The pollination and androphore structure of some Amazonian Lecythidaceae." *Biotropica* **8,** 235-241.

Prance, G. T. and Mori, S. A. (1979). "Lecythidaceae -- Part I. The actinomorphic-flowered New World Lecythidaceae (Asteranthos, Gustavia, Grias, Allantoma and Cariniana)." *Fl. Neotrop. Monogr.* **21,** 1-270.

Prance, G. T. and Mori, S. A. (1983). "Dispersal and distribution of Lecythidaceae and Chysobalanaceae." *Sonderbd. Naturwiss. Ver. Hamburg* **7,** 163-186.

Schulz, J. P. (1960). "Ecological studies on rain forest in northern Suriname." *Meded. Bot. Mus. Herb. Rijks Univ. Utrecht* **163,** 1-267.

Spruce, R. (1908). "Notes of a botanist on the Amazon and Andes." 2 vols. Wallace, A. R. (ed.). *Macmillan, London.*

Terborgh, J. (1985). "The vertical component of plant species diversity in temperate and tropical forests." *Amer. Midl. Naturalist* **126,** 760-776.

Past,

Present

and Future

The future of the tropical forests

F. ORTIZ-CRESPO

US-AID, Quito, Ecuador

There can be few things as frustrating as being a conservationist native to and living in a third world country. I should know, since I started worrying about mismanagement of my country's natural resources over 25 years ago. During this time I have discerned a few trends that may let us foresee the future, and that unfortunately gives us no reason to sit back and relax, and hope that somehow environmental issues will be resolved simply by public consensus or the sheer wisdom of our political leaders. No. Those few of us who worry about what is happening to our tropical forests, and lakes, and coasts, and rivers, and people must have an even stronger voice directed to the public and to our leaders, telling them what we think may happen. We must hope that the iron fist of reality may help us get them to listen.

One aspect of development is population growth. When I was going to primary school 30 years ago Ecuador had three million people. Now it has over ten million. A projection of this trend means 30 years from now, within our own lifespan, there will be more than 30 million Ecuadorians living in just 270 thousand square kilometers. Even now Ecuador has the distinction, if we can call it that, of being South America's most densely populated country.

Another aspect of development is the impact of technology on the renewal capacity of natural resources. When I was a child the machete and the handsaw were the tools of choice of the men pioneering agriculture in the newly opened lands. Now, a man with a power chain saw can cut a tree in a small fraction of the time needed relative to using those old-fashioned tools. Not only that, the settlers of the forest can carve planks and sell the wood with amazing ease as well, thanks to the chain saw.

Is the country getting any richer by this quantitative and qualitative augmentation of the human impact on our resource base? We certainly have a bigger pie, but also many more individual slices. Furthermore, while the pie has grown, the slices of some have grown more than the slices of others. The slices have not become any more even, but there are now a lot of slices that satisfy the aspirations of a politically powerful middle class. To accommodate population growth, Ecuador has had two choises: to redistribute agricultural land or to open up new land. We have chosen to do very little redistribution and have pushed ahead with colonization.

Every political leader has promised more "caminos vecinales" to get elected, and new roads keep penetrating into almost every piece of land available, including remote forest-clad areas west and east of the Andes.

Colonization is a strategy to scatter poverty, somebody might argue, and not a wealth-producing process. It is true that a poor farmer who takes over a 50-ha parcel of forest land is temporarily better off than a landless peasant. The pioneer has wood to build a house for him and his family, a stream to get water, free fuel from the forest, wildlife and fish to kill and eat, and initially at least soil fertility is not a problem for him. Furthermore, by cleaning land he "improves" it, so that the monetary value of his parcel goes up. The short-term economic equation gives profits for the pioneer, and this becomes the driving force that moves national and international development institutions that come in and support colonization. Unfortunately, even the 60 000 square kilometers of Napo provinces, the largest in the country, can only accommodate 120 000 farmer families with 50 hectares each. We are rapidly filling up the land with people, and obviously colonization cannot go on forever.

Furthermore, colonization conflicts with the rights of indigenous peoples, and this is nowhere more apparent than in the Ecuadorian Amazon region where the Cofan, Siona, Secoya, Quechua, Huagrani and Shuar Indians live. Protected by a better health delivery system and stripped from their traditional "uncivilized" ways, their populations are exploding in some cases at faster rates than those of other groups of Ecuadorians. Therefore, the Indians are becoming acutely aware of their need to secure land rights for now and the future, and this has forced them to become more vocal and better organized. They resent the intervention of the Agrarian Reform and Colonization Institute (IÉRAC) in favor of the non-Indian colonists, and continually press the government about establishing secure borders around Indian lands.

All of this brings us to the issue of sustainability. Is this development? How long can it last? Seen from abroad, from the perspective of the manufacturers of, say, flashlight batteries, chain saws or road graders, Ecuador is doing very well. Europe, Japan or the U.S. can sell us more of these technological wonders. Even from the perspective of the World Bank or FAO, what Ecuador is doing does not look so bad, and this is proved by the loans we get from these organizations. Again, is the short-range vision which is to blame here, ours and theirs.

The truth is, however, that this development model cannot last long and we know this as ecologists. The best agricultural land was settled long ago, and the land that is being settled now is far worse. Joshua Dickinson has noted that most Spanish cities and, indeed, Indian cities before them, in Latin America were founded and developed in places where the potential evapotranspiration rate was close to unity, or even where there was a rainfall deficit. This, he further noted, happened because it was only the land near these places that could be farmed easily, even if it had to be irrigated. Of course, technological advances now allow us to farm progressively more marginal land, but the cost increases with deficiencies in soil fertility, susceptibility to weed invasion, likelihood of plant and animal diseases, *etc*. Moreover, the costs to the individual farmer increase with time if he is located in an area of high rainfall, fragile soils and far from a market. Farming marginal land is not a sustainable proposition.

Therefore, there is little question that Third World development as illustrated by Ecuador is running in a dangerous course, along a blind alley in which true wealth is not being created. That this is done at the expense of biodiversity only compounds the problem, and must be made explicit here. For each hectare of land cleared of forest a fraction of each of thousands of species is lost, and for some species this fraction may be all there is. We are forced to conclude that not only wealth is created, but that Ecuador and the world are being decapitalized in that the planet's most complex and least understood natural ecosystems are losing their ability to self-reproduce.

There is a glimmer of hope in that development agencies in the U.S. and Europe, such as A.I.D., are beginning to see this problem, and to incorporate sustainability in their projects in the Third World. Sustainability means that we must learn to live off the interest and not the capital of our renewable resources. Sustainability means that our governments must not conceive development simply as a way to expand the agricultural frontier into the virgin forest. Sustainability cannot be achieved denying access to our resources to some sectors of our society, while granting access privileges to other sectors. Sustainability also means conserving biological and cultural diversity. Let us all join in an effort to get this message across, to our governments and to the public in our respective countries, and, whilst we cannot help being frustrated by our experience, let us not give up our hope for a better future.

Danish botanists in the tropics - tropical botany in Denmark

K. LARSEN

Aarhus University, Denmark

Tropical botany was in founded in Europe during the colonial seventeenth century by botanists, physicians, missionaries, civil servants, and army- and navy officers. It was the "Golden Age" at the University of Leiden which was, "founded during the darkest days of struggle for national independence and now enjoying the fruits of the struggle," (Morton, 1981). Leiden was for over a century the intellectual center of Europe because of its policy of tolerating all religions. It was the age of Boerhaave (1668-1695), the time when Paul Hermann (1646-1695) lived and traveled in Africa, India, and Ceylon, Rumphius (1628-1702) collected in the East Indies, and van Reede (1636-1691) produced his monumental *Hortus Malabaricus* (Heniger, 1986). Leiden, or the *Paradisus Batavus* as Hermann called it, was where the young Linnaeus spent three years from 1735 to 1738 in order to complete his schooling.

Danish botany of the seventeenth century was occupied with the floral study of the double monarchy Denmark-Norway and its northern dependencies. It was the time when the first *Flora Danica*, published in 1648 in Antwerpen, was prepared by Simon Paulli. It was a grandiose work even though not all parts were fully original. This was also the time of Danish scholars and naturalists such as Ole Borch (1626-1690) and Peder Kylling (1640-1696) who wrote *Viridarium Danicum* and had the title of Royal Botanist. The greatest of all the naturalists, and a pupil of Simon Paulli, Niels Steensen (1638-1681) founded palaeontology and chrystallography and indeed modern geology.

Early danish expeditions

In 1756 H. M. King Frederik V sent a large expedition to *Arabia Felix,* the present Yemen in the southern part of the Arabian Peninsula (Hansen, 1964). The aim was to study the language, geography and natural resources of this area. The expedition included a botanist, a philologist, an astronomer, a medical doctor, an artist, and their servant. The botanist, the young swede Peter Forsskål, who was one of Linnaeus students, died during the expedition, but his collections are among the greatest treasures in the Botanical Museum at Copenhagen (Christensen, 1918).

One of the *Flora Danica* collectors in Iceland was Johann Gerhard König (1728-85), who was born in Letland but educated in Copenhagen as apothecary and medical doctor. He was sent to the Danish colony of Tranquebar, south of Madras on the Coromandel Coast in southern India as a physician and became the first Danish plant collector in the tropics. From a wealthy friend he received funds for expeditions to India and Siam in 1779, where he was the first to collect botanical specimens, and to Ceylon in 1781. His collections were sent to the Botanical Museum in Copenhagen where large parts are still available. However, apparently all collections from Thailand were lost. König also provided Linnaeus and other botanists with herbarium specimens and made a close friendship with Roxburgh. He did not publish much, but 21 volumes of his letters and manuscript journals are kept at the British Museum in London, together with some of his specimens. König was one of the greatest plant collectors of the eighteenth century. His collections became the basis for several works, such as Roxburgh's *Plants of the Coast of Coromandel* and the Danish botanist Rottbøll's descriptions of Cyperaceae in *Descriptiones plantarum rariorum* and *Descriptiones et iconum rariores* published in Copenhagen in 1772-73. König's work as a plant collector in India was continued by Joh. R. Rottler (1749-1836) who was born in Strassbourg, France but served Denmark as a missionary in Tranquebar. He later joined the British service in Madras and collected in Ceylon and in the Ganges region. Professor Vahl from Copenhagen described numerous new species based on Rottler's specimens.

Tranquebar, could have become a solid base for Danish botany in India, but it was occupied by the British during the Napoleonic wars in which Denmark was allied with France.

An important Danish colony was the West Indian Islands which were sold to USA in 1917 and renamed the Virgin Islands. They were first explored in 1757 when Julius von Rohr, a German doctor educated in the new Linnean philosophy, was appointed surveyor. He explored the islands for 36 years until his death in 1793 which was due to a ship that took him on a planned expedition to Guiana that capsized. Von Rohr was an energetic collector and numerous of his specimens were deposited at the Botanical Museum in Copenhagen and later described as new by Martin Vahl.

Another important collector was the philologist Hans West (1758-1811) who became administrator in the West Indies. In 1793 he published a description of the St. Croix islands in which he gives the most complete list of plants from any tropical region at the time. West also collected and described algae. Large collections from the Antilles, therefore, came to Copenhagen in the last half of the eightteenth century.

The Danish government purchased a small colony on Africa's Gold Coast, now Ghana from the British in 1659. In 1850 it was resold to the British for £ 10 000. Paul Erdmann Isert (1756-1789),who was born in Brandenburg in Germany, went into Danish service and was sent as a physician to what is now Ghana and Dahomey in 1783. He died six years later. His collections were some of the very first from tropical Africa. In Copenhagen, they were examined by Vahl and Schumacher. Isert made futher collections in the West Indies not only on the Danish islands but also on Martinique and Montserrat.

Peter Thonning (1775-1848) was born in Copenhagen, he too went to the African colonies for four years as a medical doctor. He sent large collections to the Botanical Museum in Copenhagen (Hepper, 1976).

Engagements in the nineteenth century

The eighteenth century's tropical botanical exploration was mainly concentrated on the small Danish colonies in the tropics (Brøndsted, 1952-53). Much material was collected, parts of which was described, but a plan for systematic exploration was never formulated. There was not a chair for botany in Copenhagen until 1801. The first professor was Martin Vahl who was appointed 1804. Chr. Fr. Schumacher 1757-1830 was professor at the Royal Academy of Surgery and published several botanical works. Following his retirement in 1813 he worked on the West Indian and African collections.

In the first half of the nineteenth century, a number of Danish botanists and collectors went to the tropics, however, only the most important ones are dealt with in this paper. The Napoleonic wars stopped the activities in Tranquebar, but another Danish settlement at Frederiksnagore or Serampore near Calcutta became the working place for two Danish doctors.

Nathan Wulff, who later changed his name to Nathanial Wallich, was born in Copenhagen in 1786. He took his degree from the Royal Academy of Surgery in 1806 and was sent to Serampore as a physician where he became a friend of Roxburgh and went into British service in 1814. In 1815 he became assistant superintendent at the Botanical Garden, Calcutta and then superintendent from 1817 to 1846. He collected all over India and in addition, travelled to Nepal, Singapore, Penang, Assam, Mauritius, and the Cape from 1820 to 1822. From 1847 to his death in 1854 he lived in London. Wallich was an extremely energetic collector. When he visited London in 1828 he carried with him, according to a contemporary source, 20 tons of dried specimens and 23 tons of living plants. In spite of his British engagement, which he regarded as vital for his botanical exploration, he maintained close ties to Denmark at a time when the two countries were at war. He translated Danish botanical works into English to make the publications of his country known. He donated a large herbarium to the Botanical Museum in Copenhagen when the Danish Galathea Expedition visited Calcutta in 1845. Wallich published several works, including one of the most beautiful botanical books, *Plantae Asiaticae Rariores* in three large folio volumes.

Another Danish surgeon, Joachim Otto Voigt (1783-1843), became plant collector in Serampore at the same time. He was a physician for the Danish colony from 1826 until he succeeded Carey in 1834 as superintendent for the Botanical Garden at Serampore, also known as Carey's Garden, and also for a short period the Botanical Garden Calcutta. He wrote a detailed catalogue on the plants of the two botanic gardens, *Hortus Suburbanus Calcuttensis*. Bad health forced him to return to Europe in 1841. On his way back to Copenhagen, he stopped in London where he died.

In the 1840s, three large Danish expeditions ventured into the tropics, one was sponsored by the Danish goverment and the other two were private undertakings.

The largest and most expensive Danish expedition, up to that date, took place on a special occasion. After long negotiations the two Danish settlements in India were sold to Britain in 1845 for a considerable sum of money. In order to give over the sovereignty, the Danish navy ship Galathea sailed to Serampore and later to the Nicobars which remained Danish until 1868. The Galathea was equipped to circumnavigate the world and had on board a scientific staff. After the Nicobars, the ship visited the Dutch East Indies, Manila and China, from where it continued to Chile, Galapagos Islands, Cape Horn, Montevideo, Buenos Aires and Bahia. The Royal Danish Academy had chosen the young B. Kamphøvener, as the expedition´s botanist and as his assistant it had selected F. Didrichsen, a young doctor. Kamphøvener had bad health yet a few thousand collections were brought to Copenhagen including fine spirit collections. Unfortunately, there was no one to work up the material.

Frederik Michael Liebmann (1813-56) was the son of a wealthy merchant in Elsinore. He studied botany in Copenhagen under Prof. Hornemann. He became a friend of the house and later married the youngest daughter of his professor. As a student he had travelled in Germany and Norway and now planned a longer expedition to a tropical country. His choice was Mexico. He received grants from private and public funds, and because the Botanic Garden in Copenhagen was interested in adding to its collections, one of the senior gardeners was appointed to assist Liebmann. In November 1840, when Liebmann was 27 years of age, the two men sailed for Veracruz where they arrived three months later. After one year Liebmann sent the gardener back with the first collections: 50 000 specimens of dried plants, 44 boxes with living plants including 400 species, and furthermore, large zoological collection. He continued alone, worked his way from the lowlands to the highest mountains and from the rainforest to the deserts, and on his way back he visited Cuba. After two and a half years had passed, he had collected 80-90 000 dried plant specimens, numerous boxes with living plants, seeds, zoological specimens, and ethnographic and archeological samples. He literally worked day and night and was never ill. His energy was still incredible when he returned. He became professor of botany, without having obtained any academic degrees. Shortly afterwards, however, his good luck deserted him. His young wife died, and he became aware of his own illness, tuberculosis, which he did not regard as serious. But soon it became his master and he died at an age of 43. He published several works based on his Mexican collections in the proceedings of the Royal Danish Academy. His most important publications are those on the Ferns, Juncaceae, Cyperaceae, and *Begonia*.

Anders Sandøe Ørsted (1816-1872) was the son of a merchant, his uncle was the famous physicist H. C. Ørsted, the discoverer of electromagnetism. He finished his studies with a dissertation, *De regionibus marinis,* on the natural history of Øresund which is the narrow strait between the Danish island of Zealand and Sweden. He was equally interested in zoology and botany. In 1845 he sailed to Jamaica where he made collections. Later he continued to Nicaragua and from there to Costa Rica where he stayed for one year. Ørsted collected in Central America for three years and made large gatherings, even if not as large as those of Liebmann. Ørsted became one of the first to study tropical phytoplankton. In 1862 he was appointed professor of botany at Copenhagen and edited Liebmann's work on the oaks, diverting to the study of that group. Later his interests turned to mycology.

An extraordinary amateur botanist, Baron Heinrich von Eggers, was born in Schleswig, at that time a Danish town, and published a flora of St. Croix in 1876. This flora was later enlarged and translated into English as *The Flora of St. Croix and the Virgin Islands* and printed in Washington. Eggers never studied botany or any other academic disciplines. As an officer he had been a loyal supporter of Emperor Maximillian and was taken prisoner after the fall of the Mexican empire. From 1869 to 1885 he served in the Danish West Indian garrison. He had an extensive correspondence with professor Warming, who also supervised his botanical work. Eggers sent large collections to Warming including valuable spirit collections which were used for Warming's morphological and anatomical studies.

The Danish naturalist Peter Wilhelm Lund (1801-1880) was the son of a wealthy family and studied zoology in Copenhagen. When still young, he showed signs of tuberculosis and in 1825 went to Rio de Janeiro to live in a milder climate. He became fascinated by the tropical nature, including its flora. Lund returned to Denmark in 1829, but three years later decided to move back to Brazil where he undertook expeditions to the central part of the country. Here he discovered limestone caves with subfossil bones of extinct mammals. The study of these made him internationally known. He asked in Copenhagen if a young botanist could be found who could serve him as secretary. The student Eugen Warming was chosen and spent three years with Lund in Lagoa Santa, where he collected large amounts of material (Warming, 1892; Stangerup, 1981).

Eugen Warming (1841-1924) was son of a country parson on the small waddensee island of Manø (Prytz, 1981). He became the greatest Danish botanist and left his mark on botany for more than half a century. Besides his three years in Lagoa Santa from 1863 to 1866, he made an expedition in 1891-1892 to Venezuela, the Barbados, Trinidad, Puerto Rico, and the Danish West Indies. Warming, who may be regarded as the founder of tropical ecology edited *Symbolae ad floram Brasiliae cognoscendam* for 25 years and contributed treatments of numerous families to this work. In several papers he treated the family Podostemonaceae. In spite of all these connections with tropical botany, his main botanical work was connected with the flora of Denmark and Greenland. He contributed to plant geography and systematic botany in general and wrote several large textbooks for his students.

The present

In 1899, Warming sent a promising young botanist, Johannes Schmidt, to Thailand together with the zoologist Th. Mortensen. It was an expedition supported by the Carlsberg Foundation and the commercial Danish East Asiatic Company. The two naturalists spent most of their time on the island Koh Chang. After his return, Schmidt wrote his dissertation on the shoot-morphology of the Old World mangrove trees. He further edited a series of publications between 1901 and 1916, *Flora of Koh Chang*, and thus, became the founder of the modern exploration of Thailand. This could have led to the initiation of tropical botany in Denmark, but Schmidt's interest turned towards zoology and marine biology.

The first half of the 20th century was a period when Danish botanists again looked north. The tropical colonies were sold and there were no close relations to any tropical country except for Thailand, so again the work was concentrated on projects in Denmark. Large works on the Faroes and Iceland were published and an intense exploration of Greenland was initiated.

In 1957, Gunnar Seidenfaden, the Danish ambassador to Thailand , took the initiative to form a botanical expedition to that country (Larsen,

1979). On January 1, 1958, professor Thorvald Sørensen from Copenhagen and the author of this paper went to Thailand and travelled for three months throughout the country together with the director of the Forest Herbarium, Tem Smitinand. Later Bertel Hansen from the Botanical Museum in Copenhagen and his wife Birgit joined the expedition for a year. I continued the work in Thailand alone in 1961, 1962, and 1963. In 1963, the idea of the *Flora of Thailand* project was born and from this, tropical botany developed at Aarhus University. During the past 30 years, ambassador, Dr. Gunnar Seidenfaden has finished a beautifully illustrated revision of the 1000 species of orchids of Thailand. These 1000 species of orchids constitute one tenth of the total flora of Thailand. At the same time Tem Smitinand and the present author has edited 3 volumes of Flora of Thailand. Later through the initiative of Lauritz B. Holm-Nielsen, the Ecuador project (Holm-Nielsen *et al.*, 1984) was started and at present a project in the Sahel zone of tropical Africa is taking shape. All staff members in the taxonomy department at the Botanical Institute in Aarhus are working in the tropics and even if the three projects are not directly connected, we all merge our experience in educating students in this discipline.

At the Botanical Museum in Copenhagen projects in East Africa are developing.

I am convinced that tropical botany now after 300 years of trial and error has been firmly established at the two Danish universities in Copenhagen and Aarhus, and I am also convinced that there are sufficient young and energetic students who are fully able to continue the vital exploration of the rapidly vanishing tropical plant world.

Literature cited

Brøndsted, J. (1952-1953). "Vore gamle tropekolonier." 2 vols. *Westermann, København.*

Christensen, C. (1918). "Naturforskeren Pehr Forsskål (Peter Forsskål, the naturalist)." *Hagerup, København.*

Christensen, C. (1924-1926). "Den danske botaniks historie med tilhørende bibliografi." 3 vols. *Hagerup, København.*

Hansen, Th. (1964). "Arabia Felix. The danish expedition of 1761-1767." *Harper and Row, New York.*

Heniger, J. (1986). "Hendrik Adriaan van Reede tot Drakenstein (1636-1691) and Hortus Malabaricus. A contribution to the history of Dutch colonial botany." *A. A. Balkema, Rotterdam and Boston.*

Hepper, F. N. (1976). "The West African herbaria of Isert and Thonning. A taxonomic revision and an index to the IDC microfiche." *Roy.*

Bot. Gard. Kew.

Holm-Nielsen, L. B., Øllgaard, B. and Molau, U. (eds.), (1984). "Scandinavian botanical research in Ecuador." *Rep. Bot. Inst., Univ. Aarhus* **9**, 1-83.

Larsen, K. (1979). "Exploration of the flora of Thailand." pp. 125-133. *In* Larsen, K. and Holm-Nielsen, L. B. (eds.), "Tropical Botany." *Academic Press, London.*

Morton, A. G. (1981). "History of botanical science. An account of the development of botany from ancient times to the present day." *Academic Press, London.*

Prytz, S. (1981). "Warming. Botaniker og rejsende." *Bogan, Lynge.*

Stangerup, H. (1981). "Vejen til Lagoa Santa." *Gyldendal, København.*

Warming, E. (1892). "Lagoa Santa. Et bidrag til den biologiske plantegeografi." *Vid. Selsk. Skr.* **6,3**: 153-454.

Spanish floristic exploration in America: past and present

S. CASTROVIEJO

Real Jardín Botánico, Madrid, Spain

Spanish botany, after an initial strength in the eighteenth century, languished slowly over the years and, with the exception of a few outstanding individuals, reached a level of reduced presence in the international forum. The situation is fortunately beginning to change. The following presentation is a status of the history of the Spanish floristic exploration in America, the botanical expeditions, and the resulting floras.

The botanical exploration of America began immediately after its colonization. The first original Natural History ever printed was the *Sumario de Historia Natural de las Indias* by Gonzalo Fernández de Oviedo (1526). The same author later published *Historia General y Natural de las Indias* (1535) in which he described 43 fruit trees, 34 wild trees, eight medicinal herbs, and 10 herbaceous plants. Nicolas Monardes (1565-1574), a physician from Seville, published *Historia medicinal: de las cosas que se traen de nuestras Indias Occidentales que sirven en Medicina* in three parts. Acosta (1590) published *Historia Natural y moral de las Indias* in which he also described a number of species.

The first true botanical expedition was organized by Philip II in 1570. He appointed one of his physicians, Francisco Hernández as Medical Examiner, the first in the New World. Hernández departed in 1570 for New Spain, the present Mexico and neighboring countries. He worked seven years in Mexico collecting information and botanical specimens, questioning more than twenty native physicians, and testing medicines on the sick in the capital city. He was able to compile an important part of the medicinal traditions of the Aztec civilization. The result of his work was six volumes of text and ten of illustrations which he sent to Spain together with dried plants and seeds for the new gardens at Aranjuez. He died upon his return in 1578 and left his work unpublished. It was deposited in the Library of El Escorial, and was the first in a long tradition of valuable botanical works from Spain which remained unpublished.

The history of Hernández' works is long and complex, but this

TROPICAL FORESTS
ISBN 0–12–353550–6

paper will concentrate on the botanical expeditions specifically planned for the survey of the territories and the description of plants following the Linnaean methods. Several manuscripts, which remained forgotten for over 150 years, have received the attention of researchers over the last four years, generally spurred by the interest created by the celebration of the 5th Centennial of the discovery of America. Historians and botanists have met in our archives and have begun to put together clues and data which clarify many events in the floristic exploration of Tropical America which previously had no logical explanation . They are unearthing many results of these expeditions which were previously unknown in their full extent.

Expedition to the Orinoco, Venezuela, 1754-1756

This expedition was organized and sponsored by the Spanish Crown in order to establish the frontier between the Spanish and Portuguese possessions. The expedition botanist was Pehr Löfling who, at an age of 22 years, had arrived in Spain at Linnaeus' request in 1751. He collected a rich herbarium, but the whereabouts of this herbarium and a number of descriptions, plant drawings, and abundant notes on medicinal plants and ethnobotany are unknown. For two years Pehr Löfling collected plants and gathered information, mainly botanical, about the natural history of the Cumaná area near the mouth of the Orinoco. His plant descriptions for *Flora Cumanensis* are outstanding. There are two unpublished manuscripts; one is a draft with 260 descriptions, and the other is the flora proper with 211 descriptions. The early death of Löfling prevented a more complete botanical study of Cumaná, but Linnaeus published a few plant descriptions from the area in his *Iter hispanicum* (1758); therefore, *Flora Cumanensis* is the real proof of the work done by Löfling in Venezuela. The iconography of the expedition, drawn by the artists Bruno Salvador Carmona and Juan de Dios Castel, also unpublished, is an important contribution to the knowledge of American natural history. It includes 115 botanical, 81 zoological and two ethnographic drawings, and two maps. Löfling also left a sizable manuscript, still unpublished, entitled *Materia Medica Vegetal de las Provincias Americanas.*

Royal botanic expedition to the Viceroyalty of Peru, 1777- 1788

Jacques Turgot, a minister of King Louis XVI of France, wanted permission to organize a scientific expedition to recover the manuscripts of Joseph de Jussieu. The Spanish Crown replied by organizing its own expedition in which the participation of the French botanist Joseph Dombey was accepted. The young and energetic Hipólito Ruiz was appointed to head the expedition, and he was to be aided by the botanist José Pavón and the artists José Brunete and Isidoro Gálvez. They departed from Cádiz in 1777 and traveled extensively in the Andes and parts of coastal Chile and Peru until their return to Spain in 1788. They left behind a young pupil, Juan Tafalla, who had been appointed professor of botany in Lima. He would continue to explore the area and describe the new plants he found. The results of this expedition were abundant and interesting, although they included a controversy between French and Spanish botanists which would continue until the present time. The herbarium of the expedition includes 10 000 collections of which 1800 are types. One result of this expedition, *Flora Peruviana et Chilensis* by H. Ruiz and J. Pavón (1783-1810), is a well known work to all botanists studying the American flora. In 1816, upon the death of Hipólito Ruiz, the publication was discontinued after only three of eleven planned volumes had appeared. The fourth volume was published in 1957, and there is now a plan to publish the remaining ones. The flora includes 5000 descriptions and 2224 botanical and 24 200 zoological drawings. Ruiz and Pavón were aware of the difficulties of publishing a work of such a scope, and they decided to advance their new findings in their *Systema Vegetabilium Florae Peruvianae et Chilensis* and in the *Flora Peruviana et Chilensis Prodromus*. Both are necessary starting points for any taxonomic study of the flora from this region. They include describtions of 1612 new species.

A hitherto unknown result of the expedition is the *Flora Huayaquilensis* (1799-1808) by J. Tafalla. Nothing was known of this work until the Ecuadorian researcher E. Estrella, while studying the origins of botanical research in Ecuador, found the trail of the works of Juan Tafalla in Ecuador. Tafalla continued the survey of Peru when Ruiz and Pavón returned to Spain. The botanists had instructed him in the theory and practice of botanical science. Tafalla was given a Crown salary and an additional budget to allow him to cover the expenses of traveling, paying the artists, *etc.* Later, Tafalla and the Ecuadorian artist Manzanilla were asked to prepare a report on the woodlands and lumber of the *Real Audiencia* of Quito. He moved to Guayaquil and began his floristic exploration in the area. The result of this work includes descriptions of 625 species and 216 drawings of Ecuadorian plants which he, as agreed with the Royal Expedition, sent to Spain under the denomination *Flora Huayaquilensis*. When these descriptions arrived in the hands of Hipólito Ruiz, he regarded them as part of the agreed work, and they were included in his monumental *Flora Peruviana et Chilensis*. There, hidden among all the others, they remained until today. Some of them were published in the first volumes of the Flora. Because of the historic interest of this discovery and the fact that most of them are still unpublished, the Spanish Administration has agreed to publish the full work, with the manifest indication that there is no intention to create valid names under the International Code of Botanical Nomenclature.

Royal botanical expedition to the New Kingdom of Granada, 1783-1808

This expedition was sent by the Spanish Crown in 1783 on the request of Jose Celestino Mutis, a physician and botanist who was born in Cadiz and had lived in New Granada since 1760. Mutis organized around him a large and competent team of collectors and artists which allowed him to work intensively for more than 20 years in the Andean high plateau and in the Magdalena River basin. As a result of this expedition, we have a herbarium of about 20 000 specimens of which 260 are types and perhaps the most important and beautiful collection of botanical drawings in the world, including some 7000 plates in full color, most of them still unpublished. The collections also include countless papers with field notes, descriptions, observations of all kinds, and correspondence.

Some of the results are published in *Flora of the New Kingdom of Granada* (1783-1808) by J. C. Mutis. Mutis held the opinion that a good illustration was a better way to describe a plant than studying its dry skeleton; therefore, his works included illustrations made by expert artists. His herbarium contains vouchers of many of the illustrated plants. The 500 descriptions were carefully prepared by Mutis and his assistants. Unfortunately, the writing up of the work in Bogotá had barely begun when Mutis died. His nephew Sinforoso tried to continue it but was only able to arrange the papers and descriptions in order. The work remained unpublished until the 1950s when the governments of Spain and Colombia agreed to initiate a slow and supervised publication of the drawings. By 1988, 18 volumes have been published and more than 60 are planned.

Royal botanical expedition to New Spain, 1787-1803

This is the third great botanical expedition of the eightteenth century. Its director was Martín Sessé, an Aragonese physician, who was assisted by the botanists Vicente Cervantes and José Longinos Martínez. In Mexico, the botanist Jaime Sensevé and the artists Vicente de la Cerda and

Atanasio Echeverría joined them. Much later a third botanist, Juan del Castillo, also joined the expedition. Cervantes remained in Mexico City as a professor of botany. He soon selected an outstanding pupil, Mariano Mociño, who would become Sessé's best assistant after his incorporation in the expedition in 1790. José Maldonado, a classmate of Mociño, also joined the team in 1791. The expedition returned to Spain in 1803. During 16 years they surveyed the southern parts of the United States, Central America, the northwest coast up to Nootka Island, and the island of Cuba. The herbarium material they collected includes 5500 collections of which over 2000 are types. These collections are deposited in the Royal Botanic Garden, Madrid, along with the Flora of New Spain which remained unpublished in our archives for one hundred years. The magnificent collection of botanical drawings was partially lost after 1820 when it was admired by de Candolle and copied in Geneva. It was recently acquired in a dubious way by the Hunt Institute in Pittsburgh.

Among the various botanical itineraries made during the Expedition to New Spain (1787-1803) that of the Audiencia and General Captainship of Guatemala is an outstanding one. For this scientific trip the viceroy of Mexico appointed as botanist José Mariano Mociño, José Longinos Martínez as naturalist, and Vicente de la Cerda as artist. They left Mexico City in January 1795, and spent over a year traveling to the capital of New Guatemala. However, while Longinos traveled along the Atlantic coast, Mociño followed the Pacific side and collected all along the Central American isthmus. One of the results of this expedition was the *Flora of Guatemala* (1795-1803) by J. M. Mociño which is still unpublished and contains the description of 553 species in 349 genera. The 119 drawings of this flora, made by V. de la Cerda and as yet unidentified, are among some 2000 drawings from the iconography of the Expedition to New Spain.

Pacific expedition, 1789-1794

This expedition was organized to circumnavigate the world with scientific, cartographic, hydrographic, and political objectives. It was headed by Alejandro Malaspina who commanded the corvettes "Descubierta" and "Atrevida." Expedition members included the botanists Luis Neé, Antonio Pineda, and the Austrian Thaddeus Haënke. They were aided by the artists José del Pozo, Tomás de Suría, José Cardero, José Guío, Francisco Lindo, José Ravenet, and Fernando Brambilla. The expedition departed from Cádiz and visited the Atlantic coast of South America from Montevideo southwards and the entire American Pacific coast up to Alaska. Later, they traveled to the Philippines, China, New

Zealand, Australia, and several Pacific islands until their return to "el Callao." From "el Callao" they traveled further to Montevideo and Europe, ending the voyage in Cádiz. The results include a collection of 15 000 plant specimens, but notes, about 800 descriptions, and 331 botanical drawings of plants and animals mostly remain unpublished. Cavanilles, described many new species from the herbarium of the expedition.

Expedition to the Island of Cuba, 1796-1802

Under the leadership of the Count of Mopox and Jaruco, the purpose of this expedition was to survey the land and fortifications of the Bay of Guantánamo. Among the expedition members were the botanist Baltasar Manuel Boldó and the artist José Guío on his way back from the Pacific Expedition. On the island they met members of the Expedition to New Spain, who spent three years studying the flora of the Caribbean Islands before their return. Among the results of this expedition are 2000 herbarium collections and the unpublished manuscript *Flora of Cuba* (1796-1802) by B. M. Boldó and J. Estévez with 66 plant drawings by J. Guío. After having collected in the western part of the island, José Estévez joined the expedition and after the death of Boldó in 1799, undertook to arrange and complete the descriptions. The descriptions of genera and species make up a manuscript of 743 sheets. They pertain mostly to plants of La Habana and the surroundings, arranged, except for the seventh one, in the 24 Linnean classes. In fact, this manuscript together with the 66 drawings of plants by José Guío, is the first serious contribution to the flora of Cuba.

The scientific commission of the Pacific, 1862-1866

During the reign of Isabel II, the last important expedition sent by the Spanish government to America was organized. It had scientific, political, and military objectives that caused inumerable difficulties to the members of its scientific team, which was a large and highly qualified group, including even a photographer. The zoologist Marcos Jiménez de la Espada acted as scientific director and the botanist was Juan de Isern i Batlló. Juan de Isern i Batlló managed to collect a valuable herbarium of 16 000 specimens of which only a small part has been studied and published in a series entitled *Plantae Isernianeae* by José Cuatrecasas. This expedition suffered a number of misfortunes including fires, attacks from Amazon indians, and clashes between the scientists and the military which compelled the former to return to Spain on their own, after crossing on foot from the Pacific to the Atlantic coast, down the Amazon River.

Present projects

After having abandoned the floristic study of America for more than 150 years, if we exempt the works of Dr. José Cuatrecasas, Spain is now slowly retaking the initiative, and in the last ten years Spanish botanists have started to visit Venezuela, Colombia, Ecuador, Bolivia, and Paraguay. As a result of the agreement between the governments of Spain and Colombia to publish the "Mutis Flora," the Spanish Administration, through the Royal Botanic Garden, Madrid, has a permanent team of four botanists working at and together with the Institute of Natural History (National University) in Bogotá, in order to prepare the monographs and the identification of drawings for publication of the "Mutis Flora." Another field of interest is centered in Bolivia, Paraguay, and Chile. In the first two countries, work has progressed for some years now. The collections deposited in the Royal Botanic Garden are important (over 15 000 specimens), and the new findings already published are also beginning to receive recognition. On the other hand, the work in Chile is planned in a very different way, as this country has initiated the writing of a flora. Our role is, therefore, to assist in the preparation of the accounts for some genera and, above all, in the examination of the historic herbaria deposited in our institution.

Published floras

Ruiz, H. and Pavón, J. (1794). "Flora peruviana et chilensis Prodromus sive novarum generum plantarum peruvianarum et chilensium descriptiones et icones." *Gabrielis Sancha, Madrid.*

Ruiz, H. and Pavón, J. (1798-1802). "Florae peruviana et chilensis sive descriptiones et icones plantarum peruvianarum chilensium." 3 vols. *Gabrielis Sancha, Madrid.*

Ruiz, H. and Pavón, J. (1798). "Systema vegetabilium florae peruvianae et chilensis, characteres prodromi genericos differentiales, specierum omnium differentias, durationem, loca natalia, tempus florendi, nomina vernacula, vires et usos nonnullis illustrationibus interspersiscomplectens. Tomus primus." *Gabrielis Sancha, Madrid.*

Sessé, M. and Mociño, J. M. (1887-1890). "Plantae Novae hispaniae." *Mexico.*

Sessé, M. and Mociño, J. M. (1891-1896). "Flora mexicana." *Mexico.*

Diversity of minor tropical tree crops and their importance for the industrialized world

V. K. S. SHUKLA AND I. C. NIELSEN

Aarhus Oliefabrik A/S and Aarhus University, Denmark

Several underexploited or unexploited trees are found in tropical ecosystems. Some have already proved their importance for the industrialized world. The present paper deals with cocoa and its substitutes. Chocolate is associated with an imported commodity, the cocoa bean. Approximately 1.8 billion tons of cocoa beans are produced worldwide each year. Cocoa beans come from the cacao tree (*Theobroma cacao*), *Theobroma* means "food of the gods". The shrub, a native of S. America, is cultivated in West Africa, South America, Central America, and the Far East. The Ivory Coast is the leading producer, contributing about 25% of the world's production followed by Brasil. Malaysia is slowly becoming a major producer, increasing its production linearly.

The total fat content of the whole bean is around 48-49% of the dry weight. Cocoa butter (CB) is the most expensive of the major ingredients of chocolate and it is a legal requirement. Its price fluctuates unpredictably from year to year (Fig. 1). Cocoa butter is predominantly (>75%) composed of symmetrical triglycerides (Shukla *et al.*, 1983) with oleic acid in the 2-position (Table 1). It contains approximately 20% of triglycerides that are liquid at room temperature and has a melting range of 32-35°C, softening around 30-32°C. This is an essential requirement. Cocoa butter contains only trace amounts of unsymmetrical triglycerides (PPO, PSO, and SSO; see Table 1). The unique triglyceride composition, together with the extremely low levels of undesirable diglycerides, attribute to a stable crystal modification during processing.

The triglyceride composition of three different cocoa butters as determined by high performance liquid chromatography (HPLC) (Shukla *et al.*, 1983) is shown in table 1. These results show that Malaysian cocoa butter contains maximum amounts of mono-unsaturated triglycerides and minimum amounts of unsaturated triglycerides. The Brazilian cocoa butter contains minimum amounts of mono-unsaturated triglycerides and maximum amounts of unsaturated triglycerides. The solid fat contents (Shukla, 1983) of these cocoa butters are depicted in

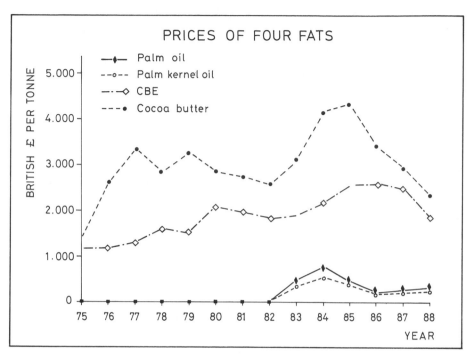

Figure 1. Price variation for four important chocolate fats; Pound Sterling per tonne.

Table 1. Triglyceride composition (mole%) of 3 cocoa butters as obtained by HPLC. Abbreviations indicate: M = Myristic acid; P = Palmitic acid; S = Stearic acid; O = Oleic acid; Li = Linoleic acid; A = Arachidic acid.

Cocoa butters	Trisaturated			Mono-unsaturated				
	PPS	PSS	POM	POP	POS	SOS	SOA	Total
Malaysian	tr.	tr.	0.4	15.2	38.5	30.3	2.8	87.2
Ghanian	tr.	tr.	0.7	17.2	34.3	26.4	1.0	79.6
Brazilian	tr.	tr.	tr.	14.4	33.7	23.3	tr.	71.4

	Di-unsaturated			Poly-unsaturated				
	POO	SOO	PLiIP	PLiS	SLiS	OOO	PLiO	SLiO	Total
Malaysian	2.0	3.1	2.1	3.3	1.2	0.3	0.6	0.2	12.8
Ghanian	3.4	4.9	2.9	4.4	2.3	0.6	1.1	0.8	20.4
Brazilian	7.1	10.4	2.5	3.9	1.7	1.0	0.6	1.3	28.5

Figure 2. There is a good correlation between the triglyceride compositions and solid fat contents of these cocoa butters. The Malaysian cocoa butter is the hardest and the Brazilian is the softest, whereas Ghanian lies in between the two. The different compositions reflect the differences between the cultivars grown in Malaysia, Brazil and Ghana. These results reveal that the quality of Brazilian cocoa butter can be improved by mixing it with the Malaysian cocoa butter.

Confectionery fats

The uncertainty in the cocoa butter supply and the volatility in the cocoa butter prices, depending on the fluctuating cocoa bean prices, has forced confectioners to seek alternatives, which may have a stabilizing influence on the price of cocoa butter. Everincreasing demand (Anon, 1984) of chocolate and chocolate-type products increases the demand for cocoa beans from year to year. However, it is difficult to predict the supply of cocoa beans.

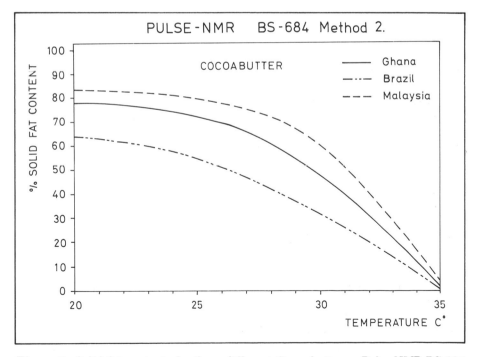

Figure 2. Solid fat contents for three different Cocoa butters. Pulse-NMR BS 684 Method 2.

This necessitates the use of economical vegetable fats to replace cocoa butter in chocolate and confectionery products. As early as 1930 attempts were made by confectioners to use fats other than cocoa butters in their formulations.   These experiments did not succeed because of the incompatibility of the fat blends used, which resulted in discoloration and bloom.   However these experiments established the need for the cocoa butter-type fats in the chocolate and confectionary industry.

The continued research in the field of confectionary science resulted in the development of fats resembling the characteristics of cocoa butter. These fats were coined the term *hard butters* (Paulicka, 1976) and were developed using palm kernel, coconut, palm and other exotic oils such as Sal (*Shorea robusta*), Shea (*Butyrospermum parkii* and Illipe (*Shorea spp.* from Borneo) as raw materials.   The processes involved in producing such fats included hydrogenation, interesterification, solvent or dry fractionation and blending.   The most elementary hard butters are manufactured by combining the processes of hydrogenation and fractionation.

Cocoa butter equivalents (CBE)

CBE are nonhydrogenated specialty fats containing the same fatty acids and symmetrical mono-unsaturated triglycerides as cocoa butter.   They are fully compatible with cocoa butter and can be mixed with cocoa butter in any ratio in chocolate formulations.

From the data presented in Table 1 it is evident that cocoa butter is a simple three component system consisting of POP, POS and SOS triglycerides and if these three triglycerides are mixed in appropriate proportions, then the resultant vegetable fat will behave as 100% cocoa butter equivalents.   Although CBE's are not produced by mixing individual triglycerides, these being very expensive to produce, this is the logic behind the whole concept of producing CBE (Faulkner, 1981; Solters, 1970).   Palm oil is fractionated to produce middle melting fraction rich in POP.   Exotic fats, such as Shea and Sal, are fractionated to get triglyceride cuts rich in POP and SOS.   A careful preparation and blending of these fractions results in a tailor made fat equivalent to cocoa butter in physical properties.   Therefore, these fats are called cocoa butter equivalents.   Cocoa butter (Shukla *et al.*, 1983) contains >75% of the symmetrical 2-oleo disaturated triglycerides (POP, POS, SOS).   In the CBE-raw material Shea (*Butyrospermum parkii*), Sal (*Shorea robusta*), Illipe (*Shorea spp.*), Kokum (*Garcinia indica*) and Mango (*Mangifera indica*) oil almost all oleic acid is also placed in the center position.   The formulation of a suitable CBE is the greatest current art in fat technology (Haumann, 1984).

The main disadvantages of using cocoa butter in chocolate

production are as follows: (i) low milk fat tolerance; (ii) lack of stability at elevated temperatures; (iii) tendency to bloom.

The principal advantages of incorporating CBE's into chocolate production are: (i) reduction in the production cost of chocolate as CBE's are cheaper than cocoa butter; (ii) stabilizing influence on fluctuating prices of cocoa butter; (iii) improvement of the milk fat tolerance; (iv) increasing resistance to storage at high temperatures and; (v) very importantly, bloom control.

Actual and potential sources of CBE.

Although CBE's are primarily produced from palm and shea oils, a number of other plant species yield fats containing an attractive triglyceride profile (see Table 2).

Shea (*Butyrospermum parkii*) is a traditional source of CBE. Shea is a medium-sized tree, 10-15m's tall, native to the southernmost parts of the Sahel and the adjacent Sudan and Guinea savanna (Maydell, 1986). Dense populations are found in the areas Guinea to Mali, Burkina Faso, Niger, N. Ivory Coast, Ghana, Togo and Benin to northern Nigeria and Cameroon. Cultivation of Shea has been tried but for various reasons Shea is not well suited for commercial cultivation. The most important reason is that the tree does not yield a crop until it is about 15 years old, reaching maximum yield after 40 years. As one of the few naturally occurring vegetable fats of these semiarid areas, cultivation of this tree should be encouraged, (Maydell, *l.c.*). No systematic collection of sheanuts takes place. Undoubtedly the volume of the crop collected is influenced both by the prices paid locally and, to some extent, by the prices obtainable for alternative crops such as groundnuts, as the collectors divert their efforts towards whatever gives the best cash return. It has not been possible to obtain reliable crop estimates or export statistics. We expect, however, a greater stability with regard to future deliveries. Based on normal harvesting conditions, Nigeria should have an export potential in excess of about 75 000 tons of sheanuts, which would reestablish that country as one of the most important suppliers.

Mowrah (*Madhuca latifolia* and *M.longifolia*) is the local name for the fat of both species. *Madhuca latifolia*, an evergreen tree,is native to Sri Lanka, S.India, the monsoon forests of the Western Ghats from Konkan and southwards. *Madhuca longifolia* is native to the Indian states of Uttar Pradesh,Madhya Pradesh,Gujarat and Andhra Pradesh (Bringi and Metha,1987). Mowrah is a forest crop with an estimated potential of 1,11 million tonnes of kernels equalling 400 000 tonnes of oil.The actual yearly collection(1976-1984) ranged from 19 000 to 30 000 tons (Bringi & Metha, *l.c.*).

Table 2. Triglyceride composition (mole%) of CBE source oils as determined by HPLC. (For abbreviations see Table 1).

| | Mono-unsaturated | | | | | | | |
	POM	POP	POS	POA	SOS	SOA	AOA	Total
Palm Oil Mid Fraction	2.3	42.7	7.8		0.7			53.5
Shea Oil		0.6	6.2		25.2	1.1		33.1
Shea stearin		1.1	9.9		65.5	3.9		80.4
Sal Oil		2.4	15.5	tr.	41.6	11.5	0.8	71.8
Mango Fat	0.9		9.5		51.7	3.7		65.8
Illipe fat (Borneo)		7.1	31.4		49.9	2.4		90.8
Allanblackia		0.2	5.1		49.1			54.4
Mangosteen		tr.	2.1		38.8	0.7		41.6
Sal stearin		tr.	12.8	tr.	63.7	17.4	o.9	94.8
Kokum fat		0.2	4.8		72.8	0.8		78.6
Mowrah fat		14.7	19.0		16.5			50.2

| | Di-unsaturated | | | | | | | |
	POO	SOO	AOO	PLiP	PLiS	SLiS	SLiA	Total
Palm Oil Mid Fraction	18.4	1.9		11.3	tr.	tr.		31.6
Shea Oil	4.5	31.3	1.2	0.4	2.2	4.4		44.0
Shea stearine	0.6	7.4			1.6	8.0		17.6
Sal Oil	4.8	10.0	3.2	2.0	3.5	tr.		23.5
Mango fat	1.2	22.7	0.6		0.9			25.4
Illipe fat (Borneo)	1.5	3.1		0.4	1.2	0.8		7.0
Allanblackia	2.2	31.7		0.3	0.5	tr.		34.7
Mangosteen	7.2	4.8			tr.	40.2		52.2
Sal stearine	0.6	1.8	0.2	0.5	0.8	0.8		4.7
Kokum fat	0.5	16.2	0.9		0.3	1.0		18.9
Mowrah fat	15.7	10.1		4.7	5.0	tr.		35.5

| | Poly-unsaturated | | | | | | Tri-saturated | | | |
	OOO	PLiO	SLiO	SLiLi	LiOO	Total	PPP	PPS	PSS	Total
Palm Oil Mid Fract.		2.9	9.3	tr.		1.2	13.4	1.5	tr.	1.5
Shea Oil	8.8	2.6	7.6	1.1	2.8	22.9				
Shea stearine	0.8		1.2			2.0		tr.	tr.	
Sal Oil	2.5	tr.	2.2			4.7			tr.	tr.
Mango fat	3.4	2.9	0.4		2.1	8.8				
Illipe fat (Borneo)	0.8	0.8	0.6			2.2				
Allanblackia	9.4	1.3	0.2			10.9				
Mangosteen	4.5	tr.	tr.		1.7	6.2				
Sal stearine	tr.	tr.							0.5	0.5
Kokum fat	2.0		0.5			2.5				
Mowrah fat	7.0	3.0	4.3			14.3				

Palm oil (*Elaeis guineensis*) another constituent of CBE, is native to the humid parts of tropical West Africa. It is now cultivated all over the world. The world production of palm oil in 1984/1985 was 7.04 million metric tons, out of which Malaysia alone provided 4.13 million metric tons. The treatment of the fruit from the moment of harvesting has great importance, for the quality of the oil and the conditions during processing, storage and further transport of the oil can all determine its suitability or otherwise for CBE production.

Shorea species are native to the south and southeast Asian area (Ashton, 1982) . The major part of the species are found in Malaya, Sumatra and Borneo, the latter being the centre of speciation. The commercial name "illipe," often used for these nuts, is a bit confusing as the same denomination is used for certain *Madhuca* (*Bassia nom. illeg.*) species of the Sapotaceae, which are native to India. "Borneo Tallow" or "Tenkawang" (Sabah) are often used as trade names for what is known as "Illipe butter". Several *Shorea spp.* yield the commercial nuts (Ashton, 1982) but as the members of the genus native to the everwet forests of S. E. Asia flower at irregular intervals (perhaps every 6-7 year) the size of a harvest varies between 2000 and 25 000 tons. Furthermore the demand also determines how much of this forest crop is marketed.

Sal (*Shorea robusta*) is a native of the Indo-Burmese region and is a dominant tree in the forests of North, East and Central India. The maturation period (seedling to fruiting) is about 25 years. The theoretical potential of Sal seed kernels, based on a forest area about 45 000 square miles would be more than 5.5 million tons. The kernels contain about 14% fat which varies considerably in quality. India can produce c. 7000-8000 tons of Sal fat but in a bad year only c. 2000 tons. The highest figure ever reached is 200 000 tons of Sal seeds amounting to 30-40 000 tons of fat.

Mango (*Mangifera indica*) probably a native of India-Burma region (Ding Hou, 1978) and now widely cultivated all over the tropics. A great deal of interest has been expressed in the use of mango kernel fat. This fat, characterized by high contents of stearic and oleic acids, has a melting point of about 40° C. The potential availability is considered promising. Production of mango fruit in India is about 10 million tons. It is estimated that this country could possibly produce about 30.000 tons of mango seed fat annually, though the technical problems of crushing the thick mesocarp still remains to be solved. The maturation period of a mango tree is about 6 years.

Guttiferae. The family Guttiferae is represented by 3 species in Table 2: *Allanblackia floribunda*, a native of the wet forests of West Africa (Bamps, 1969); Kokum (*Garcinia indica*) a native of the savanna areas of western parts of the Indian subcontinent (Ratnagiri, Maharastra, Goa and the area South of Goa); Mangosteen (*Garcinia mangostana*) a common fruit tree which is now cultivated all over tropical Asia for its sweet aril.

The results in table 2 reveal that the Kokum fat contains the highest amount of SOS triglyceride, followed by Shea and Sal stearine.

Conclusion

In summary it can be noted that investigations of alternative sources of supply are being intensified with the purpose of increasing the flexibility in the formulation of CBE's.

Table 2 shows some of the elements that can be used in the formulation.Note that species such as Kokum, Sal, and Shea have very high contents of Stearine-Oleine-Stearine (SOS) triglycerides.

Several of the species mentioned above are "wild crops" and have not yet been cultivated commercially. Experimental cultivations are needed in order to learn the cultural practices of species as *Garcinia indica, Allanblackia floribunda* and *Shorea spp.* Other species as *Mangifera indica* and *Garcinia mangostana* are widely cultivated for their fruits, but the seeds are just thrown away and harvesting and processing technologies have not yet been developed. Breeding programmes for the improvement of these species need to be carried out.

The genetic resources needed for these research programmes are found in the natural plant communities of the tropics. It is thus desireable to continue the efforts to protect at least some of the wet forests and savannas, the home of these species, which are being endangered by *Homo sapiens*.

Literature cited

Anon. (1984). "The American Candy Market." *Business Trends Analysis, Inc.* 1984: 7.

Ashton, P. S. (1982). "Dipterocarpaceae." *In* van Steenis (ed.), *Flora Malesiana ser. I. vol.* **9, 2,** 237-552.

Bamps, P. (1969). "Revision du genre Allanblackia Oliv.." *Bull. Jard. Bot. Nat. Belg.* **39,** 347-357.

Bringi, N. V. and Mehta, D. T. (1989). "Madhuca indica seed fat-Mowrah." *In* Bringi, N. V. (ed.), *"Non-Traditional oilseeds and oils in India",* Oxford and I. BH Publ. Co. PVT. Ltd., New Delhi, 56-72.

Ding Hou (1978). "Anacardiaceae." *In* van Steenis (ed.), *Flora Malesiana, ser. I, vol.* **8, 3,** 395-548.

Faulkner, R. W. (1981). "A simple and direct procedure for the evaluation of triglyceride composition of cocoa butters by high performance liquid chromatography - A comparison with the existing TLC - GC method." *Candy and Snack Industry p.* **32.**

Haumann, B. F. (1984). "Confectionery fats-for special uses." *J.Amer. Oil Chem.Soc.* **61,** 468.

Maydell, H. J. van (1986). "Trees and Shrubs of the Sahel, their characteristics and uses." *-TZ- Verlagsgesellschaft, Rossdorf*

Paulicka, F. R. (1976). "Specialty fats." *J.Amer.Oil Chem.Soc..* **53,** 421.

Shukla, V. K. S. (1983). "Studies on the chrystallization behavior of the Cocoa butter equivalents by pulse nuclear magnetic resonance-Part I." *Fette, Seifen, Anstrichmittel* **85,** 467-471.

Shukla,V.K.S.(1988). "Confectionary Fats." *Elsevier Publ. Comp., Amsterdam.*

Shukla, V. K. S., Schiøtz Nielsen, W. and Batsberger, W. (1983). "A simple and direct procedure for the evaluation of triglyceride composition of Cocoa butters by High Performance Liquid Chromatography-a comparison with the existing TLC-GC method." *Fette, Seifen, Anstrichmittel* **85,** 274-278.

Soeters,Ir. C. J. (1970). "Das physikalische Verhalten und die chemische Zusammensetzung von Schokolade,Kakaobutter und einigen Kakaobutter-Austauschfetten." *Fette, Seifen, Anstrichmittel* **72 (8),** 711-718.

Synthesis of the symposium

P. H. RAVEN

Missouri Botanical Garden, USA

This symposium has played a valuable role in helping us to understand why further studies of tropical plants are of such central importance for the development of botany. Its setting in Denmark, with its fine international tradition, has also been an inspiration to each of us for further effective collaboration between botanists of all nations in addressing these matters. In this particular field of study, we count on the continued, important participation of Aarhus University, now in its sixtieth year; and on the enthusiastic participation of its botanists, now well housed in a lovely new herbarium here on the university grounds.

It is especially suitable at this time to pay tribute to Professor Kai Larsen, who came here on July 1, 1963, establishing the Institute at that time. He studied at the University of Copenhagen with Professor T. W. Bøcher, who many of us remember and who made such wonderful contributions to biosystematics in the years following World War II, when the subject of biosystematics was becoming established. Coming from this tradition, Professor Larsen soon realized that the Mediterranean and the Canary Islands were too narrow a field to occupy him, and he subsequently devoted himself almost entirely to the study of tropical botany. It is a genuine tribute to him, and to the colleagues that he recruited, as well as to the generous support of successive university administrations, that the program here has grown into the flourishing enterprise that we see today. The program that he established has made major contributions to our knowledge of tropical plants, and will doubtless continue to make many more in the future.

During the course of his long and distinguished career, Kai Larsen has participated strongly in the development of science, not only in Scandinavia, but also internationally; these latter contributions are exemplified by his presidency of the International Association of Botanical Gardens from 1981 to 1987. Not only carrying out his own fine work, which centers in Thailand and southeast Asia, he has stimulated the development of an extensive program of collecting and research in Ecuador, thus following an excellent Scandinavian tradition of interest in Latin America. Perhaps even more importantly, Professor Larsen and the group he established here in Aarhus has done a great deal to assist in the development of the institutions of these countries, and in the training of their scientists. These cooperative efforts have been important in the formation of many centers, which will continue to contribute to the solution of the problems that concern us all far into the future. It is appropriate, therefore, to congratulate him and his wife for all of their significant accomplishments, and to wish them the very best for the years to come.

The background

In summarizing the major currents of this symposium, I will attempt to say something about where we find ourselves in 1988, and where I think we should be heading in the future. Tropical flowering plants amount to perhaps 170 000 or 180 000 species. About half of those are in Latin America, another third in Asia, and about a sixth in tropical Africa, for example, we do not know for either region how many kinds of plants more remain to be recognized. Guessing, we may suppose that perhaps 10% of the species of tropical plants remain to be found, and perhaps 20% remain to be named, although many already exist in herbaria. Obviously, a great many names that are currently applied to tropical plants will fall into synonymy as studies continue.

A remarkable feature of global biological diversity is that half of the plants in Latin America occur in just the three countries of Colombia, Ecuador, and Peru, which is an area for scale about a third the size of the United States. That would amount to about 45 000 species of plants, or a sixth of the world total. There might be many more species in these countries, which are incredibly rich biologically and also poorly known. Unfortunately, there are probably only a few hundreds of us capable of dealing with the classification of tropical plants professionally, and one of the most urgent tasks we face is to increase our numbers.

In order to evaluate this urgency further, I offer the following background. When Professor Larsen was completing his graduate study at Copenhagen in 1952, half as many people - about 2.6 billion - lived in the world as now - 5.2 billion. A doubling of the global population in this short time has certainly had an enormous negative impact on all of the things we are talking about. In the early 1950's, about 1 billion people lived in countries that are partly tropical, excluding China, and now there are about 2.7 billion. Within the tropics, therefore, human populations have nearly tripled. During this period of unprecedented human growth, the efforts of this Institute, and of all the rest of us who are dealing with tropical plants, have been and clearly will remain highly important and urgent. Beyond the ecological pressures associated with large and rapidly growing human populations, the maldistribution of wealth is a very important factor driving the destruction of forests in the tropics. Most tropical countries have a clearly defined policy in the area of family planning, and are striving for stability in this area. Their populations include so many young people, however, that the effort is extraordinarily difficult. While they do so, however, their poverty, and the national and international political system within which they live, is exacerbating the problem greatly. Nonetheless, if efforts to level off the global human population are sustained, it might reach stability by the middle of the next century at about 10 billion people.

Meanwhile, the one-quarter of the global population that lives in industrialized countries like those of Europe and the United States controls about 85% of the world's wealth, and consumes about 80 to 95% of the goods that contribute to the quality of life. Particularly important for us to consider as we meet here in this fine educational institution is the fact that three-quarters of the people in the world - those who live in developing countries - have only about 6% of the world's scientists and technologists, while the industrial countries have the remaining 94%. This is one of the major reasons that it is difficult for people living in developing countries to improve their productive base, and constitutes a problem that we must address cooperatively, and with urgency.

At any event, all people are participating in the major changes that are taking place in the atmosphere, in the loss of topsoil, in the destruction of the forests, and in the rapid elimination of biological diversity. The global economic system now operates in such a way that large numbers of poor people are permanently destroying the productive capacity of forests everywhere, while funds are being transferred rapidly at the rate of tens of billions of dollars a year from poor, developing countries to rich industrial ones. The resulting instability has many profound consequences for human beings, but it will also result in the loss of many plants, animals, and microorganisms on which human progress could otherwise be based in the future, especially when threats to the environment becomes less intense as the global population of human beings stabilizes. At present, we are consuming, diverting, or wasting some 40% of terrestrial photosynthetic productivity, and the future can be as good even as the present, much less better, only if we increase our efficiency - the sustainibility of the systems that we utilize - greatly.

Extinction rates

What is the current rate of extinction? Probably 2% of the remaining tropical lowland evergreen forest is being destroyed each year; for example, 200 000 square kilometers of the Brazilian Amazon alone were seen to be burning, in some 170 000 seperate fires, in the summer of 1987. The human population is growing rapidly, especially in the tropics, with more than 80 million people - virtually the equivalent of a new Mexico - being added each year. Although certain areas, such as the interior of the Guyanas, the northern and western Brazilian Amazon, and perhaps, and this is less certain, the central Zaire basin in Africa, may last longer, because of their relatively low populations and less intense utilization at present, most areas of tropical lowland evergreen forest will be gone, destroyed, or cut up into small patches in areas too steep or too wet to cultivate within the next 20 years or so. Other types of tropical forest are, for the most part, already decimated and in far worse condition that the tropical lowland evergreen forest we are discussing here; for example, the deciduous forest along the Pacific Coast of southern Mexico and Central America, which occupied an area about twice the size of France at the time of Columbus, now is represented by remnants amounting to less than 2% of its original extent.

The theory of island biogeography, which relates area to species number, indicates a logaritmic relationship between the two. A tenfold increase in area is, in general, necessary to double the number of species within a region of uniform habitat, whether the areas concerned be islands or mainlands regions. Correspondingly, a tenfold decrease in area can be expected to place half of the original species that inhabited that area in danger of extinction. Obviously, the loss will not be instantaneous, and its rate depends, among other factors, on the size and configuration of the patches that remain; but the general relationship is clear. In small patches, not only are numbers of individuals reduced, so that the possibility of loss by chance is enhanced, but also there is a danger of inbreeding and an exposure to altered environmental conditions, especially around the margins of the patches. Human beings may conveniently hunt through them for commodities of interest, and the chances of survival for most organisms is markedly decreased.

When, therefore, we hear that the area of natural vegetation in Sri Lanka is now 9% of its original extent (Sumithraarachchi, this volume), we can conclude that half the species that were there before the destruction of forest began, are now endangered. Consequently, it is not surprising to find that so many of the plants of Sri Lanka are now known as a few individuals, and so exist already only in botanical gardens. Similar relationships are already true in a number of tropical areas, some of them rich in endemic species, such as Madagascar, Hawaii, and western Ecuador. These areas, recently characterized as "hot spots" of extinction by Norman Myers, should command out urgent attention.

Taking the overall picture, we can estimate the probable rate of extinction for plants over the next few decades. I estimate that just under half of the world's plant species occur in the tropics outside of the three large blocks of tropical lowland evergreen forest that were enumerated above. Parks and reserves may be established in the remaining areas, but whether many of them can be maintained in the face of the global economic forces to which I have just alluded is problematical; certainly, this could be accomplished only by the effective operation of an international consortium that served to fund them sufficiently. At any rate, if nearly half of the world's plant species do occur in areas in which the vegetation is going to be completely destroyed or reduced to small patches over the next 20 years or so, we might assume that half of those, or a quarter of the world total, some 50 000 to 60 000 species, are at risk of extinction during that period of time. How many of them actually survive will depend, in large measure, on human activities, on *ex situ* preservation in botanical gardens and seed banks, on reserves, and on the degree of internationalism that is exhibited by the industrial nations of the world. At any rate, it is clear that each of us can make a real difference in the face of this problem, and that if the global population docs stabilize, it will be some of the very plants that are in danger of extinction that will be most desired to re-vegetate devastated areas. Knowledge will be fundamental in making the best choises in this tumultuous era.

In studying the tropics, it is important to remember that they do provide baseline conditions for many biological trends, and that they need to be evaluated for their relevance to many others. For example, Ashton's (this symposium) analysis of species richness took up Tillman's hypothesis, a hypothesis published within the past two years, and one that badly needs evaluation in relatively undisturbed tropical vegetation. The very principles that underlie the analysis presented by Hartshorn, Oldeman, Bruenig and Huang, Swaine, and Gentry in this symposium - features of composition, dynamics and functioning - are the same principles that might be applied to its better management if they are understood sufficiently and in time. In other words, such information is badly needed both for human progress - the construction of stable, productive systems - and for its scientific implications. In many caces, we do not know whether a particular situation in the tropics differs for the corresponding one in temperate regions, and only further analysis will help us to determine the answer.

In a more theoretical field, hypotheses about speciation in tropical plants, such as those presented by Geesink and Kornet, Chen, Gentry, Dransfield, Andersson, Luteyn, or Polhill, depend mainly on gross observations of variation in morphological patterns, often bolstered by field observations of pollination systems or other more or less evident ecological features.   Charles Darvin or Alfred Russel Wallace could have made essentially similar observations, given sufficient time, and yet they are basic to our understanding of tropical plants. What is necessary is that the principles of modern biosystematics also be applied to some selected groups of tropical plants, so that we can know, for example, whether speciation in *Swartzia* or *Calliandra* is like that in *Acacia* or *Quercus* or by some miracle actually does correspond to the "biological species concepts" that were developed from studies of birds half a century ago.   We need hybridization experiments involving at least a few genera of tropical trees and lianas, so that we can understand unique evolutionary principles that may apply to them, or to fit them in the whole emerging evolutionary synthesis about plants.   More detailed analyses have been made of the evolutionary patterns in a few genera of tropical herbs, and one excellent analysis of *Asplenium* has been offered by Iwatsuki in this symposium; but data for tropical trees and lianas, so characteristic of the forests we have been mainly emphasizing here, are mostly lacking.   When evolutionary hypotheses have been offered, they have been based, in general, on a few observations of some feature of the breeding system, or a broad-based survey of cytology, for example, and not usually on the kind of detailed analysis that could yield even more significant findings.

In relation to these deeper and more extensive studies of selected groups of plants, we need the application of modern cladistic methods to the analysis of some groups of tropical plants, so that we can appreciate their patterns of evolution more precisely.   We need biochemical studies, involving the careful analysis of macromolecules, so that we can appreciate taxonomic distances and the mechanisms of evolution more profoundly in some tropical plants.   In short, although we must complete the cataloguing and "general evolutionary pattern" phase of our analysis, for evident reasons, we must also learn even more about some groups to enrich the theoretical foundations of plant evolution more broadly.

Unique in the tropics is the sheer richness of communities in terms of their biological diversity.   Although we know little about the details of the functioning of plant populations and communities even in Denmark, we know next to nothing about these features for any tropical community.   One of the reasons we are always so interested in the all-too-scarce historical analyses of tropical vegetation, such as that presented by Irion (this volume), is that we want to understand how the vegetation that we see today, with all of its diversity, came into being. The many excellent analyses of vegetation in different parts of the tropics presented here, including those of Balslev and Renner, Huber, Junk, Salo, and Spichiger and Ramella, are of interest both because of what they tell us about particular regions, what they reflect of the history of those regions in a dynamic concept, and the way in which they provide a basis for further, more detailed studies.

A particularly remarkably feature of tropical vegetation, and one in which it exceeds other types of vegetation characteristic of nontropical regions, is its diversity in form. Doubtless related to the richness of tropical vegetation, this diversity, elegantly presented in posters by Ulloa, Oldemans, and Barthélémy and his associates in this symposium, is related to plant architecture. It is no accident that our modern conceptions of plant architecture began with a consideration of tropical plants, because tropical plants are so much more diverse in these respects. Those studies, in turn, have helped us to understand much better the ways in which temperate plants grow and adapt to their surroundings - a model for the general scientific utility of tropical studies.

Among the other biological features that I wish to take up here are epiphytes, very characteristic of tropical vegetation. Although I do not believe that anyone emphasized orchids specificially, they constitute about one in eight species of vascular plants worldwide, and perhaps one in six in the tropics. Most of the tropical species are epiphytes, so that this one family alone surely constitutes a characteristic feature of tropical vegetation, and one that we must understand if we are going to try to comprehend the whole. Understanding just why there are so many kinds of orchids, and how they fit together in tropical communities, becomes an important general problem; to illustrate this point, there are far more kinds of orchids in Latin America than there are species of vascular plants in Europe.

Special problems

One can scarcely think about tropical vegetation these days without considering the problem of refugia. Half a century or longer ago, it was fashionable to search for Pleistocene refugia in formely glaciated areas - centers of survival from which plants and animals had spread, of to which they had remained confined, following the withdrawal of the ice. Some were identified, and some later confirmed by records of fossil pollen. Today in the tropics, distributions of plants and animals have been analyzed and used as evidence to postulate the existence of similar refugia - in this case, forest refugia that survived the dry periods that accompanied the glacial maxima at low latitudes. Since no one doubts that the forest contracted during these dry cycles, the real question is, to what extent are the refugia responsible for the patterns that we see at the present day. On that point, we remain far from certain.

For that matter, what about the tropical lowland forest itself? On the one hand, it is said to be very ancient; on the other, very young. Only a careful analysis of the fossil record, and a precise charting of past climates, will provide an accurate answer to this question. It is of central importance, however, in understanding both the dynamics of the forest itself, and the evolutionary and ecological patterns that underlie the biological richness that exists in tropical regions today. To begin to date the appearance of tropical lowland moist forest, it would first be necessary to evolve some more precise definitions than we have available at present. When would we say that such a forest, clearly derived from pre-existing vegetation, began to exist in the past? Large-leaved fossils undeniably have occured at much higher latitudes than those where plants with comparable leaves occur today, but what does that mean about the nature of the communities where they grew? The once-popular notion that something like tropical lowland evergreen forest once occured everywhere, and has now contracted to its present area, is clearly erroneous: but what is the true answer? Certainly the worldwide temperature gradient that extends from the Equator to the poles has steepened during the past 15 million years or so, as many modern plant communities came into existence in something approaching their contemporary form: the lowland tropical forest obviously changed significantly during this time also. When we understand better the details of the way in which this change occured, we shall be able better to deal with the contemporary forest that we and our fellow human beings are destroying so rapidly.

A few words about speciation. Speciation in its modern form is a word that largely was invented in the heyday of the biological definition of species, which came about under the influence of Dobzhansky, Mayr, and others starting about 50 years ago. Contemporary experimental evidence, however, has shown that plants exist in local populations of limited size, so that the concept of gene pools affecting entire, widespread species in some mystical way and holding them together simply is not supported. The populations of a widespread species in southern Mexico are completely out of contact genetically with those in Paraguay, and their evolution is independent; we must therefore conclude that they retain a somewhat uniform appearance because of their common evolution, the integration of their genome, and because they are continuing to play similar ecological roles, or be subjected to similar selective forces. In order to understand better the kinds of units that we call species, we must perform genetic experiments to find out whether they are capable of interbreeding, and to find improved ways of analyzing their history, and especially their adaption to their environment. Merely repeating outmoded species concepts, or applying them on the basis of relatively superficial observations to the situation found in particular groups, may be appropriate at a descriptive level, but it can make no valid contribution to theory.

The fact that J. D. Hooker once wrote that no hybrids existed in the British flora - it was all a matter of knowing the plants well enough to escape error - while now we recognize some 2000 hybrid combinations there, ought to be a warning to those who assume that they can simply look at morphological variation in tropical, or any other plants, and accurately assess the degree to which the species are hybridizing, or the extent to which past hybridization has affected their evolution. *Juniperus* subg. *Sabina* in central and eastern North America provides another example of this sort - a group in which Edgar Anderson concluded that introgressive hybridization was the dominant force in shaping the observed pattern of variation, whereas B. L. Turner and his associates concluded, using biochemical techniques, that there was no hybridization at all. Which view is correct is not the point here: the examples are provided merely to indicate that situations in nature are often not simple, and to urge those who are studying at least some groups of tropical plants to use the more precise methods of analysis that are available now, so that a few examples may be known in greater depth and perhaps shed some light on the whole.

At the same time, I would like to stress that it is a matter of importance, and remains both valid and useful to continue to describe species as morphological entities, because that is the key to being able to understand the composition of forests - an important activity while there are still forests to describe - and to save them, or at least some of the elements in them. The great task of taxonomy is yet far from completed.

The age of species, genera, and other taxa remains a fascinating subject, in relation to evolutionary rates. For example, large trees look like they took a long time to evolve. How long did they actually take, however, to reach their present form? Macromolecular methods now available can assist greatly to resolving questions of that sort if they are properly applied. Regardless of how bizarre a plant may appear, it doesn't follow that it is very old. For example, *Espeletia*, which forms such a characteristic feature of the *páramos* of northern South America, does not appear to have great antiquity in the family Asteraceae, which itself may not be ancient. The point here is, however, that the problem can be solved if we have the will to do it. There is less and less point in simply offering speculative views about these situations.

Many areas of plant biology have scarcely been investigated in the tropics. One of the most obvious is plant physiological ecology, which clearly has great importance in determining the ranges and patterns of adaption and evolution in tropical plants, as it does for plants everywhere. Another area would be the study of mycorrhizae, for example. What role do the peculiar mycorrhizae have in the success of Ericaceae as epiphytes? Could some connections be forged between different groups of biologists, such as the excellent group in population genetics here at Aarhus and this taxonomy institute, with positive results for both? These and many other problems need to be approached, on a selective basis, while there is still time. If we do so, we shall be able to apply the solutions constructively to important problems of ecological sustainability in an intelligent manner.

In the 25 years that this Institute has existed, about a third of the tropical lowland forest that existed then has been cut. That gives us a good idea of what the next quarter century holds, and of the challenge that faces us. In attacking this problem, we have tools that were unimaginable a short time ago, and we should use them extensively. In the face of the problem that exists, we must all devote ourselves to learning about those magnificent tropical forests that we have been considering here. In doing so, we are inspired by the accomplishments that have been attained by our colleagues here in Aarhus, and offer special congratulations again to Professor Kai Larsen for all that he has made possible.

Index to plant names